W9-BCC-933

STRATEGIES AND TACTICS
IN ORGANIC SYNTHESIS

STRATEGIES AND TACTICS IN ORGANIC SYNTHESIS

Edited by

Thomas Lindberg
G. D. Searle Research and Development
Skokie, Illinois

1984

ACADEMIC PRESS, INC.

(Harcourt Brace Jovanovich, Publishers)

Orlando   San Diego   San Francisco   New York   London
Toronto   Montreal   Sydney   Tokyo   São Paulo

ACADEMIC PRESS, INC.
Orlando, Florida 32887

United Kingdom Edition published by
ACADEMIC PRESS, INC. (LONDON) LTD.
24/28 Oval Road, London NW1 7DX

Library of Congress Cataloging in Publication Data

Main entry under title:

Strategies and tactics in organic synthesis.

    Includes bibliographies and index.
    1. Chemistry, Organic--Synthesis. I. Lindberg,
Thomas, Date   •
QD262.S85   1984   574'.2   83-15674
ISBN 0-12-450280-6

PRINTED IN THE UNITED STATES OF AMERICA

84 85 86 87   9 8 7 6 5 4 3 2 1

CONTENTS

3. A PROSTAGLANDIN SYNTHESIS

Josef Fried

4. SYNTHESIS OF INDOLE ALKALOIDS

Philip Magnus

5. SYNTHESIS OF TYLONOLIDE, THE AGLYCONE OF TYLOSIN

William P. Jackson, Linda D.-L. Lu Chang, Barbara Imperiali, William Choy, Hiromi Tobita, and Satoru Masamune

8. THE SYNTHESIS OF FOMANNOSIN AND ILLUDOL

M. F. Semmelhack

9. EVOLUTION OF A SYNTHETIC STRATEGY: TOTAL SYNTHESIS OF JATROPHONE

Amos B. Smith III

10. ON THE STEREOCHEMISTRY OF NUCLEOPHILIC ADDITIONS TO TETRAHYDROPYRIDINIUM SALTS: A POWERFUL HEURISTIC PRINCIPLE FOR THE STEREORATIONALE DESIGN OF ALKALOID SYNTHESES

Robert V. Stevens

11. A NONBIOMIMETIC APPROACH TO THE TOTAL SYNTHESIS OF STEROIDS: THE TRANSITION METAL-CATALYZED CYCLIZATION OF ALKENES AND ALKYNES

K. Peter C. Vollhardt

12. EVOLUTION OF A STRATEGY FOR TOTAL SYNTHESIS OF STREPTONIGRIN

Steven M. Weinreb

CONTRIBUTORS

Numbers in parentheses indicate the pages on which the authors' contributions begin.

VIRGIL BOEKELHEIDE *(1), Department of Chemistry, University of Oregon, Eugene, Oregon 97403*

WILLIAM CHOY *(123), Department of Chemistry, Massachusetts Institute of Technology, Cambridge, Massachusetts 02139*

RICK L. DANHEISER *(21), Department of Chemistry, Massachusetts Institute of Technology, Cambridge, Massachusetts 02139*

JOSEF FRIED *(71), Department of Chemistry, University of Chicago, Chicago, Illinois 60637*

BARBARA IMPERIALI *(123), Department of Chemistry, Massachusetts Institute of Technology, Cambridge, Massachusetts 02139*

WILLIAM P. JACKSON *(123), Department of Chemistry, Massachusetts Institute of Technology, Cambridge, Massachusetts 02139*

LINDA D.-L. LU CHANG *(123), The Schering Corporation, Bloomfield, New Jersey 07003*

PHILIP MAGNUS *(83), Department of Chemistry, Indiana University, Bloomington, Indiana 47405*

SATORU MASAMUNE *(123), Department of Chemistry, Massachusetts Institute of Technology, Cambridge, Massachusetts 02139*

K. C. NICOLAOU *(155), Department of Chemistry, University of Pennsylvania, Philadelphia, Pennsylvania 19104*

LEO A. PAQUETTE *(175), Evans Chemical Laboratories, Department of Chemistry, The Ohio State University, Columbus, Ohio 43210*

N. A. PETASIS *(155), Department of Chemistry, University of Pennsylvania, Philadelphia, Pennsylvania 19104*

M. F. SEMMELHACK *(201), Department of Chemistry, Princeton University, Princeton, New Jersey 08544*

AMOS B. SMITH III *(223), Department of Chemistry, Laboratory for Research on the Structure of Matter, and the Monell Chemical Senses Center, University of Pennsylvania, Philadelphia, Pennsylvania 19104*

ROBERT V. STEVENS *(275), Department of Chemistry and Biochemistry, University of California, Los Angeles, Los Angeles, California 90024*

HIROMI TOBITA *(123), Department of Chemistry, Massachusetts Institute of Technology, Cambridge, Massachusetts 02139*

K. PETER C. VOLLHARDT *(299), Department of Chemistry, University of California, Berkeley, Berkeley, California 94720*

STEVEN M. WEINREB *(325), Department of Chemistry, Pennsylvania State University, University Park, Pennsylvania 16802*

JAMES D. WHITE *(347), Department of Chemistry, Oregon State University, Corvallis, Oregon 97331*

PREFACE

The inspiration for this book came from an article by I. Ernest on R. B. Woodward's prostaglandin synthesis.[1] In his paper Ernest describes the trials and tribulations that had to be endured before success was finally attained. As Ernest states in his introduction ". . . the sober and dispassionate form of today's scientific publications does not leave much room to express or even suggest the creative motivation and atmosphere in which ideas originate and are further developed." At that time there were no books in which chemists described their syntheses in the way that Ernest did in his paper. I felt that such a book would be especially valuable to students learning organic synthesis. To try and remedy this situation I asked a group of outstanding chemists to give a ". . . sincere and more or less complete account of the chronological development of ideas and experimentation which finally led to the solution of the problem."

Many syntheses only appear as terse communications in journals. Very rarely do chemists discuss the blind alleys and dead ends that were encountered in a synthesis. This is unfortunate for the student who wants to learn about synthesis. I think many students have the mistaken impression that organic chemists conceive a brilliant "paper" synthesis in 1 hour and hand it over to their graduate students who see it through to completion without any problems or difficulties. However, in almost every synthesis there are problems to be overcome and obstacles to be surmounted. The outstanding chemists in this book have done an excellent job in describing the strategies and tactics that they have used in synthesis. One can easily see that the road from a "paper" synthesis to the final product is a long and difficult one. I believe that students and

[1]Ernest, I. (1976). *Agnew. Chem. Int. Ed. Engl.* **15,** No.4, 207.

chemists will find these accounts of synthesis to be interesting and informative.

Finally, I would like to express my sincere appreciation to the contributors, for without their efforts there would be no book.

Thomas Lindberg

Chapter 1

THEME AND VARIATIONS:
A SYNTHESIS OF SUPERPHANE[*,†]

Virgil Boekelheide

Department of Chemistry
University of Oregon
Eugene, Oregon

I. Introduction, Goals, and Synthetic Philosophy

The word cyclophane designates the general class of bridged aromatic compounds. The field was pioneered by Cram and Steinberg,[1] who first synthesized [2.2]paracyclophane (**1**) and who invented the nomenclature used for these compounds. The numbers in the brackets indicate how many

---

* I have long felt that painting and the composition of music, more than most other academic disciplines, have much in common with the conceptional aspects of scientific research. The intellectual ferment leading to the creation of a painting or a musical composition seems quite akin to the intellectual ferment leading to the creation of a new idea, a new concept, or a new principle in science. Just as the style of eighteenth century music is replete with compositions that are a theme with variations, so also are scientific ideas often exploited by variations to discover the limits and generality of the idea. This aside is offered in explanation of the title.

† We thank the National Science Foundation for their generous financial support of the work described in this review.

1

bridges there are and how many bridging atoms in each bridge. Although the prefixes ortho-, meta-, and para- continue to be used, the positions of bridging are more commonly indicated by numbers in parentheses following the brackets. Thus, [2.2]paracyclophane can equally as well be designated as [2.2](1,4)cyclophane. Where all of the bridges have the same number of bridging atoms, the number of bridges may be indicated by a subscript number. For example, structure **5** is [2$_3$](1,3,5)cyclophane.

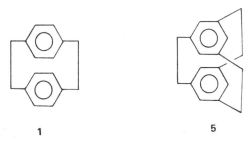

1   5

From the very beginning of his work, Cram appreciated the great value such rigid, caged structures would have in providing insight into questions of ring strain, bond stretching, bond angle distortion, aromatic ring deformation, and π–π orbital interactions. The extremes for such behavior are to be found in examples where the aromatic rings are forced together face-to-face in the closest proximity possible. Thus the multibridged cyclophanes having two bridging atoms in each bridge are of special interest. There are only 12 such possible [2$_n$]cyclophanes with the same substitution pattern in each deck. These are shown in Scheme 1.

At the time we began the work under discussion only compounds **1, 2, 3, 5** and **6** shown in Scheme 1 were known. It was important to devise syntheses for the remaining members of the series so that a complete correlation could be made of the variations in physical properties with variations in geometry, particularly in the distance between aromatic decks. It was already well appreciated that molecules such as [2.2]paracyclophane exhibit a strong, between decks, π–π orbital interaction, and how much this interaction would be intensified as the distance between decks was shortened was of prime interest. For this purpose the molecule most desired was [2$_6$]-(1,2,3,4,5,6)-cyclophane (**12**), bearing the trivial name superphane.[2] Superphane is the ultimate in bridging in the series, is the most symmetrical, and should have the aromatic rings forced in closer proximity than for any other member.

Before discussing possible synthetic routes to the multibridged [2$_n$]cyclophanes and superphane, some basic points in the philosophy of designing any synthesis need to be restated. A synthesis should be designed to be efficient, short, and, if possible, display either novel chemistry or new applications of known chemistry. Because our purpose in making these molecules was to

examine their physical and chemical properties, we had an additional requirement: the synthesis must be convenient and practical enough to provide
sufficient quantities for studying these properties. Specifically, we felt that a
good synthesis should be capable of providing at least 1 g of the final product.

This latter requirement of designing a synthesis to provide the final product
in adequate quantities, whether it be for biological testing, studying its chemical and physical properties, or preparing analogs, is frequently neglected in
the design of syntheses and deserves greater emphasis. All too often in the
competition for syntheses of natural products or novel structures, success is
judged by who is the first to accomplish the goal rather than whether a useful
synthesis has been developed. Developing new syntheses that fail to provide
the target molecule in useful quantity may have heuristic value, but modern

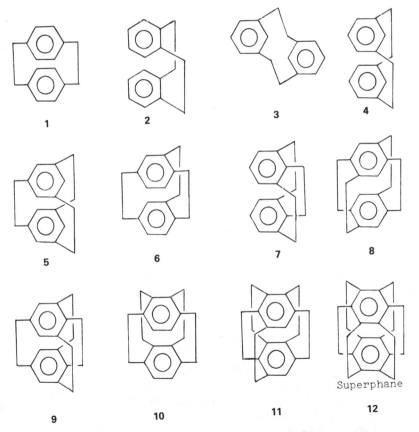

SCHEME 1.   Possible symmetrical [2$_n$]cyclophanes. (Although 3 and 4 appear to be conformational isomers, the energy barrier for interconversion is sufficiently high that the separate structures can be isolated.)

synthetic chemistry should now have the ability to advance beyond that point. One of the requirements for any significant new synthesis should be that it provide the desired product in adequate quantity to satisfy the avowed purposes for which the synthesis was undertaken.

Also, before designing the synthesis of a particular molecule, the status of the field needs to be examined to ascertain whether the goal can be reached efficiently by simple extension of known methods or whether a completely new approach is needed. The first two decades of cyclophane syntheses were dominated by two methods: (1) the Wurtz coupling reaction [3] and (2) the 1,6-elimination of p-methylbenzylammonium hydroxides.[4] Around 1970 the dithiacyclophane route to cyclophanes was introduced and quickly became an important general method for making all types of cyclophanes.[5-7] Then, in 1972, Hopf introduced a Diels–Alder method for making polysubstituted [2.2]paracyclophanes, which had important advantages of convenience and adaptability for preparing large quantities.[8]

Thus valuable methods for synthesizing cyclophanes were already at hand. However, to apply these to the synthesis of the highly bridged cyclophanes looked to be a cumbersome and tedious task, involving many steps for the introduction of each bridge beyond those of [2.2]paracyclophane itself.

II. Theme

The present account, describing a trail of research leading to the synthesis of superphane, begins in the fall of 1974 with a remarkable graduate student, Richard T. Gray. At the time, our interest in a synthesis of superphane was somewhat remote.

Gray had begun his doctoral research on a project directed toward a synthesis of cyclophanes having cyclooctatetraene units as decks, a project not yet accomplished but still of much interest. My suggestion to Gray regarding a possible route to the desired cyclooctatetraene cyclophane involved first synthesizing [2₄](1,2,4,5)cyclophane (**8**) as a precursor. Based on our experience in developing the dithiacyclophane route to cyclophanes, it seemed that this synthetic method would be appropriate for a synthesis of [2₄](1,2,4,5)-cyclophane. In fact, as shown in Scheme 2, Gray's first successful synthesis of [2₄](1,2,4,5)cyclophane (**8**) followed this approach.[9]

While pursuing experimentally the synthetic route shown in Scheme 2, Gray, as is typical for a good graduate student, gave serious thought to how to improve on his mentor's proposal. As a result he conceived the idea of a one-step synthesis of [2₄](1,2,4,5)cyclophane (**8**), as outlined in Scheme 3. The exploitation of Gray's idea led to some false starts and some disappointments, as will be pointed out, but eventually to a new, general method for preparing

multibridged cyclophanes and ultimately to a successful synthesis of superphane.

Gray's idea was based on two assumptions: (1) that *o*-methylbenzyl
chlorides thermally eliminate hydrogen chloride to give an equilibrium
mixture of an *o*-xylylene and a benzocyclobutene and (2) that the equilibrium
mixture of *o*-xylylene and benzocyclobutene undergoes thermal dimerization
to dibenzocyclooctadienes. An early observation by Hart and Fish had shown
that trichloromethylpentamethylbenzene loses hydrogen chloride on mild

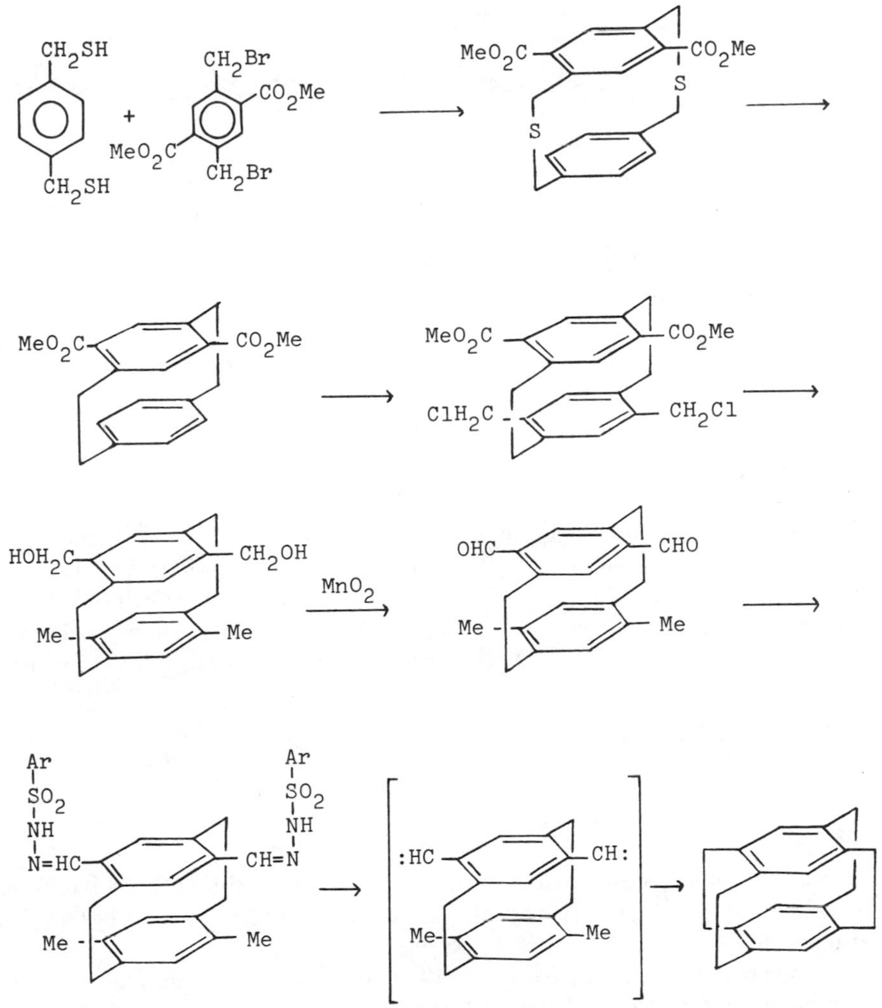

SCHEME 2

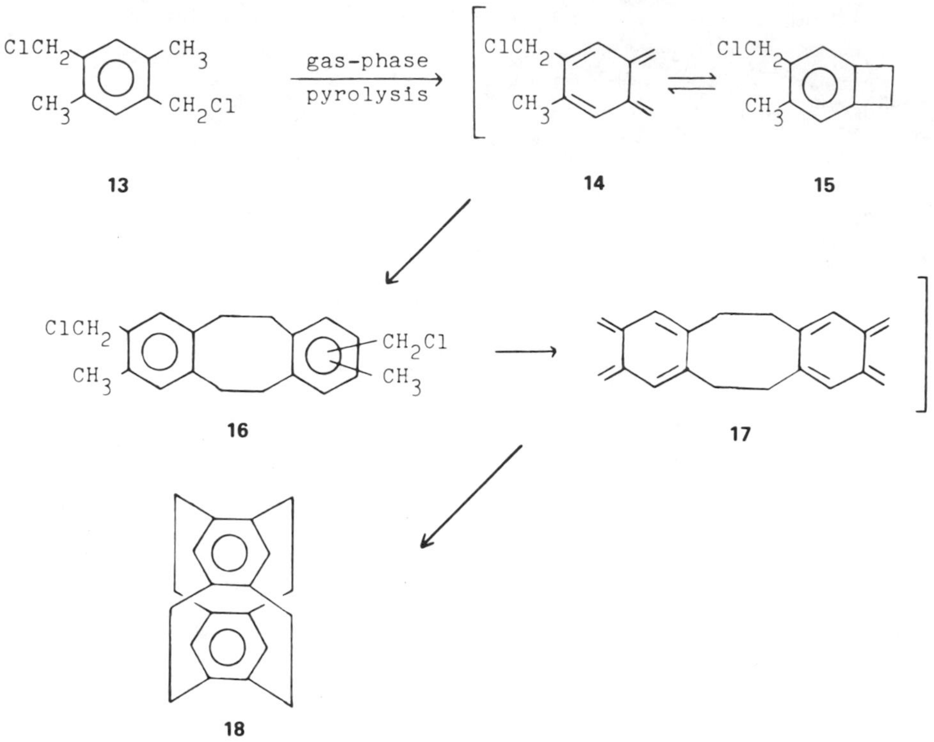

13   14   15

16   17

18

SCHEME 3

heating to give the corresponding benzocyclobutene derivative.[10,11] Furthermore Loudon *et al.* had shown that *o*-methylbenzyl chloride (**19**) itself is
smoothly converted to benzocyclobutene (**20**), and there were indications
from their study that benzocyclobutene underwent thermal dimerization to
dibenzocyclooctadiene (**21**).[12] This latter finding was also supported by the

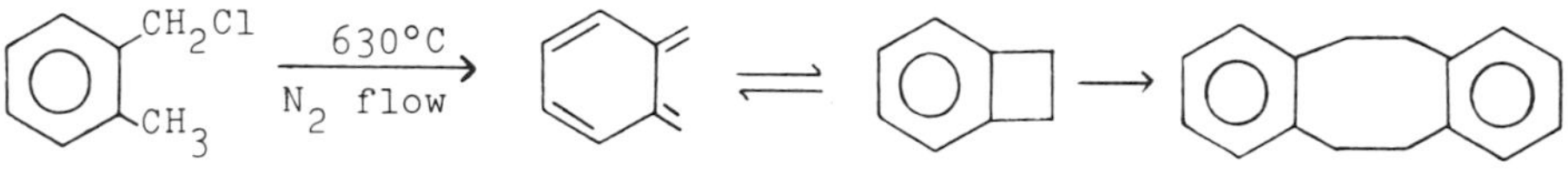

19   20   21

earlier work of Cava and Deana on the pyrolysis of 1,3-dihydroisothianaph-
thene 2,2-dioxide.[13] Thus there was reasonable precedent for both of Gray's
assumptions.

However, the gas-phase pyrolysis of **13** at 700°C and 10^{-3} mm Hg pressure
gave none of the desired [2₄](1,2,4,5)cyclophane (**8**) but, instead, benzo-
[1,2;4,5]dicyclobutene (**22**).[14] The easy formation of **22** was interesting but,

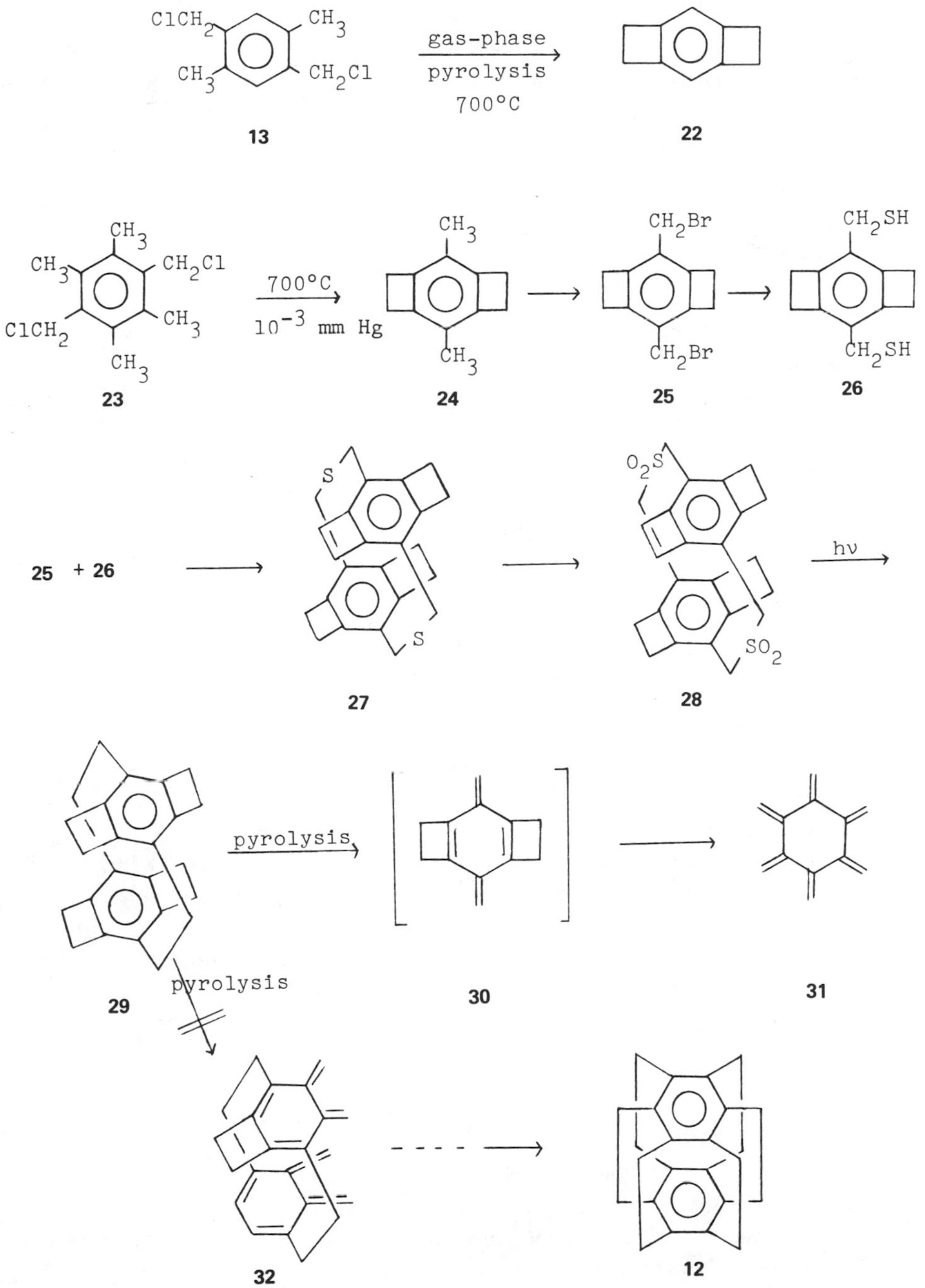

SCHEME 4

at the time, it was not obvious how it could serve as a precursor for [2₄]-(1,2,4,5)cyclophane. Therefore, Gray returned to the original dithiacyclophane route and successfully prepared [2₄](1,2,4,5)cyclophane (**8**), as well as making a careful study of its physical and chemical properties.[15]

The idea of preparing multibridged cyclophanes by thermal dimerization of benzocyclobutenes, though, still seemed worthy of investigation. In an attempt to test this idea experimentally, we prepared the cyclophane derivative **29** by the route shown in Scheme 4.[16]

However, gas-phase pyrolysis of **29** did not give the desired [2₆](1,2,3,4,5,6)-cyclophane (**33**), but rather gave hexaradialene (**31**). Apparently, the strained [2.2]paracyclophane **29** prefers carbon–carbon cleavage to a *p*-xylylene rather than to an *o*-xylylene.

Hexaradialene (**31**) is an interesting molecule in its own right. We,[16] as well as Schiess and Heitzmann,[17] were able to show that hexaradialene can be prepared conveniently in good yield by the gas-phase pyrolysis of 2,4,6-tris(chloromethyl)mesitylene (**33**).

33   **31**

We then returned to Gray's original idea but decided to try to accomplish the desired goal in a stepwise fashion rather than in a single-pot reaction.[18] As shown in Scheme 5, gas-phase pyrolysis of 2,5-dimethylbenzyl chloride (**34**) proceeded in good yield to give 4-methylbenzocyclobutene (**35**) which, on thermal dimerization in a boiling diethyl phthalate solution, gave **36** as a mixture of isomers. Chloromethylation of **36** then gave **37**, again as a mixture of isomers. No attempt was made to separate the isomers, though, because both on pyrolysis would be expected to form the same *o*-xylylene intermediate **38**. In fact, pyrolysis of **37** gave a mixture of [2₄](1,2,4,5)cyclophane (**8**) plus the cyclobutene analog **39**.[18] Later, it was possible to find conditions for the pyrolysis that effected the conversion of **37** to [2₄](1,2,4,5)cyclophane (**8**) cleanly in 20% yield with essentially no formation of the unwanted cyclobutene analog **39**.[19]

Thus Gray's basic idea was shown to be eminently sound. When two benzocyclobutene moieties are held in suitable proximity by stable bonds, thermal dimerization occurs on gas-phase pyrolysis to give multibridged cyclophanes (Scheme 5).

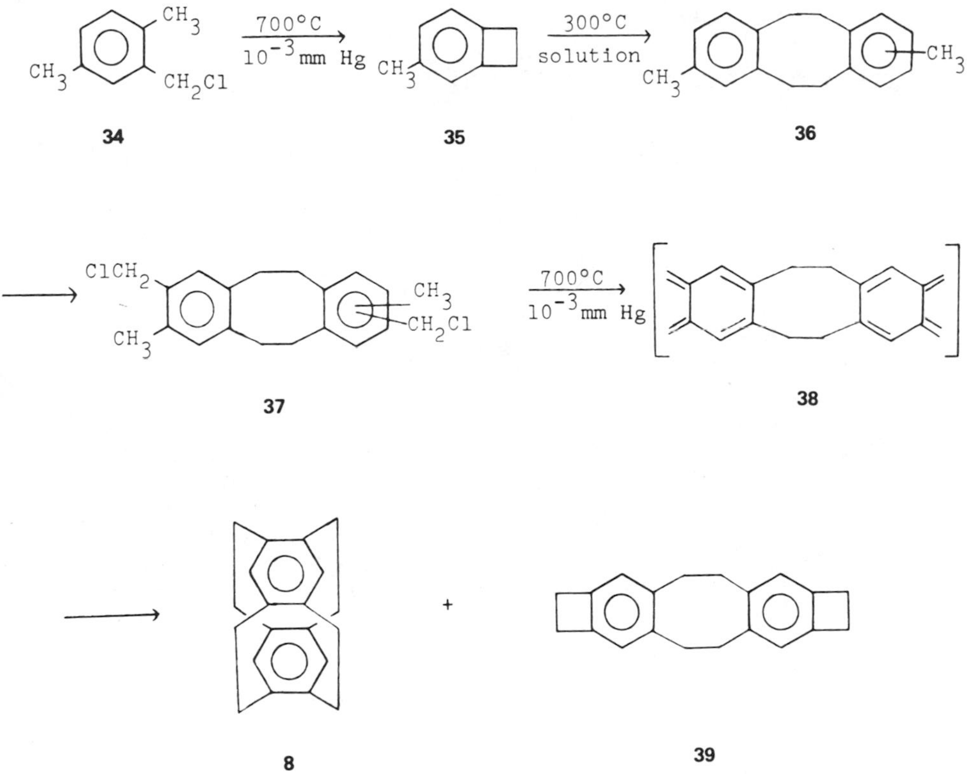

SCHEME 5

III. Variations

With the establishment of the dimerization of benzocyclobutenes as a synthetic route for preparing multibridged cyclophanes, it was of interest to explore the generality of the method, its limits, and its possible variations. The first variation was a synthesis of the known [2₃](1,2,4)cyclophane (**6**), first prepared by Truesdale and Cram[20,21] and later in a very convenient fashion by Trampfe, Hopf, and Menke.[22] Our synthesis is shown in Scheme 6.[23] Subsequently, a synthesis very similar to ours was reported by Aalbersberg and Vollhardt.[24]

In an analogous way, we also prepared the previously unknown [2₃](1,2,3)-cyclophane (**7**), as is likewise shown in Scheme 6.[25]

Our attempt to prepare superphane (**12**), as shown in Scheme 4, had been thwarted by the fact that thermal bond rupture of the intermediate [2.2]para-cyclophane (**29**) occurred preferentially to a *p*-xylene intermediate **30** rather

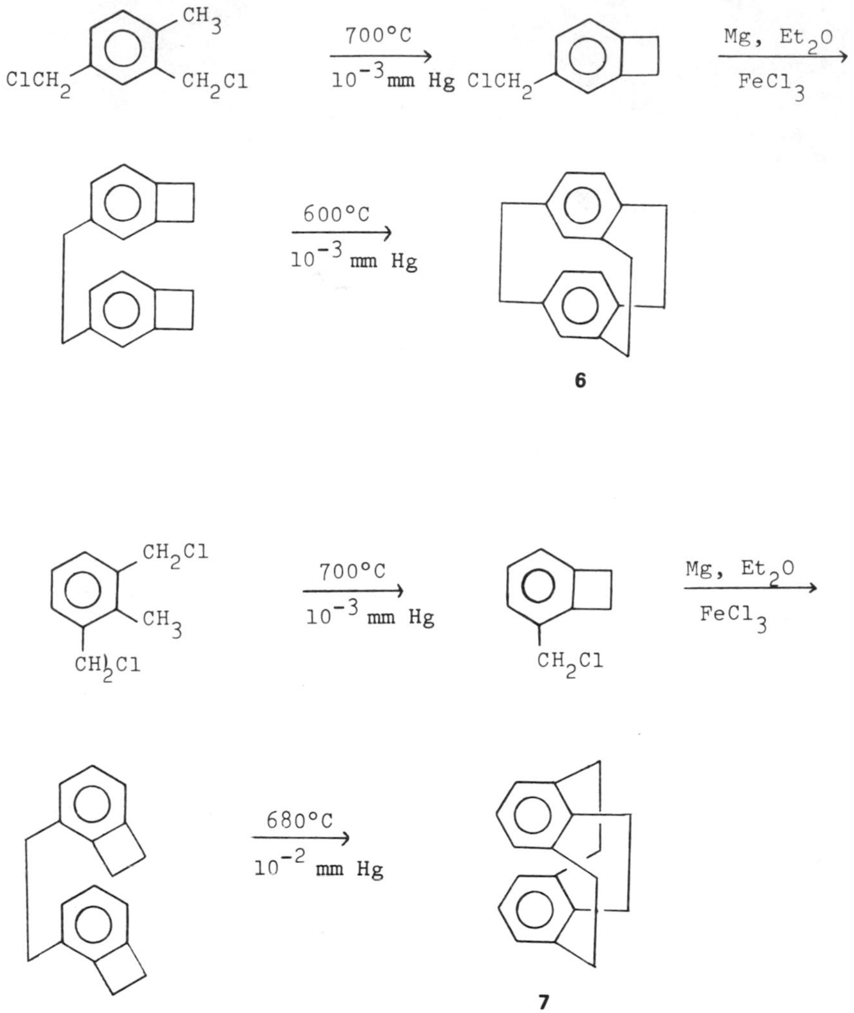

SCHEME 6

than the desired *o*-xylene intermediate **32**. It now seemed that the idea of a thermal bond isomerization of two benzo[1,2;4,5]dicyclobutene moieties held face to face was sound, if the bonding holding the benzo[1,2;4,5]dicyclobutenes was thermally stable. Paulo Schirch, a Brazilian graduate student, then took up the idea of doing a synthesis analogous to those shown in Scheme 6 but where the benzocyclobutene moiety was replaced by a benzo[1,2;4,5]-dicyclobutene unit.

For this purpose a convenient synthesis of an appropriately substituted benzo[1,2;4,5]dicyclobutene was needed. We first investigated the gas-phase pyrolysis of 3,5-bis(chloromethyl)-2,6-dimethylbromobenzene (**42**), which proceeded smoothly in 53% yield to give 3-bromobenzo[1,2;4,5]dicyclobutene (**43**).[26] Replacement of the bromine by cyanide to give **44** was successful, using the von Braun reaction, but the yield was poor and the experimental procedure tedious. Even so, the cyano derivative **44** was readily converted to the corresponding aldehyde **45** and undoubtedly the further steps in this route could have been successfully accomplished to prepare $[2_5](1,2,3,4,5)$-cyclophane (**11**).

42   **43**   **44** : R = CN

45 : R = CH=O

However, as indicated earlier, one requirement for a successful synthesis is that it be sufficiently convenient to provide adequate quantities of the final product. This route did not look promising in this respect. Therefore, we turned to a different approach.

Our successful solution is shown in Scheme 7.[26] Chloromethylation of methyl 2,6-dimethylbenzoate (**46**) gave **47** in a convenient, high-yield reaction. Presumably, the methyl groups at the 2- and 6-positions prevent the ester from being coplanar with the aromatic ring, and so there is very little deactivation of the ring by the ester group toward further electrophilic substitution. Gas-phase pyrolysis of **47** then readily gave 3-carbomethoxybenzo-[1,2;4,5]dicyclobutene (**48**). The further steps of converting **48** to the desired dimer **51** were accomplished following the general procedures as outlined previously in Scheme 6. The gas-phase pyrolysis of **51** occurred smoothly to give $[2_5](1,2,3,4,5)$cyclophane (**11**).

The final step in the synthesis of $[2_5](1,2,3,4,5)$cyclophane (**11**), as shown in Scheme 7, is remarkable in that it introduces four new bridges in a yield of 69%. Although there is no experimental evidence available as yet to define the reaction path, we presume that the conversion of **51** to **11** proceeds via intermediates such as those shown in the brackets. The overall yield for the six-step synthesis of $[2_5](1,2,3,4,5)$cyclophane, as shown in Scheme 7, is 31%, and it has been carried out on a scale providing 5 g of $[2_5](1,2,3,4,5)$cyclophane per run.

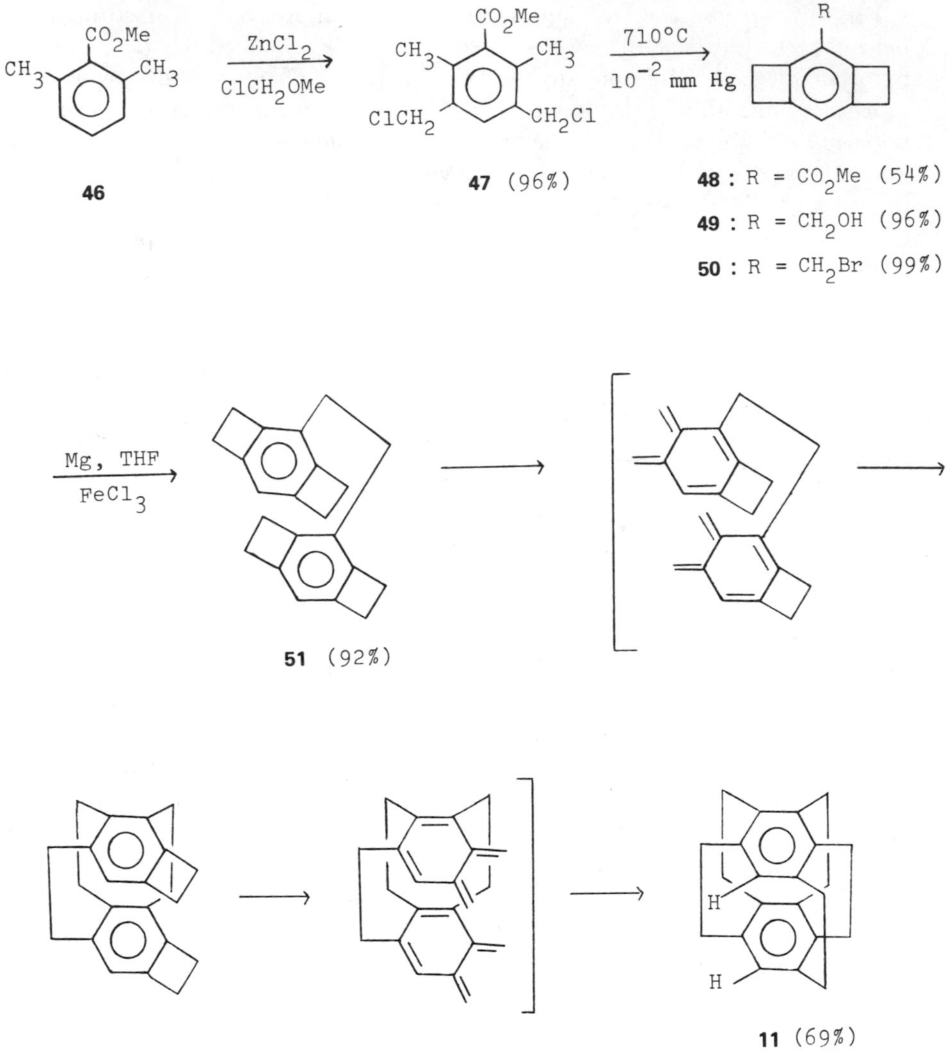
CO2Me
CH3
CH3
ZnCl2
ClCH2OMe
CH3
CH3
ClCH2
CH2Cl
710°C
10-2 mm Hg
R
46
47 (96%)
48 : R = CO2Me (54%)
49 : R = CH2OH (96%)
50 : R = CH2Br (99%)
Mg, THF
FeCl3
51 (92%)
H
H
11 (69%)

SCHEME 7

IV. A Synthesis of Superphane

With the firm establishment of the thermal formation and dimerization of benzocyclobutenes as a method for preparing multibridged cyclophanes, our attention turned to the most interesting member of the series, $[2_6]$-(1,2,3,4,5,6)cyclophane (**12**), for which the trivial name superphane has been coined.[2] Superphane represents the ultimate in bridging for a $[2_n]$cyclophane, is highly symmetrical (D_{6h}), and should have its benzene-ring decks both more planar and in closer proximity than for any other member. Its properties should describe the extreme for a π–π orbital interaction. A final reason for undertaking its synthesis, of course, was that the superphane molecule is simply aesthetically pleasing.

Because each two-step sequence of (1) gas-phase pyrolysis of o-methylbenzyl chlorides to form benzocyclobutenes and (2) thermal dimerization of benzo-cyclobutenes introduces two cyclophane bridges, a synthesis of the six-bridged superphane requires repeating the two-step sequence three times. Of the various possible ways of doing this we chose to explore the route shown in Scheme 8.[27]

Before discussing the details of the route presented in Scheme 8, a final requirement for success in any multistep, difficult synthesis must be stated. This is the skill of the experimentalist. Most experienced research directors are painfully aware of beautifully designed syntheses they have conceived, whose successful completion failed because of inadequate skill on the part of the experimentalist. The synthesis of superphane, as outlined in Scheme 8, was undertaken by Yasuo Sekine, a postdoctoral fellow, and it is a tribute to his experimental skill, determination, and resourcefulness that the synthesis of superphane was accomplished as readily as it was.

At the time we began our synthetic work shown in Scheme 8, $[2_4](1,2,3,4)$-cyclophanes were still unknown. Thus our synthesis of 4,13-dimethyl$[2_4]$-$(1,2,3,4)$cyclophane (**60**) was a first example of this class of compounds. However, simultaneously with our studies, Kleinschroth and Hopf prepared the parent $[2_4](1,2,3,4)$cyclophane (**10**) by an independent synthesis.[28]

The reagents and yields presented in Scheme 8 are those eventually found to be most successful. As shown, the synthesis required 10 steps and proceeded in an overall yield of 4% on a scale giving approximately 1 g per run. Access to this quantity of material has permitted a fairly extensive study of the physical and chemical properties of superphane.[28-31]

From the initial experimental studies there were two significant weaknesses in Scheme 8. One was the fact that formylation of **43** by the Rieche procedure gave a mixture of products **55** and **56** in yields of 49 and 29%. Fortunately, the desired compound **44** was the predominant isomer. As is often the case, careful attention to the exact details of the reaction conditions made it possible to improve both the overall yield and the relative amount of **55**. Because **55** and **56** are easily separated, improvement in the procedure to give **55** in 60% yield together with **56** in 20% yield made this a satisfactory step.

A more serious problem was the dimerization of **53** to give **54**. Traditionally, the dimerization of benzocyclobutenes has been carried out at 300°C in boiling diethyl phthalate following a procedure first devised by Cava and Deana.[13]

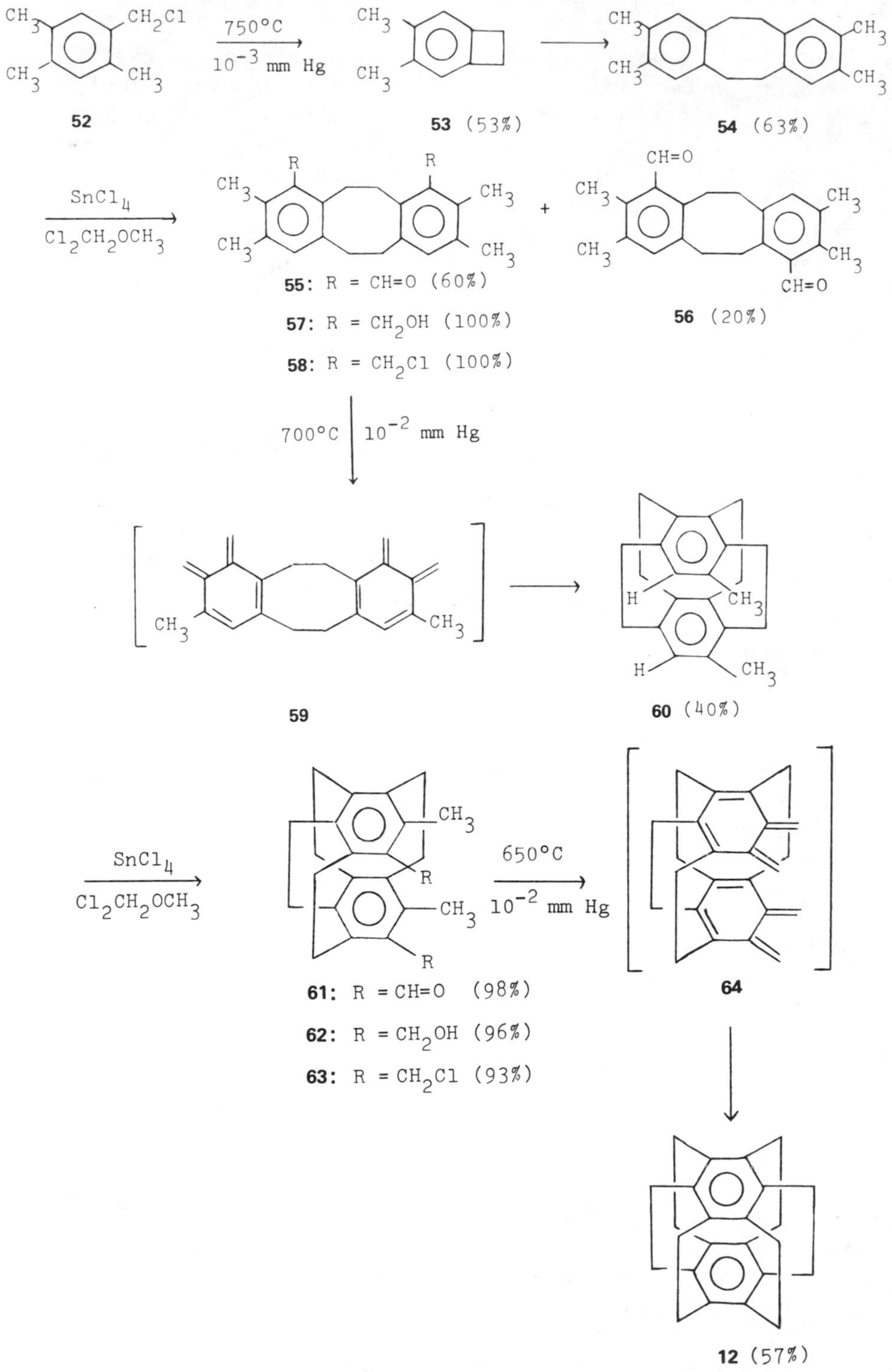

52

750°C
10⁻³ mm Hg

53 (53%)

54 (63%)

SnCl₄
Cl₂CH₂OCH₃

55: R = CH=O (60%)
57: R = CH₂OH (100%)
58: R = CH₂Cl (100%)

56 (20%)

700°C | 10⁻² mm Hg

59

60 (40%)

SnCl₄
Cl₂CH₂OCH₃

61: R = CH=O (98%)
62: R = CH₂OH (96%)
63: R = CH₂Cl (93%)

650°C
10⁻² mm Hg

64

12 (57%)

However, the isolation of the dimer **54** from the diethyl phthalate was a slow and tedious process and, when **54** finally was isolated and purified, its yield was only 23%. We then turned to a gas-phase, thermal dimerization of **53**, using a nitrogen-flow system at atmospheric pressure with a vertical pyrolysis furnace. By adjusting the nitrogen flow rate and the rate of introduction of **53**, we were able to find conditions favoring a bimolecular reaction of **53** in the gas phase. This made possible a clean conversion of **53** to **54** in 63% yield.

It is common practice in describing thermal reactions of benzocyclobutene to write *o*-xylylene derivatives as intermediates, and we have followed this practice in the reaction schemes we have presented here. However, there is no compelling evidence requiring such *o*-xylylenes as intermediates and, quite possibly, an aromatic diradical is the correct intermediate.[32]

The completion of a synthesis of superphane established once again the general usefulness of the two-step, thermal formation and dimerization of benzocyclobutenes as a synthetic method for preparing multibridged cyclophanes. Of the twelve categories of [2$_n$]cyclophanes listed in Scheme 1, six have now been synthesized by this technique, including the most highly bridged ones.

13   **22** (53%)   **65**

66   **17**

8 (~10%)

In view of the usefulness of Gray's basic idea, the question of why his original proposal of a one-step synthesis of [2₄](1,2,4,5)cyclophane (8) had not been successful seemed worthy of reexamination. The obvious explanation seems to be that the conditions used in his experiment were those best suited for thermal elimination of hydrogen chloride from *o*-methylbenzyl chlorides to give benzocyclobutenes. These conditions, though, of high-temperature, gas-phase pyrolysis at low pressure would not be expected to provide a sufficient number of bimolecular collisions to promote the second step of dimerization. Thus it seemed possible that a two-step synthesis of [2₄](1,2,4,5)cyclophane (8) could be devised if appropriate conditions were provided for each step.

This has proved to be the case.[32] Repetition of the pyrolysis of 2,5-bis-(chloromethyl)-1,4-dimethylbenzene (13) at 700°C and 10^{-2} mm Hg pressure gave benzo[1,2;4,5]dicyclobutene (22) in 53% yield, just as Gray had found. A second pyrolysis of benzo[1,2;4,5]dicyclobutene (10), using the nitrogen-flow system at atmospheric pressure to allow control of the concentration in the hot zone of the furnace and so increase the number of bimolecular collisions, led directly to [2₄](1,2,4,5)cyclophane (8), presumably via the intermediates 65, 66, and 17, as shown.

Under the conditions explored thus far, the yield of [2₄](1,2,4,5)cyclophane (8) is only about 10%. In addition to 8, which is a dimer of 22, there are formed higher molecular weight products including trimers, tetramers, pentamers, and polymers. These higher molecular weight products appear to represent novel structures of interest in their own right and are being examined further.[33]

V. Further Variations

There are, of course, many other variations possible on this basic theme of thermal formation and dimerization of benzocyclobutenes. Extension to other aromatic hydrocarbons is one obvious possibility. Another is the extension of the method to aromatic heterocycles. As shown in Scheme 9, 4,13-diaza[2₄](1,2,4,5)cyclophane (79) and 4,16-diaza[2₄](1,2,4,5)cyclophane (78) have been readily prepared in this way.[34]

The dimerization of the pyridinocyclobutene derivative 70 by nitrogen-flow, gas-phase pyrolysis proceeds in good overall yield (48%) giving both possible isomers, 72 and 75, in essentially equivalent amounts. This is somewhat surprising because an *o*-xylylene intermediate, such as 71, would have been expected to show a greater selectivity in its dimerization. This behavior is again more readily explained if the intermediate is in fact a pyridine diradical.

Although the syntheses of 78 and 79 illustrate the generality of the method, the reasons for preparing these compounds were more profound than that.

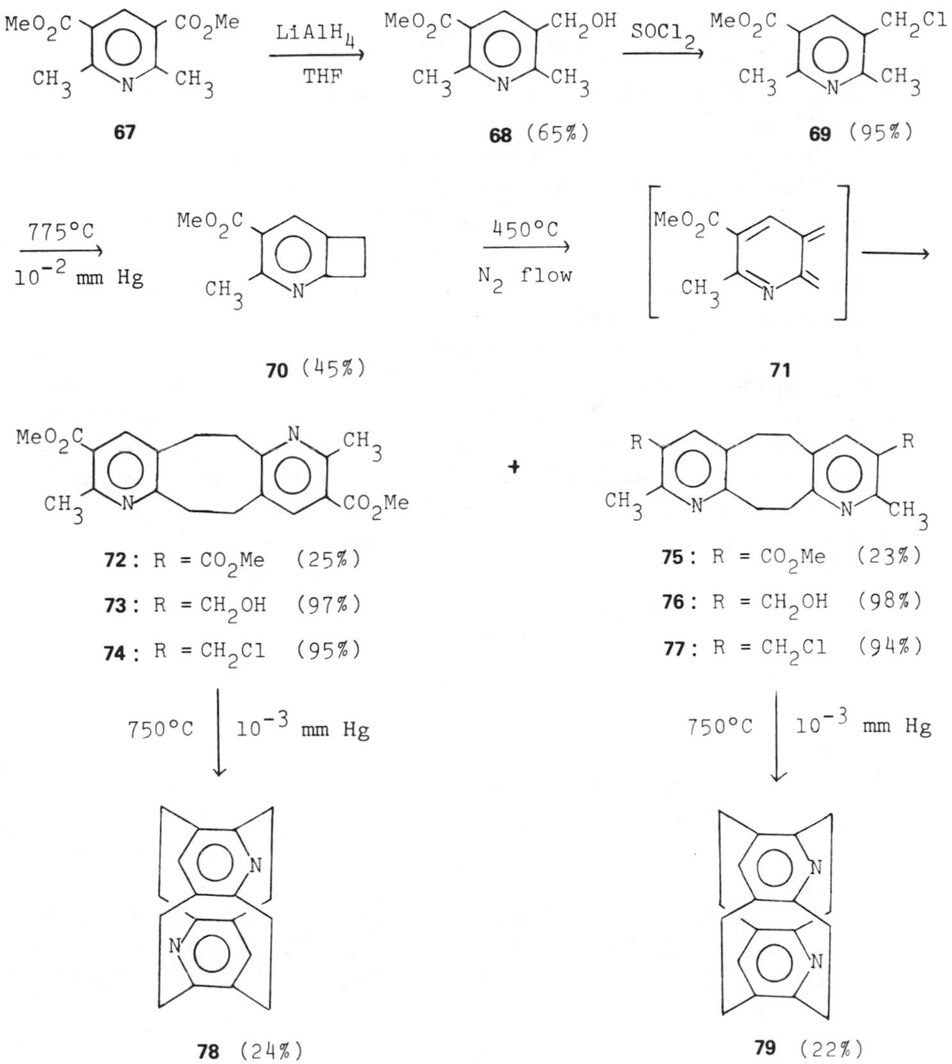

67 $\xrightarrow[\text{THF}]{\text{LiAlH}_4}$ **68** (65%) $\xrightarrow{\text{SOCl}_2}$ **69** (95%)

$\xrightarrow[10^{-2} \text{ mm Hg}]{775°C}$ **70** (45%) $\xrightarrow[N_2 \text{ flow}]{450°C}$ **71** $\longrightarrow$

72: R = CO$_2$Me   (25%)

73: R = CH$_2$OH   (97%)

74: R = CH$_2$Cl   (95%)

\+

75: R = CO$_2$Me   (23%)

76: R = CH$_2$OH   (98%)

77: R = CH$_2$Cl   (94%)

$\xrightarrow[10^{-3} \text{ mm Hg}]{750°C}$

$\xrightarrow[10^{-3} \text{ mm Hg}]{750°C}$

78 (24%)

79 (22%)

Scheme 9

We were interested in studying not only the π–π orbital interaction of two pyridine rings held in very close proximity but also the interaction between the nitrogen-lone pair electrons in 4,13-diaza[2$_4$](1,2,4,5)cyclophane (**79**). Photoelectron spectral studies have now shown that the bands due to electron ejection from the lone pair on nitrogen are split by 0.8 eV in the case of 4,13-diaza[2$_4$](1,2,4,5)cyclophane (**79**), whereas no such split can be detected in the case of the 4,16-diaza[2$_4$](1,2,4,5)cyclophane (**78**).[35]

This remarkable difference in properties is also reflected in the difference in basicity of the two isomers,[32] the pK_a values for the 4,13-diaza isomer **79** are $pK_{a1} = 7.71$ and $pK_{a2} = 2.81$, whereas the values for the 4,16-diaza isomer **78** are $pK_{a1} = 7.23$ and $pK_{a2} = 5.70$.[36]

Another variation is that involving heteroanalogs of benzocyclobutene. Preliminary investigations of their behavior have been carried out with the oxygen, nitrogen, and sulfur analogs.[37,38] For these heteroanalogs, the pyrolysis is best carried out with the corresponding benzyl alcohols rather than the benzyl chlorides. This is shown for the sulfur analog, which most nearly resembles its hydrocarbon counterpart both in its formation and dimerization.[37] Thus benzo[b]thiete (**82**) is formed cleanly and in good yield by gas-phase pyrolysis of the readily available *o*-mercaptobenzyl alcohol (**80**), presumably via the intermediate **81**. Apparently, valence tautomerization of benzo[b]thiete (**82**) to **81** occurs very easily. Simply heating **82** in boiling xylene converts it in very high yield to the corresponding dimer **83**.

In summary, pursuit of a novel idea has led to the development of a general synthetic method of which a synthesis of superphane is only one example.

REFERENCES

1. D. J. Cram and H. Steinberg, *J. Am. Chem. Soc.* **73**, 5691–5704 (1951).
2. Y. Sekine, M. Brown, and V. Boekelheide, *J. Am. Chem. Soc.* **101**, 3126 (1979).
3. M. Pellegrin, *Rec. Trav. Chim. Pays-Bas* **18**, 458 (1899).
4. H. E. Winberg, F. S. Fawcett, W. E. Mochel, and C. W. Theobald, *J. Am. Chem. Soc.* **82**, 1428–1435 (1960).

5. F. Vögtle, *Angew. Chem.* **80,** 258 (1969); *Angew Chem., Int. Ed. Engl.* **8,** 274 (1969); *Chem. Ber.* **102,** 3077–3081 (1969).

6. R. H. Mitchell and V. Boekelheide, *J. Am. Chem. Soc.* **92,** 3510–3512 (1970); *J. Chem. Soc., Chem. Commun.* pp. 1555–1557 (1970); *Tetrahedron Lett.* pp. 1197–1202 (1970).

7. R. H. Mitchell and V. Boekelheide, *J. Am. Chem. Soc.* **96,** 1547–1557 (1974).

8. H. Hopf, *Angew. Chem.* **84,** 471–472 (1972); *Angew. Chem., Int. Ed. Engl.* **11,** 419–420 (1972).

9. R. Gray and V. Boekelheide, *Angew Chem.* **87,** 138 (1975); *Angew. Chem., Int. Ed. Engl.* **14,** 107–108 (1975).

10. H. Hart and R. W. Fish, *J. Am. Chem. Soc.* **82,** 749–751 (1960).

11. H. Hart, J. A. Hartlage, R. W. Fish, and R. R. Rafos, *J. Org. Chem.* **31,** 2244–2250 (1966).

12. A. G. Loudon, A. Maccoll, and S. K. Wong, *J. Am. Chem. Soc.* **91,** 7577–7580 (1969).

13. M. P. Cava and A. A. Deana, *J. Am. Chem. Soc.* **81,** 4266–4268 (1959).

14. R. Gray, L. G. Harruff, J. Krymowski, J. Peterson, and V. Boekelheide, *J. Am. Chem. Soc.* **100,** 2892–2893 (1978).

15. R. Gray and V. Boekelheide, *J. Am. Chem. Soc.* **101,** 2128–2136 (1979).

16. L. G. Harruff, M. Brown, and V. Boekelheide, *J. Am. Chem. Soc.* **100,** 2893–2894 (1978).

17. P. Schiess and M., Heitzmann, *Helv. Chim. Acta* **61,** 844–847 (1978).

18. V. Boekelheide and G., Ewing, *Tetrahedron Lett.* pp. 4245–4248 (1978).

19. V. Boekelheide and E. D., Laganis, unpublished work.

20. E. A. Truesdale and D. J. Cram, *J. Am. Chem. Soc.* **95,** 5825–5827 (1973).

21. E. A. Truesdale and D. J. Cram, *J. Org. Chem.* **45,** 3974–3981 (1980).

22. S. Trampe, H. Hopf, and Menke, K., *Chem. Ber.* **110,** 371–372 (1977).

23. G. D. Ewing and V. Boekelheide, *Chem. Commun.* pp. 207–208 (1979).

24. W. G. L. Aalbersberg and K. P. C. Vollhardt, *Tetrahedron Lett.* pp. 1939–1942 (1979).

25. B. Neuschwander and V. Boekelheide, *Isr. J. Chem.* **20,** 288–290 (1980).

26. P. Schirch and V. Boekelheide, *J. Am. Chem. Soc.* **103,** 6873–6878 (1981).

27. Y. Sekine and V. Boekelheide, *J. Am. Chem. Soc.* **103,** 1777–1785 (1981).

28. J. Kleinschroth and H. Hopf, *Angew. Chem.,* **91,** 336 (1979); *Angew. Chem., Int. Ed. Engl.* **18,** 329 (1979).

29. B. Kovac, M. Mohraz, E. Heilbronner, V. Boekelheide, and H. Hopf, *J. Am. Chem. Soc.* **102,** 4314–4324 (1980).

30. P. H. Scudder, V. Boekelheide, D. Cornutt, and H. Hopf, *Spectrochim. Acta, Part A* **37A,** 425–435 (1981).

31. R. G. Finke, R. H. Voegeli, E. D. Laganis, and V. Boekelheide, *Organometallics* **2,** 347 (1983).

32. B. E. Eaton, E. D. Laganis, and V. Boekelheide, *Proc. Natl. Acad. Sci. U.S.A.* **78,** 6564–6566 (1981).

33. V. Boekelheide and B. E. Eaton, unpublished work.

34. H.-C. Kang and V. Boekelheide, *Angew. Chem.* **93,** 587 (1981); *Angew. Chem., Int. Ed. Engl.* **20,** 571–572 (1981).

35. Y. Zhong-Zhi, E. Heilbronner, H.-C. Kang, and V. Boekelheide, *Helv. Chim. Acta* **64,** 2029–2035 (1981).

36. We thank K. Jäkel of the Ciba-Geigy Corporation for making these measurements.

37. Y.-L. Mao and V. Boekelheide, *Proc. Natl. Acad. Sci. U.S.A.* **77,** 1732–1735 (1980).

38. Y.-L. Mao and V. Boekelheide, *J. Org. Chem.* **45,** 1547–1548 (1980).

Chapter 2

THE TOTAL SYNTHESIS
OF GIBBERELLIC ACID

Rick L. Danheiser

Department of Chemistry
Massachusetts Institute of Technology
Cambridge, Massachusetts

Toward the end of 1978 a pair of communications from the laboratory of
E. J. Corey[1,2] at Harvard University appeared in the *Journal of the American
Chemical Society* announcing the realization of "one of the more intriguing and
salient objectives in the area of organic synthesis." Extensive investigations
spanning more than a decade of research had culminated with the first total

synthesis of the plant hormone gibberellic acid. As a graduate student at Harvard it was my privilege to participate in the final 5 years of this landmark effort, during which period studies on the ultimately successful route to gibberellic acid were initiated, and at last brought to fruition. Here then is a definitive account of this investigation: one of the more celebrated achievements in the annals of modern synthetic organic chemistry.

I. Introduction

Cultivators of rice throughout the world have long feared a peculiar disease of the rice seedling, which is characterized by a sudden and dramatic elongation of the stems of afflicted plants. Infected seedlings often attain heights twice that of healthy plants before they turn brown, wilt, and finally die. Historically this malady has had particularly serious economic and social consequences in Japan, where it is known as the *baka-nae* ("foolishly overgrown seedling") disease.

Early studies carried out in Japan identified a secondary metabolite of the fungus *Gibberella fujikuroi* as the causative agent of the *baka-nae* disease. Following the Second World War, extensive investigations were undertaken in Japan, at Imperial Chemical Industries, Ltd., and at the U.S. Department of Agriculture to elucidate the structure of the major active principle of the fungal toxin.[3-6] By 1959 the planar structure of "gibberellic acid" (gibberellin A_3–"GA_3") was in hand,[7] and remaining stereochemical ambiguities were resolved by X-ray diffraction studies carried out independently in Glasgow[8] and at Harvard.[9] Structure **1** thus represents the absolute configuration of gibberellic acid.

1

Gibberellic acid is the most widely distributed and one of the most structurally complex members of the gibberellins—a family of more than 60 diterpenoid acids now recognized to constitute a fundamental class of phytohormones exercising a wide range of physiological functions in all green plants. The extraordinary biological effects of the gibberellins have been widely exploited to the benefit of agriculture. Gibberellic acid is produced industrially by stirred fermentation of *G. fujikuroi* and has been applied commercially to stimulate the growth of sugarcane and to assist in the malting of barley and the production of a variety of fruits and vegetables.

II. The Synthetic Challenge

The striking biological properties and novel structural features of the gibberellins attracted the attention of organic chemists during the late 1950s. Serious and intensive efforts commenced with the elucidation of the structure of gibberellic acid in 1963. Although by 1977 the literature had witnessed over 150 papers (from more than 25 different research groups) relating to gibberellin synthesis, the total synthesis of the most prominent member of the class, gibberellic acid itself, had not yet been realized.

Why has gibberellic acid proved to be such an elusive synthetic target? A detailed examination of the structure of the molecule will prove valuable in appreciating the special problems peculiar to the synthesis of the gibberellins.

Admittedly, the tetracyclic carbon skeleton of gibberellic acid presents no insurmountable synthetic challenge. Perhaps worthy of note is the presence in the gibberellin skeleton of the unusual bicyclo[3.2.1]octane unit, which has stimulated the design of many ingenious methods for the construction of such systems (see below).

The structure of gibberellic acid includes eight chiral centers. In reality we need only concern ourselves with six of these, because the lactone and D-ring bridges* are each constrained to be cis to the molecular framework. The B and C rings of gibberellic acid join in cis fusion: normally the more stable arrangement in simple hydrindanes. Unfortunately, in the BCD tricyclic system of gibberellic acid the cis-fused isomer appears to be considerably more strained than the trans, and this point has important consequences in the chemistry of molecules incorporating this system.

The C-6 carboxyl substituent in gibberellic acid possesses the pseudo-equatorial β configuration, and it is useful to note that in several gibberellin degradation products a mixture of epimers at this position may be equilibrated entirely to the natural configuration. On the other hand, the hydroxyl function at C-3 has the contrathermodynamic β (axial) orientation, and reduction of a keto function at this position leads predominantly to the unnatural equatorial α-alcohol. All in all, a certain lack of "stereochemical communication" between the left and right portions of the molecule serves to complicate the design of highly stereocontrolled gibberellin syntheses.

Ultimately, the formidable character of the gibberellic acid problem must be attributed to the unique juxtaposition of several reactive functional systems

* In this account the following designations are used:

within the structure of the molecule: a "singularly diabolical placement and density of functionality."[1] Consider first the notoriously sensitive allylic tertiary alcohol moiety of the C and D rings. Exposure of this system to any conditions that develop significant electron deficiency at C-16 initiates a rearrangement of the C-12—C-13 bond, resulting in the formation of a new bicyclo[3.2.1]octane with inverted D-ring configuration.

The reactivity of the gibberellin A-ring functionality rivals that of the C and D rings. Upon standing in distilled water at pH 7, gibberellic acid slowly undergoes a self-induced acid-catalyzed trans elimination of the lactone bridge, affording gibberellenic acid (2).[10] Treatment of gibberellic acid with dilute mineral acid or boiling water leads to a decarboxylative elimination of water and CO_2, yielding allogibberic acid (3), which possesses an aromatic A ring and inverted stereochemistry at C-9.[11] This epimerization, which occurs via protonation of a C-9—C-10 unsaturated derivative, leads to the thermodynamically favored trans B–C ring fusion. The C-4—C-10 lactone bridge of gibberellic acid experiences an allylic transposition $(1 \rightarrow 4)$ when exposed to cold dilute aqueous alkali.[12] Finally, gibberellins containing saturated A rings (e.g., GA_1, 5) undergo a remarkably facile epimerization of the C-3 hydroxyl upon treatment with base;[12] retroaldol cleavage of the C-3—C-4 bond leads to the aldehyde 6, which recyclizes to the more stable, equatorial alcohol.[13]

Thus, as the second decade of synthetic efforts began, the total synthesis of GA_3 remained an elusive goal. To be sure, the synthesis of several simpler gibberellins and degradation products had been achieved, but these routes in some instances required in excess of 50 steps and depended on the availability of relay compounds.[14] As yet the multiple reactive functional systems of gibberellic acid had confounded all synthetic assaults on the parent compound itself.

III. The Synthetic Strategy

Where does one begin in performing an antithetic analysis on so complex a synthetic target as gibberellic acid? The strategy that would eventuate in the first total synthesis of GA_3 began to take shape in 1970.[15] Well aware of the perverse antics that characterize the A-ring functionality of gibberellic acid, we elected to postpone the elaboration of this sensitive system until the final stages of the synthesis. Several considerations suggested that the diene acid system in 8 might serve as an attractive precursor to the gibberellic acid A ring. First of all, this compound was anticipated to be readily available through degradation of gibberellic acid, thus allowing us to examine the last steps of the synthesis without consuming valuable synthetic material. This expectation was confirmed experimentally[16]: treatment of the monotosylate derivative of methyl gibberellate with sodium bromide in hexamethylphosphoramide produced a mixture of epimeric bromides (**10**), which afforded the desired key intermediate **8** (henceforth referred to as the "Corey–Carney acid") upon exposure to excess activated zinc in ethanol.

We regarded the Corey–Carney acid as a particularly versatile intermediate for gibberellin synthesis. Our plans for the elaboration of the A ring revolved around an electrophilically induced lactonization involving the carboxyl group and the diene system. Although the regiochemical course of such a process could not be predicted with certainty, routes leading to gibberellic acid could be conceived for each of the most likely γ-lactone products. Note that in this strategy the neighboring C-4 carboxyl group serves an essential role in controlling the stereochemical course of electrophilic additions to the diene system, thus allowing the stereospecific introduction of the A-ring substituents at C-3 and C-10.

Continuing our retrosynthetic analysis, we considered several routes to the Corey–Carney acid (**8**) that would utilize highly stereocontrolled pericyclic reactions. For example, one such process would involve the intramolecular cycloaddition of the acetylenic ester **13**. The intramolecular Diels–Alder cyclization of this intermediate would lead, after alkylation with methyl iodide, to the lactone **11**, whose transformation to the Corey–Carney acid would simply require oxidation and epimerization of the appendage at C-6.

This intramolecular Diels–Alder reaction is really the pivotal step in our synthetic strategy: in addition to actually constructing the A ring, it establishes the crucial stereochemistry at C-5 and C-4 and therefore (indirectly, see above) at C-3 and C-10 as well. An interesting feature of this strategy is that the C-6 appendage in **13** must possess the unnatural α configuration! In the concerted cycloaddition step, the dienophilic acetylene bond is then constrained to react with the α face of the diene system, resulting in the desired stereochemistry at C-5. The presence of the additional lactone ring in the enolate derived from **12** then "warps" the molecule so that only the β "convex" face of the anion is exposed to alkylating agent (**12 → 11**). Finally, as mentioned above, the correct stereochemistry at C-6 can later be established by epimerization of an appropriate carbonyl derivative to the thermodynamically favored β orientation.

At this stage of our strategic planning we envisioned two alternative approaches to the synthesis of the key cycloaddition precursor **14**. In both routes the D ring is completed by coupling carbons 13 and 16, so that the stereochemistry of this bridge is determined by the orientation of the angular substituent in the cis-fused BC intermediates **16** and **20**. One of the proposed strategies would thus proceed through the cis-decalone **16**, which itself could be prepared in stereocontrolled fashion via an intermolecular Diels–Alder reaction employing a suitably dienophilic cyclohexene derivative. This approach would later require the contraction of ring B; our second strategy would employ the tricyclic intermediate **19** and require the regiospecific attachment of two appendages to the B ring. We shall defer to later sections further discussion of the means by which we envisioned executing these B-ring transformations. Let it suffice to say here that the conversion of both **15** and **19** to the key cycloaddition precursor **14** proved feasible in the end.

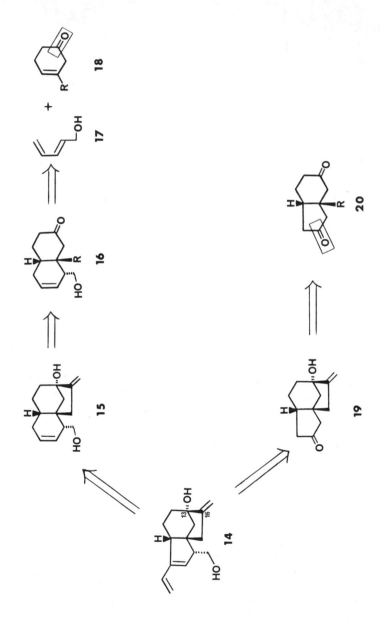

IV. Preliminary Studies: Stereocontrolled Elaboration of the A Ring

Before embarking on what augured to be a long and arduous synthetic journey, it seemed wise to us first to establish the viability of the Corey–Carney acid as a key intermediate for the elaboration of the gibberellic acid A ring.[16] An ample supply of the Corey Carney acid was therefore produced by degradation of methyl gibberellate[17] (according to the procedure detailed in the previous section), and a systematic investigation of the reactions of this substance with various electrophilic species was undertaken. Not unexpectedly, most electrophiles attacked the more accessible C-2—C-3 double bond in **8**, and the resulting cationic intermediates were intercepted by the carboxyl group at the less sterically hindered C-2 position. For example, careful oxidation of the Corey–Carney acid with *m*-chloroperbenzoic acid in methylene chloride at −25°C furnished the hydroxy lactone **21** in 76% yield. Although

this compound possesses the requisite C-3 β-hydroxyl group, it unfortunately also contains the C-4 → C-2 lactone bridge found in "isogibberellic acid" (**4**), the product previously generated from the rearrangement of GA$_3$ in aqueous base (see above). We therefore now required a method for effecting the reverse allylic transposition of the A-ring lactone bridge.

This appeared to be a feasible transformation. For example, saponification of **21** and iodolactonization of the resulting carboxylic acid would generate the kinetically favored γ-lactone **23**; reductive vicinal elimination of the elements of IOH would then introduce the C-1—C-2 double bond, and provide methyl gibberellate. In the event, hydrolysis of **21** was achieved by exposure to excess ethanolic sodium hydroxide, and the resulting acid was treated with 1.1 equivalents of iodine in a mixture of aqueous sodium bicarbonate, tetrahydrofuran, and methylene chloride to generate the crystalline iodo lactone **23** in 60% overall yield. Not surprisingly, several attempts to effect bromolactonization of **22** led to the competitive rearrangement of the CD ring system.

In order to effect the necessary vicinal elimination, it next proved essential to derivatize the C-2 hydroxyl in order to render it a better leaving group than the lactone bridge. This was most easily achieved by converting it to the corresponding trifluoroacetic ester (**24**). The conversion of **23** to methyl

21   **22**

23 : R = H
24 : R = COCF$_3$

1

25 : R = COCF$_3$
26 : R = H

gibberellate was thus accomplished in one flask by the following sequence of reactions: (a) trifluoroacetylation with excess trifluoroacetic anhydride in pyridine–tetrahydrofuran at 0°C for 1 h, (b) elimination by addition of excess zinc dust (0°C, 1 h), and (c) hydrolysis of the resulting product (**25**) by addition of 10% aqueous sodium bicarbonate solution (25°C, 14 h). Methyl gibberellate was obtained in 90% overall yield from **23** in this fashion, and was then converted to GA$_3$ by treatment with lithium n-propylmercaptide according to the procedure of Johnson and Bartlett.[18] The viability of the Corey–Carney acid as a key intermediate in our strategy was thus unequivocally established.

V. Model Studies: The Intramolecular Diels–Alder Strategy Is Tested

Crucial to our plans for the synthesis of the Corey–Carney acid was the intramolecular [4 + 2] cycloaddition **13** → **12**. As we pondered our next move in 1972, we noted that remarkably few examples of the *intramolecular* version of the Diels–Alder reaction had been reported. We therefore deemed

it advisable to examine the feasibility of this pivotal reaction in our strategy
before proceeding further with the total synthesis of gibberellic acid.[19] We
chose as our objective the model compound **27**, whose structure incorporates
the AB ring system of the Corey–Carney acid.

    27   **28**   **29**

Thus, the monocyclic cycloaddition precursor **36** was assembled without
incident, employing the following sequence. Condensation of diethyl sodiomalonate and diethyl 1,1-cyclopropanedicarboxylate in refluxing ethanol
for 12 h afforded 2,5-diethyl-1-cyclopentanone dicarboxylate (**31**) in more
than twice the previously reported yield.[20] This interesting one-pot transformation proceeds via the homoconjugate addition of malonate to **30**,
Dieckmann cyclization of the resulting tetraester, and decarbethoxylation.
Treatment of **31** with 1.2 equivalents of sodium borohydride in ethanol
($-24°C$, 17.5 h) then furnished the alcohol **32** in 90% yield.

  30   **31**   **32**

The transformation of **32** to the unsaturated diester **33** was facilitated by
prior conversion of the hydroxyl to the corresponding methanesulfonate
ester. Thus, exposure of **32** to methanesulfonyl chloride and diazabicyclo-
[5.4.0]undec-5-ene in benzene (0.5 h at 0°C, 5 h at room temperature)
afforded **33**[21] in more than 90% yield after distillation. Reduction of the
unsaturated diester with diisobutylaluminum hydride in toluene (0°C, 2.5 h)
or alane in ether (room temperature, 4 h) then furnished the diol **34**. Selective
oxidation of the allylic hydroxyl of **34** was accomplished using manganese

      32   **33**

34 : R = CH_2OH
35 : R = CHO
36 : R = $CH=CH_2$

dioxide in methylene chloride, and the resulting α,β-unsaturated aldehyde was transformed to **36** by treatment with 3 equivalents of methylenetriphenylphosphorane in tetrahydrofuran. The dienol **36** was thus obtained in 45% overall yield from **33**.

The esterification of this alcohol with propiolic acid proved unusually challenging. Reactions of derivatives of propiolic acid are often complicated by their proclivity toward polymerization, generally initiated by the conjugate addition of nucleophilic species. Propiolic anhydride is reportedly[22] a rather intractable substance (". . . . *es ist eine widerwärtig und stechend riechende Flüssigkeit von begrenzter Haltbarkeit . . . . bei der Destillation aber nur dann Neigung zu explosions-artiger Zersetzung zeigt, wenn sie gewisse Verunreinigungen enthält*"*) and the preparation and characterization of the similarly unstable propiolyl chloride was not achieved until a year after our investigation![23] We consequently focused our attention on esterification methods employing reactive propiolate derivatives that could be generated *in situ* under very mild conditions.

A systematic investigation of a wide variety of coupling reagents showed the dicyclohexylcarbodiimide (DCC)-mediated procedure[24] to be uniquely effective for the required esterification. Treatment of the dienol **36** with propiolic acid and 1.2 equivalents of DCC in methylene chloride at 0°C thus afforded the desired ester **29** in 80–85% yield after chromatographic purification. The stage was now set for the first real test of our intramolecular Diels–Alder approach to the construction of the gibberellin A ring.

When heated in benzene solution in a sealed tube at 150°C for 22 h, the acetylenic ester smoothly underwent the desired intramolecular Diels–Alder reaction to provide a single cycloadduct **28**. This unstable compound was best converted to **37** without purification because of its propensity toward aro-

matization. Alkylation was readily accomplished by adding **28** to a solution of 1.05 equivalents of lithium isopropylcyclohexylamide and 3 equivalents of HMPT in tetrahydrofuran at −78°C, and then treating the resulting mixture with excess methyl iodide.[25] Isolation and purification of the product by chromatography afforded a single crystalline lactone in 72% overall yield from **29**.

* It is a repulsive and acrid-smelling liquid of limited stability . . . it shows a tendency to explode on distillation if it contains certain impurities.

This substance exhibited spectral characteristics in complete accord with the stereochemistry predicted from steric considerations discussed earlier. Thus, as anticipated, *intramolecularity* ensures that the Diels–Alder reaction produces exclusively the cycloadduct with the desired regiochemistry and stereochemistry.

At this point, with characteristic prescience, Professor Corey suggested that we examine an alternative Diels–Alder reaction employing the β-chloroacrylyl ester **38**. Cyclization of this compound and subsequent dehydrochlorination would provide a "back-up" route to the key lactone **28**.

Unfortunately, we could obtain only moderate yields of **38** by employing the DCC method or by reaction of dienol **36** with (*E*)-2-chloroacrylyl chloride in the presence of various amines. Satisfactory results were eventually achieved by treating the lithium alkoxide derived from **36** with 1.25 equivalents of the acid chloride in tetrahydrofuran at −78 to −40°C. The desired ester was obtained in 83% yield in this manner. Upon heating in benzene at 160°C for 34 h, this compound then cyclized to afford the endo Diels–Alder cycloadduct **39** in 85% yield. Finally, treatment of **39** with 2 equivalents of lithium isopropylcyclohexylamide at −78°C, followed by excess methyl iodide, furnished the same lactone (**37**) produced previously via cyclization of the propiolyl ester.

Armed thus with two routes to the key intermediate **37**, we next turned our attention to the conversion of this lactone to the Corey–Carney acid model compound (**27**). In essence, this transformation simply requires hydrolysis of the lactone, oxidation of the B-ring hydroxymethyl group, and epimerization of this substituent to the thermodynamically favored β configuration. In reality this transformation proved far from simple. Indeed, exposure of **37** to hydroxide did cleanly effect hydrolysis of the lactone; however, we were then unable to achieve the oxidation of the hydroxymethyl group in the resulting carboxylate salt without also inducing substantial relactonization. We consequently devised an alternative scheme enlisting the diene system of ring A to internally protect the carboxyl group during the required operations on ring B.

An aqueous solution of the potassium salt obtained by saponification of **37** with potassium hydroxide in anhydrous methanol was neutralized to pH 7 at 0°C with carbon dioxide and then treated with 1 equivalent of aqueous potassium tribromide solution at −10°C for 1 min. Two bromo lactones were

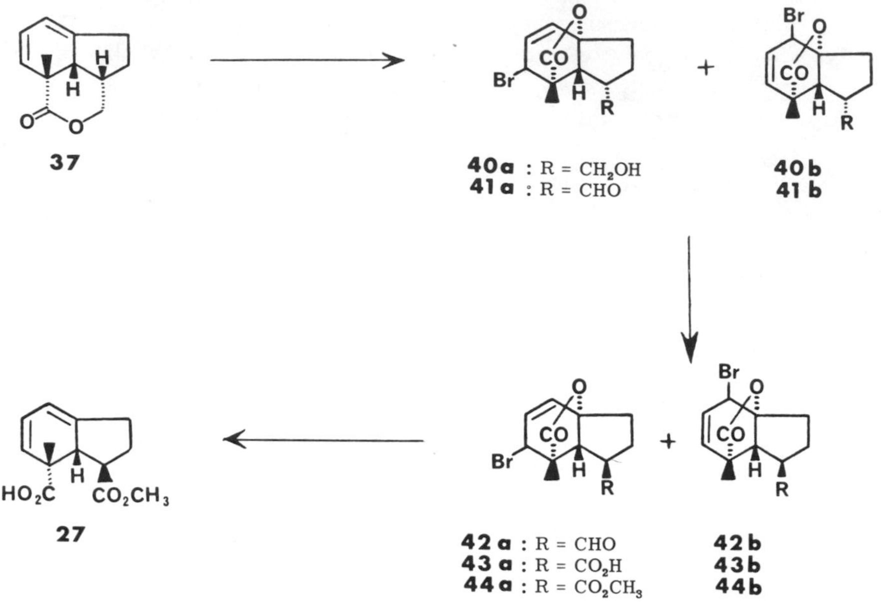

37

40a : R = CH$_2$OH
41a : R = CHO

40b
41b

42a : R = CHO
43a : R = CO$_2$H
44a : R = CO$_2$CH$_3$

42b
43b
44b

27

isolated and identified as **40a** and **40b** on the basis of their IR and NMR spectra. Oxidation of this mixture with Collins reagent in methylene chloride at −25°C then gave the aldehydes **41a–41b**, which epimerized to **42a–42b** during chromatographic purification on silica gel. The overall yield of epimerized aldehydes from **37** ranged from 45 to 56%. Oxidation of **42a–42b** with Jones reagent and esterification of the crude acids with ethereal diazomethane next gave the methyl esters **44a–44b**. Finally, exposure of these esters to excess activated zinc in anhydrous ethanol (room temperature, 2 h) regenerated the A-ring diene acid system, providing the crystalline Corey–Carney acid model compound **27** in 63% overall yield from **42a–42b**. A stereocontrolled method for the synthesis of the A and B rings of gibberellic acid was finally in hand.

VI. Model Studies: A Method for the Construction of the C and D Rings

It should not be assumed from the preceding discussion that the right half of the gibberellin molecule was entirely ignored during this preliminary phase of our investigations. In fact, reconnaisance of the CD region of gibberellic acid had begun in 1969, and within 2 years not one, but two efficient routes to the bicyclo[3.2.1]octane unit of GA$_3$ were in hand.[26,27] As alluded to earlier, the synthesis of the BCD tricyclic unit of gibberellic acid poses a formidable synthetic challenge as a consequence of the unexpectedly high degree of steric strain associated with this system. Thus, on more than one occasion the attempted application of well-established routes to bicyclo[3.2.1]-octanes in the construction of the BCD tricyclic system had proved totally futile.[28] In fact, as we began our investigation, the only really promising strategy for the synthesis of the gibberellin D ring was the elegant method of Stork, involving the reductive coupling of acetylenic ketones as exemplified in the transformation **45** → **46**.[28a]

The first D-ring strategy devised at Harvard was an outgrowth of the pioneering research on the application of organocopper compounds in synthesis, which was then reaching a climax in the Corey laboratory. In a variety of model compounds, such as **47**, it was observed that exposure to several equivalents of di-*n*-butylcopperlithium in ether at low temperature resulted in efficient cyclization to a bicyclo[3.2.1]octane derivative replete with the gibberellin tertiary allylic alcohol functionality.[26]

Further studies in our laboratory during this period focused on the intramolecular reductive cyclization of keto aldehydes such as **49**. Oxidation and Wittig methylenation of the resulting pinacolic coupling product would provide an efficient alternative approach to the gibberellic acid BCD tricyclic unit. However, the lack of any precedent for *intramolecular* pinacolic coupling reactions of this type prompted us to examine the feasibility of this strategy, using the model system **49**.[27]

With some effort (see below) a stereoselective route to this model substrate was devised, and we were soon after able to demonstrate that the proposed coupling reaction could indeed be achieved by the action of various reducing agents, among which magnesium amalgam in tetrahydrofuran proved most efficacious. Crucial to the success of this process is the inclusion of 2 equivalents of dimethyldichlorosilane in the reaction medium. In the absence of this silylating agent, the magnesium alkoxides generated during the course of cyclization catalyze the internal aldol cyclization of the keto aldehyde substrate, so that in fact the ketol **52** becomes the preponderant product of the reaction.[29]

52

VII. The Hydrindane Approach

By the autumn of 1973 we had laid a solid groundwork for each of the most crucial stages of our proposed synthesis, and the prospects for the expeditious and successful implementation of our strategy appeared to be excellent. The feasibility of our intramolecular Diels–Alder strategy for the development of the AB-ring system was securely established; we had clearly demonstrated the viability of a scheme for the elaboration of the sensitive A-ring functionality; and we now had at our disposal two efficient methods by which the strained CD-ring network could be reproduced. The time had arrived at last to hurdle the parapet and commence the actual synthetic assault on gibberellic acid.

We chose as our first objective the tricyclic ketone **19**. Our plan for the synthesis of this key intermediate centered on the application of our pinacolic coupling strategy to a keto aldehyde of the type represented by **53**, which

14   **19**   **53**   **54**

would in turn be assembled by the stereocontrolled conjugate addition of an acetaldehyde equivalent to an appropriate hydrindenone derivative. Functionalization of indene would provide ready access to **54**.

The oxidation of indene with 40% peracetic acid in the presence of sodium carbonate thus placed in our hands an ample supply of indene oxide (55). Addition of this epoxide in tetrahydrofuran–*tert*-butanol to a solution of lithium in liquid ammonia brought about the selective cleavage of the benzylic oxirane bond with concomitant Birch reduction of the aromatic

56 : R = H
57 : R = CH$_2$Ph

ring. Distillation provided us with the diene alcohol 56 in 70% overall yield from indene. Next, the hydroxyl group in 56 was protected as the benzyl ether derivative, and the more accessible double bond of the resulting diene was selectively hydroborated using disiamylborane in tetrahydrofuran. Now it only remained to oxidize this homoallylic alcohol and then to isomerize the double bond into conjugation, and our initial target, a hydrindenone of type 54, would be in hand.

In fact, the oxidation of 58 to the desired β,γ-unsaturated ketone was more difficult than anticipated, as the product enone proved to be unstable in the presence of the more common oxidizing agents. Fortunately, we soon discovered that this transformation could be accomplished without complication by using the recently introduced "Corey–Kim reagent."[30] Thus exposure of 58 to the complex formed from N-chlorosuccinimide and dimethyl sulfide in toluene at $-25°C$, followed by brief treatment with triethylamine, cleanly afforded the desired ketone (59).

A variety of catalysts were examined for the critical isomerization step 59 → 60. In fact, we approached this stage of the synthesis with considerable circumspection, alert to the danger that our product (60) might easily suffer elimination to the dienone 61. In the event, acids and most basic catalysts

were totally unsatisfactory for this transformation, and it was only after considerable experimentation that we found that potassium carbonate in methanol effected the desired conjugation with minimal elimination, provided

that the reaction was carried out at $-30°C$. A 60–65% yield of the beautifully crystalline α,β-enone could be obtained in this manner.

We next directed our attention to the elaboration of the D-ring bridge.
Our strategy called for the addition of a lithium diorganocuprate to enone **60**
to establish the cis B–C ring fusion and simultaneously introduce the remaining two carbons of the D ring. The addition of lithium dialkylcuprates to
enones is well known to occur via the antiparallel approach of the reagent to
the less hindered face of the conjugated system; in the case of hydrindenones,
this results in the exclusive formation of a cis-fused adduct.[31]

As a versatile two-carbon precursor to the D ring, the vinyl group appeared
best to meet our requirements. In fact, in earlier model studies we had already
demonstrated the utility of "divinylcopperlithium" as an acetaldehyde equivalent for the synthesis of keto aldehyde **49**.[27] In a similar fashion, the more
complex enone **60** smoothly combined with this reagent at $-78°C$ in a mixture
of diethyl ether and diisopropyl sulfide to afford the angularly substituted
hydrindanone **62** in 90% yield. As anticipated, only the cis-fused isomer was
detected as a product of this reaction.

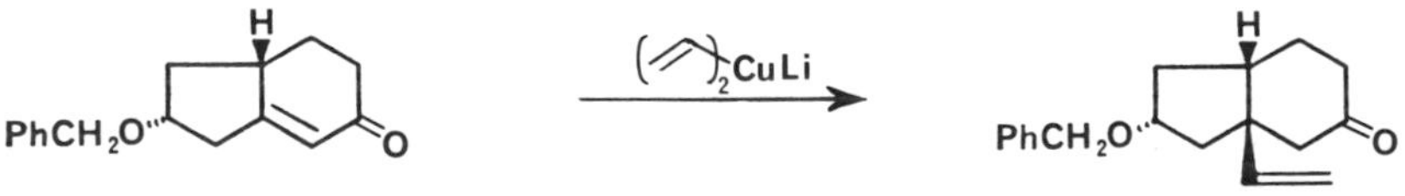

60   **62**

At this point we considered several alternate schemes for accomplishing
the next stage of the synthesis, the oxidation of the vinyl appendage to an
angular $-CH_2CHO$ unit. Two facts with important tactical consequences
emerged from our initial experiments. First, the carbonyl group in **62** exhibited
the greater tendency to combine with electrophiles in comparison to the vinyl
substituent. In particular, reaction of **62** with peracids and with various
hydroborating agents proceeded preferentially at the keto function. Clearly,
protection of this carbonyl would be necessary in order to permit the required
operations on the carbon–carbon double bond.

In selecting a suitable protective group for the ketone, it became necessary
for us to take into consideration a second important point. Our objective,
the keto aldehyde **66**, was observed to be unusually sensitive to both acid and
base, readily undergoing an internal aldol closure analogous to the cyclization
49 → **52** discussed earlier. Consequently, it was not surprising to find that in
this case the most popular carbonyl protective group, the ethylene ketal,
could not be hydrolyzed without further inducing the destruction of the desired
product (that is, **66**) by aldol reaction.

From these considerations evolved the following scheme. Exposure of **62**
to 2-mercaptoethanol in the presence of boron trifluoride etherate furnished

the hemithioketal **63** as a mixture of diastereomers. Regioselective hydroboration of the vinyl group was achieved using disiamylborane in tetrahydrofuran, and the resulting alcohols (**64**) were then oxidized with Collins reagent in methylene chloride at −20°C. Finally, deketalization of **65** occurred instantly, and without aldol cyclization, upon treatment with mercuric chloride and calcium carbonate in aqueous acetonitrile. The keto aldehyde **66** was available in more than 50% overall yield from **62** by employing this sequence.

Thus we arrived at the first pivotal stage in our synthetic strategy: the intramolecular pinacolic coupling reaction that would establish the D ring of gibberellic acid. As rehearsal, we once again repeated the magnesium amalgam-promoted cyclization of a sample of model keto aldehyde **49** and, as usual, obtained the desired stereoisomeric pinacols in 75% combined yield. Then we subjected the "real" keto aldehyde (**66**) to the same protocol. No reaction occurred! Our key intermediate resisted coupling and, under forcing conditions, only displayed a disconcerting tendency to slowly undergo the unprofitable aldol side reaction. In some obscure way the simple substitution of ring B with a benzyloxy group completely inhibited the desired ring closure.

Thereupon we embarked on an odyssey into pinacolic coupling chemistry which would eventually encompass the examination of well over one hundred and fifty different coupling procedures. A decisive turning point in these studies came with our realization that certain low-valent transition metals possess a unique capacity to bring about the desired coupling reaction. The first intimation of this process should be credited to Sharpless, who in 1972 reported that certain low-valent tungsten reagents have the ability to induce the intermolecular reductive condensation of aromatic ketones and aldehydes.[32] Unfortunately, the Sharpless reagent $(WCl_6-n\text{-BuLi})$ does not efficiently couple *aliphatic* carbonyl compounds and hence proved ineffective for the cyclization of keto aldehyde **66**. We therefore undertook a vigorous and systematic investigation of the utility of a wide variety of related low-valent transition metal reagents for the desired pinacolic coupling reaction.

It was at this point that we became aware of the felicitous discovery in the separate laboratories of Mukaiyama and McMurry of the ability of low-valent titanium species to bring about the reductive coupling of carbonyl compounds.[33-35] A few experiments quickly showed that the McMurry reagent $(TiCl_3-LiAlH_4)$ in particular held some promise as a means by which the cyclization of keto aldehyde **66** might be accomplished. Although the McMurry reaction normally results in the reductive coupling of carbonyl compounds to form olefins, in our intramolecular case deoxygenation of the pinacolic intermediate is obviously impeded by the strain associated with the resulting bridgehead double bond.

After considerable further experimentation, we finally arrived at optimal conditions for effecting the required intramolecular reductive coupling:[36]

exposure of **66** to the low-valent titanium species generated by the combination of 6 equivalents of monocyclopentadienyltitanium trichloride with 4.5 equivalents of lithium aluminum hydride in tetrahydrofuran at 50°C provided the crystalline pinacols **67** in 45–50% yield. In 12 steps (beginning with indene) we had assembled a functionalized, tricyclic BCD precursor to gibberellic acid.

Actually, **67** does not represent the first tricyclic intermediate to be prepared in our laboratory. Some time before, when the prospects for the successful execution of the crucial pinacolic coupling appeared most disheartening, we had initiated exploratory studies on an alternative strategy for the synthesis of the key tricyclic BCD intermediate **19**. This alternate approach to the gibberellin D ring was to involve the base-promoted cyclization of a suitable hydrindanone, in which an electron-withdrawing group would activate the angular appendage for anion formation and also serve to facilitate the subsequent introduction of the requisite exo methylene at C-16.

Our initial results did not prove encouraging. We began this investigation by examining the cyclization of the model series hydrindanones **68** and **71**. In the case of the diketone, both base-catalyzed and enamine-mediated aldol cyclizations led exclusively to bicyclo[4.3.0]octane derivatives. For example, the reaction of **68**[37,38] with lithium ethoxide in ethanol at room temperature

produced a mixture of the enone **69** and ketol **70** in 80% yield. Nor did the nitrile **71** undergo the desired mode of ring closure. When exposed to lithium *tert*-butoxide in 10:1 ether–*tert*-butanol at −35°C, the cyano ketone cyclized to form **72** in nearly quantitative yield.

After some deliberation we reached the conclusion that a change in tactics was called for. It had become evident that the desired mode of cyclization was probably disfavored in a thermodynamic sense, as a consequence of the strain attending the desired bicyclo[3.2.1]octane relative to the alternative bicyclo-[4.3.0] system. Clearly then, the success of our strategy would require the deployment of a W group not itself susceptible to nucleophilic attack, and more potent in its electron-withdrawing capacities than the hydrindanone carbonyl.

The nitro group impressed us as almost unique in fulfilling these requirements, and we quickly set about devising a route to the requisite nitro ketone intermediate **73**. To this end the previously described alcohol **64** was converted to the corresponding tosylate derivative, which was then exposed to the action of anhydrous sodium iodide in refluxing acetone. Fortuitously, these conditions effected the cleavage of the hemithioketal protective group in addition to accomplishing the desired nucleophilic substitution. Finally, displacement of iodide with sodium nitrite in dimethyl sulfoxide provided us with the desired cyclization substrate **73**, in nearly 40% overall yield from the vinyl ketone **62**.

We were not too surprised to find next that the outcome of the base-promoted cyclizations of nitro ketone **73** were uncommonly sensitive to the identity of the base and the nature of the reaction medium. With amine bases such as DBU a maximum of only 60% of the thermodynamically disfavored cyclization product could be generated. However, by employing lithium alkoxides in nonpolar media at relatively low temperature, complete cyclization could in fact be achieved. The formation of a stable lithium chelate derivative of **74** under these conditions is no doubt essential to the success of the reaction. Thus, exposure of **73** to a slight excess of lithium ethoxide in a 4:1 mixture of diethyl ether and ethanol at −40°C for 30 min provided **74** as a single crystalline diastereomer in 86% yield after purification by recrystallization.

The next stage of our plan for the elaboration of the D ring called for the replacement of the nitro substituent by an exo methylene, presumably via the intermediacy of a ketone derivative such as **76**. It quickly became apparent that protection of the bridgehead hydroxyl would be crucial for the success of this operation. The nitro alcohol system present in **74** proved exquisitely sensitive to both acid and base, readily reverting to **73** via retroaldol cleavage, a transformation which occurred to a significant extent even upon the attempted purification of this substance by chromatography. Unfortunately, the unusual lability of **74** also worked to frustrate our attempts to contrive some means of protection for the recalcitrant hydroxyl group. Neither acetylation, benzylation, nor the conversion of **74** to an acetal derivative could be achieved without substantial retroaldolization. In the end the acetate derivative **75**

was finally prepared through recourse to the novel, reactive acylating agent trimethylsilylketene.[39,40] Reaction of the alcohol **74** with 1.2 equivalents of this stable ketene in carbon tetrachloride at 0°C gave, after brief treatment with potassium fluoride in methanol,[41] the desired nitro acetate **75** in nearly quantitative yield. Reduction of **75** was then achieved by reaction of the corresponding nitronate anion with titanous chloride as described by Mc-Murry,[42] and the resulting ketone **76** was finally converted to **77** by Wittig methylenation in dimethyl sulfoxide. However, the overall yield for the transformation of **74** to **77** was only 40%, and all efforts to further improve the efficiency of this sequence proved fruitless.

We now had at our disposal two methods for the synthesis of a tricyclic gibberellin intermediate. Unfortunately, the viability of each of these approaches was compromised by key steps that proceeded in only modest yield. Whether either of these routes was efficient enough to support a total synthesis of the projected magnitude appeared to be at best problematical.

An important strategic decision was thus reached. The hydrindane approach would, at least for the time being, be abandoned. We would apply all our resources and all our efforts to an alternative, potentially more efficient strategy, which we had been simultaneously investigating during the preceding 12 months: the hydronaphthalene approach.*

VIII.   The Hydronaphthalene Approach:
(1) The Advance to Chandra's Dione

In the closing months of 1974 we committed all our forces to the hydronaphthalene approach and launched a new synthetic offensive to gibberellic acid along this pathway. The essential features of the hydronaphthalene strategy are set forth in the following scheme. Clearly, the feasibility of this plan would hinge on our ability to execute efficiently the contraction of ring B, preferably via the regiocontrolled aldol cyclization of the key dialdehyde intermediate **79**. This would be a risky operation. It did not escape our notice that a number of destructive side reactions could conceivably intervene at the stage of the crucial aldol transformation and that the desired product might itself prove unstable either to elimination or, with equally detrimental consequences, to epimerization at the appendage at C-6. Nonetheless, we

---

* The further investigation of the hydrindane approaches discussed in this section was never resumed. However, in 1979 Corey and Smith completed an efficient synthesis of the same tricyclic ketone (**19**) that had been the objective of these earlier studies.[43] This intermediate was also successfully transformed to the cycloaddition precursor **14**, which we had by that time prepared and carried on to GA$_3$, using the hydronaphthalene approach described in the following sections of this account. Also noteworthy is the report in 1982 by Corey and Munroe of an astonishingly efficient nine-step synthesis of the tricyclic ketone **19**.[44] Finally, it should also be noted that four ingenious stereocontrolled routes to the same tricyclic ketone have been developed independently in the laboratories of Professor Gilbert Stork at Columbia.[45]

prepared to boldly forge ahead with this approach, convinced that these potential hazards could be circumvented, and that this very attractive and expeditious route would in the end prove to be feasible.

Our plans for the formation of the gibberellin D ring called for another
application of the pinacolic coupling methodology which we had by this time
perfected in the course of our studies on the hydrindane approach to GA_3.
We further envisaged that an intermolecular Diels–Alder reaction could
provide a highly stereocontrolled route to the requisite keto aldehyde intermediate **81**. For example, addition of *trans*-2,4-pentadienol to a dienophilic
cyclohexene derivative (**18**) would generate a hydronaphthalene (**82**) with

precisely the desired stereochemical disposition of substituents. In this transformation the suprafacial course of the Diels–Alder reaction dictates that
only the cis-fused cycloadduct can be formed, and the desired orientation of the
hydroxymethyl appendage is a consequence of the well-known Alder endo
rule.

Several considerations contributed to our selection of the quinone **84** as
the dienophile component for the proposed cycloaddition. First of all, this

"cyclohexene derivative" obviously possesses considerable dienophilic reactivity and also incorporates the requisite latent carbonyl function at the indicated position. Numerous examples[46] have established the reactivity of related quinones as dienophilic partners in Diels–Alder additions, and the application of the Diels–Alder reaction of methoxybenzoquinones to decalone synthesis, first recognized by Woodward,[47] is also well documented.

The substituents incorporated in quinone 84 also provide for the eventual elaboration of the requisite acetaldehyde appendage and, in addition, serve to direct the regiochemical orientation of the cycloaddition. The regiochemical course of the Diels–Alder reaction of substituted quinones has been clarified by several recent studies,[48] and the desired intermediate (85) is predicted to be the predominant cycloadduct generated from the combination of 84 with trans-2,4-pentadienol. Thus, the diene combines with the less electron-rich, alkyl-substituted double bond of the quinone, and the diene hydroxymethyl group assumes a position in the cycloadduct "ortho" to the more electron-deficient quinone carbonyl. This regioselectivity may be viewed as a consequence of a dipolar asymmetric transition state in which the formation of one new bond in the cycloaddition process occurs partially in advance of the other.

Our synthesis of the key bicyclic intermediate 85 (hereafter referred to as "Chandra's dione"*) commenced with o-eugenol (86). We planned to generate the requisite benzyloxyethyl appendage from the allyl group of this compound by employing the sequence: oxidative cleavage, reduction, and benzylation.[49] The phenolic hydroxyl was first acetylated to permit these

operations on the side chain. Johnson–Lemieux oxidation[50] of o-eugenol acetate proceeded smoothly in aqueous tetrahydrofuran to furnish the desired

* After Dr. S. Chandrasekaran, the first member of our group to prepare this substance.

aldehyde; however, subsequent reduction with sodium borohydride in
ethanol afforded the phenol **88** rather than the expected alcohol. Unable to
contrive some means of suppressing this transesterification, we fell back and
modified our tactics along the following lines.

Saponification of **88** afforded the corresponding diol, which was selectively
benzylated at the less sterically encumbered primary hydroxyl to provide the
desired quinone precursor **93**. However, the low yield attending this operation
checked our progress along this otherwise promising route.

Next we examined the protection of the phenolic hydroxyl group in **86** as
various acetal derivatives. For example, the tetrahydropyranyl ether of
o-eugenol was prepared, but proved to be unstable to the Johnson–Lemieux
oxidation conditions. Success was finally achieved through recourse to the
β-methoxyethoxymethyl (MEM) ether group, which had recently been
developed in our laboratory in connection with another aspect of the gibberellic acid problem (see below). Thus exposure of the sodium salt of *o*-eugenol
to 1.6 equivalents of MEM chloride in tetrahydrofuran afforded **89** in 86%

yield following distillation. Oxidative cleavage of this olefin was then accomplished by treatment with 3 equivalents of sodium periodate and a catalytic
amount of osmium tetroxide in aqueous tetrahydrofuran, and the resulting
unstable aldehyde was immediately subjected to reduction with sodium borohydride in ethanol at 0°C. Next, the alcohol **91** was converted without purification to the corresponding benzyl ether, and finally selective cleavage of the
MEM ether group in **92** was effected, using 1.2 equivalents of trifluoroacetic
acid in methylene chloride. The desired quinone precursor (**93**) was obtained
in 74% overall yield from **89** by employing this sequence.

It is a basic tenet of organic synthesis that in the more extended multistage
syntheses, it is frequently logistical considerations that play a decisive role in
determining the ultimate success of the entire synthetic venture. In this connection it is therefore worth noting that in the aforementioned sequence up

to 200 g of *o*-eugenol could be protected in a single operation, and that the resulting product was routinely transformed to **93** in 100-g lots, and without the purification of intermediates along the way.

At this stage we were ready to undertake the oxidation of **93** to the requisite quinone dienophile **84**. Although initially we accomplished this step using 2 equivalents of Fremy's salt {potassium nitrosodisulfonate, $K_2[ON(SO_3)_2]^{51}$} in aqueous methanol, the laborious operations associated with the preparation of this reagent eventually prompted us to examine alternative oxidizing agents for effecting this transformation. The most convenient procedure for achieving this oxidation thus involved stirring a dimethylformamide solution of phenol **93** with 0.08 equivalent of bis(salicylidene)ethylenediiminocobalt-(II) ("salcomine," **94**)[52] under an atmosphere of molecular oxygen for 4 days.

The reaction mixture was then filtered with the aid of ether and acetone through a pad consisting of alternate layers of sand and anhydrous sodium sulfate; concentration of the filtrate afforded the desired quinone **84** as splendid, bright yellow crystals in 65–87% yield.

Thus we arrived at what we viewed as the first of the five pivotal steps in the hydronaphthalene strategy: the Diels–Alder addition of quinone **84** to 2,4-pentadienol. The importance of this intermolecular cycloaddition is

derived from the fact that this reaction establishes the cis junction of what are destined to become rings B and C of gibberellic acid and thereby also provides for the subsequent stereospecific elaboration of the D-ring bridge. Furthermore, this pivotal operation also determines the relative configuration of the hydroxymethyl substituent joined at what will later develop as C-6 of ring B.

It is a cornerstone of our strategy that this stereocenter serves (in the temporary guise of the "unnatural" epimer) as a *stereochemical mediator*, facilitating the communication of stereochemical information from the C and D rings to the developing A-ring region of our synthetic intermediates.

We therefore considered it especially auspicious that this crucial transformation proceeded successfully and with notable efficiency in the very first trial. Upon heating in benzene at reflux for 30 h, our quinone combined with *trans*-2,4-pentadienol[53,54] to afford a single crystalline Diels–Alder adduct in more than 90% yield. Spectroscopic examination of this material identified it as Chandra's dione (**85**), isomerically pure within our limits of detection. Chemical transformations performed on intermediates derived from **85** later in the synthesis unambiguously confirmed the regio- and stereochemical structural assignment.

IX. The Hydronaphthalene Approach: (2) Operations on the C Ring

Ring C of our Diels–Alder adduct develops as the next theater of action in the hydronaphthalene approach. As our stockpiles of Chandra's dione began to mount (eventually to total several hundred grams), we turned our attention to expunging the extraneous functionality in the cycloadduct **85** so as to bring about its transformation to our next key objective, the keto aldehyde **102**.

$$ \textbf{85} \qquad\qquad\qquad\qquad \textbf{102} $$

Our first route to **102** involved an adaptation (with certain technical refinements) of tactics introduced by Woodward for the execution of a similar transformation in the total synthesis of cholesterol.[47] The primary alcohol contained in Chandra's dione was first protected as the tetrahydropyranyl ether **95** in order to permit the required operations on ring C. Exposure of **85** to 1.1 equivalents of dihydropyran in the presence of a catalytic amount of *p*-toluenesulfonic acid thus furnished the desired THP derivative in nearly quantitative yield. This intermediate was then transformed to the desired keto aldehyde in five steps. First, reduction of **95** with diisobutylaluminum hydride in toluene[55] produced the diol **96**. It will be noticed that the structure of this intermediate incorporates a vinylogous hemiacetal moiety, a functional system which upon hydrolysis is readily transformed into an α,β-unsaturated ketone. In the case of **96**, we found this process to be facilitated by the prior

95   **96**   **97**

activation of the hydroxyl leaving group as the corresponding methanesul-fonate ester. Thus, **96** was first treated with a mixture of methanesulfonyl chloride and triethylamine and then subjected in the same flask to hydrolysis using aqueous tetrahydrofuran buffered with 2,6-lutidine. The rather hindered remaining hydroxyl group was then acetylated employing acetic anhydride and 4-dimethylaminopyridine.[56]

Exposure of the resulting enone (**97**) to the action of excess lithium metal in liquid ammonia then effected the reductive expulsion of the α-keto acetoxy group, the conjugate reduction of the α,β-unsaturated ketone system, and the hydrogenolysis of the benzyl ether protective group, all in a single "one-pot" operation to afford the saturated ketone **99**. Collins oxidation finally provided keto aldehyde **81**.

97   **99**   **81**

Our initial enthusiasm for this relatively direct approach began to wane, however, as we undertook to employ it for the assembly of a sizeable quantity of the keto aldehyde **81**. In short, the sequence suffered from undependable, variable yields, which proved treacherously erratic when we attempted to advance more than 25 g of Chandra's dione along this route in a single run. We had little choice but to reevaluate our tactics and institute a new investigation in search of more secure avenues of approach to the requisite keto aldehyde. Ultimately, we found it prudent to utilize the following somewhat longer route (seven steps), which was able reliably and reproducibly to support the transformation of Chandra's dione to keto aldehyde **81** in greater than 40% overall yield.

Thus, selective reduction of the more electrophilic carbonyl group in dione **95** produced **100** as a single alcohol epimer. Corroboration for the stereochemistry assigned to this intermediate was provided by its quantitative conversion to the cyclic ether **101** upon exposure to *N*-bromosuccinimide in tetrahydrofuran. Examination of molecular models clearly indicates that only the cis-fused, α-hydroxy isomer of **100** possesses the capacity to undergo this transformation.

The hydrolysis of **96** to **97** had proved to be the most refractory step in our earlier, "short route" to keto aldehyde **81**. Further experimentation now showed that the efficiency and reproducibility of this key transformation could be considerably improved if the second, "spectator" hydroxyl group in **96** was first protected as an acetal or ether derivative. For example, reaction of **100** with chloromethyl methyl ether and diisopropylethylamine in methylene chloride at reflux provided the α-methoxymethylenoxy ketone **102**, reduction of which then afforded the vinylogous hemiacetal **103**. The stage was set for us once again to essay the key hydrolysis step.

In the event, our usual hydrolysis conditions proved totally unsatisfactory, producing the desired enone as the main component of a very complex mixture of products. After extensive experimentation, we eventually arrived at the following optimal procedure for achieving this transformation. A concentrated solution of **103** in tetrahydrofuran was first treated with methanesulfonyl chloride and triethylamine at −60°C for 15 min. To the resulting mesylate, in the same flask, was then slowly added a saturated aqueous solution of potassium bicarbonate, and the reaction mixture was then allowed gradually to warm to 0°C. The desired enone (**104**) was obtained in 65–77% overall yield from Chandra's dione in this fashion.

Although **104** could be directly converted to the diol **106** via the action of lithium in liquid ammonia, the following two stage reduction proved more manageable in large-scale operations. First, reaction of **104** with exactly 1 equivalent of hydrogen, using a 5% rhodium-on-carbon catalyst in tetrahydrofuran, accomplished the selective reduction of the enone double bond[57] to provide the saturated ketone **105**. Further reduction with excess lithium metal in a mixture of liquid ammonia, *tert*-butyl alcohol, and tetrahydrofuran then furnished the diols **106** in ∼50% overall yield from Chandra's dione.

At this point in the investigation we chose to divert a small portion of the diol **106** from the main route in order to provide further corroboration for the structure assigned to this intermediate. The alert reader will have noticed that at the stage of enone **104** a potentiality had developed for the loss of stereochemical integrity at the B–C ring junction. The following experiment was designed to confirm the persistence of the desired cis-ring fusion, and also to demonstrate unambiguously that the hydroxymethyl appendage in ring B possessed the required α orientation.

Thus, reaction of the diol **106** with 1 equivalent of methyl chloroformate in a mixture of pyridine and tetrahydrofuran selectively generated the carbonate derivative of the primary alcohol, which was then conveniently converted to **108** by Collins oxidation. Exposure of the resulting ketone to *p*-toluenesulfonic acid in methanol produced the internal ketal **109**. Only the keto alcohol intermediate with the desired stereochemistry can cyclize in this fashion.

106 : R = H
107 : R = CO₂Me

The unusual sensitivity toward aldol cyclization of the hydrindane series keto aldehydes **49** and **66** presaged the difficulties we would encounter in effecting the oxidation of diol **106** to **81**. Although the desired transformation could be accomplished by means of the Collins reagent, conventional extractive isolation procedures were totally unacceptable. The following procedure, employing essentially anhydrous workup conditions, ultimately proved satisfactory.

Oxidation of **106** was first carried out with excess Collins reagent in the presence of dry, acid-washed Celite in methylene chloride between −25 and −45°C. The resulting mixture was then treated with powdered sodium bisulfate monohydrate at −10°C for 30 min, diluted with diethyl ether, and filtered. Concentration provided almost pure keto aldehyde **81**, used in the next step without further purification.

X. The Hydronaphthalene Approach: (3) The D Ring

With the successful conclusion of the C-ring campaign we now arrived at the second of the five pivotal steps in the hydronaphthalene approach: the intramolecular pinacolic coupling reaction.

Once again we found that only low-valent titanium reagents possessed the capability to bring about this key transformation. For large scale work the following procedure proved superior with respect to both efficacy and convenience.[58] First, the keto aldehyde **81** was gradually added to the black suspension of Ti⁰ generated by the reaction of 8.5 equivalents of titanous chloride with 24 equivalents of potassium metal in tetrahydrofuran. The resulting mixture was stirred at room temperature for 2.5 h, cooled to 0°C, and very cautiously treated dropwise with anhydrous methanol and aqueous potassium carbonate solution, then filtered through a mixture of sand and

Celite, and finally subjected to an extractive workup employing a 4 : 1 mixture of ether and methylene chloride. Purification of the crude coupling product was accomplished in 10-g batches by preparative HPLC, which provided the pure cis (40%) and trans (10%) pinacols as well as a small amount (5–10%) of the diol **106**. This side product could of course be oxidized to **81** and thus recycled in the sequence.

It may be noted in passing that this key pinacolic coupling reaction was a particularly interesting reaction to carry out on a preparative scale. Experienced chemists will appreciate that the execution and subsequent quenching of reactions involving 30–50 g of finely divided potassium metal can provide moments of great drama, as well as a certain amount of piquant stimulation for the experimentalist.

To this point we had made relatively rapid progress along the hydronaphthalene route to GA_3. It was against the next stage of our synthetic advance, the elaboration of the D-ring allylic alcohol system, that gibberellic acid would for the first time fully mobilize its formidable defensive resources.

The elaboration of our pinacolic intermediate to the gibberellin allylic alcohol functional system seems to be a deceptively simple proposition. In principle, it would appear that this transformation could even be accomplished in as few as two steps, via oxidation of the pinacol-type intermediate **111** and Wittig methylenation of the resulting α-ketol (**112**). In reality, however, the extraordinary sensitivity of the D-ring system toward cleavage and rearrangement complicates both of these normally straightforward operations. For example, carbon–carbon bond cleavage (**111** → **115** or **116**) predominates

115 : R = H
116 : R = OH

upon attempted oxidation of the vicinal diol system using a variety of oxidizing agents,[59] including the Jones and Collins reagents, N-bromosuccinimide, *tert*-butyl hypochlorite, various Pfitzner–Moffatt-type reagents, and silver carbonate on Celite. It is an especially frustrating feature of these reactions that under most oxidation conditions a keto alcohol is in fact formed; however, this compound turns out not to be the desired ketol **112** but rather the *rearranged* system **117**, which is formed by the oxidative cleavage of **111** and subsequent aldol cyclization of the resulting unstable keto aldehyde.

Once obtained, the desired oxidation product **112** itself proves to be a rather intractable substance. For example, the very facile base-catalyzed α-ketol rearrangement of keto alcohols of this type requires that the tertiary hydroxyl group be protected prior to Wittig methylenation. This is a particularly insidious rearrangement: its overall effect is simply to invert the configuration of the D-ring bridge (**112** → **114**)! Needless to say, the lability of the α-ketol system also complicates attempts to contrive some means of protection for the tertiary hydroxyl group.

Earlier investigations in several laboratories had demonstrated that it is possible to accomplish the required D-ring functional manipulations by using relatively circuitous, multistage routes.[59] We considered it worthwhile, however, to invest some effort in the development of new technology that would allow the more direct execution of the desired transformations. These studies in fact proved to be quite rewarding and eventually resulted in the introduction of two new synthetic methods of considerable general utility.

We first considered the problem of the oxidation of the vicinal diol system. To minimize glycol cleavage, it seemed clear that it would be best to effect this reaction under relatively neutral conditions using nonpolar reaction media, and we therefore focused our attention on the class of oxidations proceeding via the intermediacy of alkoxysulfonium species. Several "Corey–Kim" reagents were thus developed, which we found also to be effective for the selective oxidation of sensitive *sec,tert*-1,2-diol derivatives.[30,60] For example, a model system of type **111** (namely **50**) could be oxidized to the corresponding α ketol in 72% yield employing the complex prepared by the combination of dimethyl sulfoxide and chlorine. Unfortunately, this reagent also reacts with olefins and therefore could not be applied for the oxidation of our gibberellin intermediate **110**.

We thereupon undertook an extensive examination of various other oxidation procedures also involving dimethylalkoxysulfonium intermediates. Eventually we found that the desired oxidation was most efficiently achieved by employing the following variation of a method introduced by Swern.[61] First, the pinacol **110** was added to a suspension of the complex generated by the reaction of 7 equivalents of dimethyl sulfoxide and 3.5 equivalents of trichloroacetic anhydride in methylene chloride at −60°C. After 45 min,

$110 \quad \xrightarrow[\substack{DMSO \\ Et_3N}]{(CCl_3CO)_2O} \quad 118 \quad \xrightarrow{MEMCl} \quad 119$

3.5 equivalents of triethylamine was added, and the reaction mixture was allowed to warm to room temperature. The desired ketol **118** was formed in excellent yield without any detectable D-ring cleavage.

We now confronted the problem of the protection of the tertiary hydroxyl group. A careful inspection of the armamentarium of alcohol protection methods led us to conclude that no protective group then available could satisfy the stringent requirements of our synthetic plan. For example, we required a protective group that could be introduced under near neutral conditions in order to avoid promoting the acid- or base-catalyzed rearrangement of our sensitive α ketol (**118**). Note also that this protection step would be further complicated by the relative steric inaccessibility of the tertiary bridgehead hydroxyl group. Naturally, the protective group selected would then have to be sufficiently stable to survive subsequent reaction conditions— in particular strongly basic media and the acidic conditions envisioned for the selective hydrolysis of the B-ring THP ether. Finally, it would in the end be necessary to be able to selectively excise the protective group from our molecule, without wreaking havoc on the very sensitive A-ring functionality, and also without thereby destroying the D-ring allylic alcohol generated in the deprotection step.

The MEM ether protective group was specifically designed to satisfy these exacting requirements.[63] The several oxygens in the methoxyethoxymethyl chain have the capacity to coordinate to Lewis acids, such as zinc bromide, and thereby facilitate the cleavage of this protective group under unusually mild conditions. On the other hand, the MEM group is sufficiently stable to protic acid to resist cleavage when exposed to the normal THP ether hydrolysis procedure. Finally, the introduction of this protective group can be accomplished under relatively mild conditions. Thus, treatment of **118** with 3 equivalents of MEM chloride and 10 equivalents of diisopropylethylamine in methylene chloride at reflux furnished the desired MEM ether (**119**) in 55–68% overall yield from the pinacolic coupling product.

XI. The Hydronaphthalene Approach:
(4) Subjugation of the B Ring

With some foreboding we now advanced to the third of the five pivotal steps in the hydronaphthalene strategy: the internal aldol reaction that would establish ring B. Although our route to the tricyclic ketone **119** was relatively efficient (25 g of **119** could be prepared from 150 g of *o*-eugenol), we chose to employ a more easily synthesized model compound for our initial examination of the crucial ring contraction. The cyclic ether **122** seemed ideally suited for this purpose.

The required model compound was conveniently prepared by using the following route. Exposure of *trans*-2,4-pentadienol to 1.1 equivalents of dihydropyran and a catalytic amount of *p*-toluenesulfonic acid in methylene chloride produced the corresponding THP ether, which smoothly combined with maleic anhydride in benzene at reflux to furnish the Diels–Alder adduct **120**. Reduction of this anhydride with excess lithium aluminum hydride in tetrahydrofuran then gave the diol **121**. Finally, gradual addition of 1.9 equivalents of *p*-toluenesulfonyl chloride to a solution of **121** in pyridine at 65°C afforded the desired cyclic ether in greater than 60% overall yield from 2,4-pentadienol.

It will be recalled that our plan for the contraction of ring B called for the oxidative cleavage of the cyclohexene double bond, followed by regiocontrolled internal aldol cyclization of the resulting dialdehyde. Unfortunately, the convenient Johnson–Lemieux method could not be applied for the oxidative cleavage of **122** because of the propensity of the dialdehyde product to form cyclic acetals upon exposure to the protic solvents required in this procedure. Fission of the six-membered ring was therefore accomplished employing the following mild, two-stage method. The double bond of **122**

was first hydroxylated using osmium tetroxide in a mixture of ether and pyridine; treatment of the resulting diol with 1 equivalent of lead tetraacetate in benzene at 5°C then effected the instantaneous cleavage of **123** to afford the desired dialdehyde. The stage was now set for us to essay the formidable intramolecular aldol cyclization.

Why did we approach this crucial step with such apprehension? In this pivotal reaction we require that the new carbon–carbon bond be formed via regioselective attack on the more sterically hindered carbonyl group. Clearly the possibility exists that this transformation could be complicated by the intervention of an alternative mode of ring closure that would produce the hydroxy aldehyde **127**. We anticipated, however, that under basic conditions this aldol cyclization would lead via irreversible dehydration to the desired conjugated aldehyde **125**, since the reversibly formed alternative product cannot undergo a similar elimination.

Unfortunately, there exists in this reaction even further potential for molecular mayhem. Under aldol cyclization conditions, the enolization of the α,β-unsaturated aldehyde product can initiate a variety of calamitous processes, including, for example, the elimination of the tetrahydropyranyloxy group (**125** → **126**), and the epimerization of the B-ring side chain.

In the event, treatment of the model dialdehyde **124** with a wide range of alkoxide and tertiary amine bases afforded, in every case, a complex mixture of products, which included all three of the aldehydes **125–127**. It was emphatically clear that we would have to devise new tactics in order to bring about the desired transformation.

After much further consideration of the problem, we concluded that Knoevenagel-type condensation conditions would probably provide us with the most promising avenue of attack. Woodward had previously employed a 1:1 mixture of piperidine and acetic acid for the construction of the cholesterol D ring,[47] and related Knoevenagel conditions had been successfully applied to the cyclization of a variety of 1,6-dialdehydes.[64] We anticipated that under these conditions the aldol cyclization of **124** would proceed via the thermodynamically favored enamine derivative of the less sterically hindered aldehyde (see **128**) and would thus lead regioselectively to the formation of the desired aldol product.

This proved to be the case. Exposure of the model dialdehyde to 1 equivalent each of piperidine and acetic acid in benzene produced the desired aldehyde **125**, contaminated by only a small amount of the elimination product **126**. In fact, we soon after discovered that even less of this side product formed if the less basic amine morpholine (pK_A of conjugate acid, 8.3) was substituted for piperidine (pK_A 11.1).

In subsequent studies we investigated the utility of more than 25 amine–carboxylic acid systems as catalysts for the aldol cyclization of **124**. The effects of catalyst concentration and stoichiometry, choice of solvent, and reaction time and temperature were all systematically examined. Eventually we found that a combination of dibenzylamine (pK_A 8.0) and trifluoroacetic acid provided optimal results. Thus, treatment of dialdehyde **124** with 0.3 equivalent of preformed dibenzylammonium trifluoroacetate in tetrahydrofuran at room temperature for three hours afforded the desired α,β-unsaturated aldehyde **125** in 78% overall yield from the olefin **122**. Neither the elimination product **126** nor the alternative aldol product **127** was observed to form under these conditions.

It was most gratifying to find that this ring contraction methodology could be applied with equal success to our tricyclic gibberellin intermediate **119**. Thus, treatment of a solution of **119** in aqueous acetone with 0.05 equivalent of osmium tetroxide and 1.3 equivalents of N-methylmorpholine N-oxide[65] at room temperature for 80 h furnished a single cis diol (**130**) in 89% yield after chromatography.

In one run of this catalytic osmylation step, we were mystified to find that a small amount of starting material completely resisted hydroxylation, even when treated with additional osmium tetroxide. NMR examination of this material subsequently revealed it to be not the unreacted olefin **119**, but rather the corresponding B-ring *saturated* compound (**129**), resulting from

129

overreduction during the earlier hydrogenation step **104** → **105**. Although **129** has obviously lost the ability to function as an intermediate in our synthesis, this "eunuch ketone" (as it became known in our group) did eventually serve us as an excellent model compound for studies on the esterification of the B-ring hydroxymethyl group.

To effect the desired ring contraction, the diol **130** was first cleaved by exposure to 1.02 equivalents of lead tetraacetate in benzene at 5°C for 30 min, and the resulting sensitive dialdehyde was then cyclized to **132** by treatment with 0.2 equivalent of dibenzylammonium trifluoroacetate in benzene at 50°C for 1 h. The α,β-unsaturated aldehyde **132** was thus obtained in 64% overall yield from **119**, although the actual efficiency of this process is actually somewhat higher since **132** was observed to undergo partial decomposition

upon chromatography. In practice, therefore, this intermediate was generally subjected to Wittig methylenation without prior purification. Reaction of **132** with 5 equivalents of methylenetriphenylphosphorane in a 2:1 mixture of tetrahydrofuran and hexamethylphosphoramide at reflux for 3.5 h thus furnished the desired triene **133** in 44% overall yield from the cis diol **130**. The battle for the B ring was won.

XII. The Hydronaphthalene Approach: (5) The Triumph of the Intramolecular Diels–Alder Strategy

Now we approached the climactic phase of our approach to the total synthesis of gibberellic acid. Exposure of the THP ether **133** to a 3:1:1

mixture of acetic acid, tetrahydrofuran, and water at 35°C for 40 h smoothly provided the key B-ring alcohol **134**. As expected, no hydrolysis of the D-ring MEM ether protective group was detected under these conditions.

With the preparation of this tricyclic intermediate we had now reached a particularly significant point in our synthetic journey. It will be noticed that the left portion of the tricyclic alcohol **134** exactly corresponds to the monocyclic diene alcohol **36**, whose synthesis and conversion to a gibberellin AB-ring model compound we had already thoroughly investigated in our earlier preliminary studies. From this point on, every remaining transformation in our synthetic plan had previously been accomplished employing analogous compounds in the AB-ring model series. It was thus with renewed enthusiasm and optimism, and the highest possible morale, that we attacked the next stage of the synthesis: the esterification of **134** to produce cycloaddition substrates **135** and **136**.

The conversion of **134** to its β-chloroacrylyl ester proceeded without incident according to the procedure perfected in the monocyclic model series. Thus, exposure of the lithium salt of **134** to 1.55 equivalents of trans-2-chloroacrylyl chloride in tetrahydrofuran at −40°C for 30 min provided the desired ester in 72% overall yield from the tetrahydropyranyl ether **133**. Unfortunately, the formation of the propiolate derivative of **134** proved much more difficult than the analogous esterification of the somewhat less sterically encumbered model compound **36**. Ultimately we were able to obtain the desired propiolate ester (**135**) in a modest 56% yield by the reaction of the lithium salt of **134** with propiolic anhydride at low temperature in dimethoxyethane. The delicate anhydride was generated *in situ* for this reaction by the treatment of anhydrous potassium propiolate with 0.5 equivalent of oxalyl chloride in the presence of a catalytic amount of dimethylformamide.

Of the five pivotal steps in the total synthesis of gibberellic acid, certainly no transformation was as crucial to our strategy as the intramolecular Diels–Alder reaction now confronting us. Although our carefully contrived model studies had seemingly established the feasibility of this process, we could not help but approach this key step with some trepidation. All too clearly we recalled the fact that the key pinacolic coupling reaction had also been thoroughly tested in exhaustive model studies prior to its initial disastrous failure when applied to the "real" gibberellin intermediates **66** and **81**.

The first test of the pivotal intramolecular Diels–Alder reaction was now at hand. A benzene solution of the β-chloroacrylyl ester **136** was heated in a sealed tube to 160°C for 36 h in the hope of effecting the clean and efficient intramolecular cycloaddition observed for the model ester **38** under identical conditions. *None of the Diels–Alder adduct was produced!* To our consternation, this treatment resulted in the complete conversion of **136** to an uncharacterizable tar.

Appalled by this development, and rapidly approaching a state of panic, we retreated from the laboratory to ponder the unexpected disparate behavior of the model ester **38** and our tricyclic analog **136**. We thereupon evolved the reasonable hypothesis that the observed polymerization of the tricyclic ester involved the decomposition of the sensitive allylic ether system of the D ring, perhaps catalyzed by the elimination of HCl from the β-chloroacrylyl ester function. The Diels–Alder cyclization was therefore immediately attempted in the presence of calcium carbonate as an insoluble acid scavenger.

This produced tar-coated chalk. Undismayed, we examined other, more innocuous acid scavengers and instituted the additional precaution of thoroughly degassing the reaction mixture through several freeze–thaw cycles at high vacuum. In this fashion dramatic success was finally realized. Heating a degassed solution of **136** in benzene in the presence of 100 equivalents of propylene oxide at 160°C for 45 h accomplished the desired intramolecular Diels–Alder reaction and provided the crystalline lactone **137** (a single diastereomer) in 55% yield after recrystallization. NMR spectral analysis fully supported the stereochemical assignment, which was ultimately confirmed by the conversion of this intermediate to gibberellic acid.[66]

136   137   138 : R = MEM
                                                                               11 : R = H

The next stages of the synthesis of gibberellic acid proceeded smoothly under the conditions previously employed in the model series. Elimination of chloride and methylation at C-4 of the A ring was achieved in 75% yield by treating the cycloadduct **137** with 2.2 equivalents of lithium isopropylcyclo-hexylamide in a mixture of tetrahydrofuran and hexamethylphosphoramide at −78°C, followed by quenching with excess methyl iodide. Exposure of the resulting pentacyclic lactone to 25 equivalents of zinc bromide in a 15:5:1 mixture of chloroform, ether, and nitromethane at room temperature for 3 h then effected the selective cleavage of the MEM ether group to afford the key intermediate **11** in 70% yield.

For reasons that will become clear in the next section, we later chose this pentacyclic lactone as the most suitable intermediate to resolve in our route to GA$_3$. The resolution was readily achieved by the chromatographic separation of diastereomeric carbamate derivatives of the D-ring tertiary hydroxyl group. Thus, exposure of **11** to excess phosgene and 3 equivalents of 4-di-methylaminopyridine in methylene chloride gave the corresponding chloro-formate, which was treated without purification with (−)-α-phenylethylamine to provide the desired diastereomeric carbamates in 95% yield. After separation by silica gel chromatography, the less polar urethane was then cleaved in 95% yield according to the procedure of Pirkle[67] by treatment with 5 equivalents of triethylamine and 3 equivalents of trichlorosilane in benzene at room temperature. An optically active synthetic intermediate incorporating the complete carbon skeleton of gibberellic acid was now in hand.

XIII. The Hydronaphthalene Approach:
(6) Stalemate at the A Ring

And so it was that we finally turned our attention to the last pivotal stage in the total synthesis of gibberellic acid: the elaboration of the A- and B-ring functional systems. Our immediate objective was, of course, the Corey–Carney acid (**8**), whose efficient transformation to gibberellic acid had already been accomplished in preliminary studies as described in Section IV of this account.

Thus, to complete the total synthesis of GA$_3$, all that remained was the hydrolysis of the lactone ring, the oxidation of the C-6 hydroxymethyl group, and finally the epimerization of this appendage to the thermodynamically favored, natural β configuration.

In two successful model investigations we had already laid the groundwork for this last stage of the synthesis. The tricyclic lactone **37** (corresponding to the left portion of **11**) had been efficiently converted to the desired Corey–Carney acid model compound by a route that featured the internal protection of the A-ring functionality as a bromo lactone in order to permit the required operations on ring B. This study, it will be recalled, was described in detail in Section V of this account. We subsequently undertook a second model study employing the tetracyclic model compound **139** to ensure that the desired transformations could be effected in the presence of the sensitive D-ring allylic alcohol system. Reduction of the Corey–Carney acid with 4 equivalents of diisobutylaluminum hydride in toluene at $-40°C$ thus furnished the required model compound **139** in 72% yield; notice that except for the configuration of the C-6 hydroxymethyl group, this alcohol is identical to the product expected from the hydrolysis of our key gibberellin intermediate **11**. The conversion of **139** back to the Corey–Carney acid then proceeded without incident in 50% overall yield, employing the methodology previously developed in the bicyclic model series. In this case, however, we chose to protect the A-ring functionality as an iodo lactone because of the previously observed sensitivity of the D-ring allylic alcohol system to brominating agents.

All of the remaining steps in our synthetic plan had now been optimized, using model systems that incorporated both the A- and D-ring functionalities of the envisaged synthetic intermediates. Confident that we had provided for every possible contingency, we believed that the triumphant completion of the total synthesis of gibberellic acid was finally at hand.

That was naive. We were totally unprepared for the special surprise our foe had held in reserve for the final confrontation. All our attempts to generate a halo lactone derivative of type **140** from the pentacyclic lactone **11** were thwarted by the irrepressible propensity of the B-ring C-7 hydroxyl to attack the neighboring C-4 carboxyl group, thus reforming the original δ-lactone system, and concomitantly initiating further destructive transformations of the fragile A-ring functionality. All efforts to suppress this translactonization proved futile: relative to the bicyclic model series, the C-4 and C-6 substituents in these tetracyclic intermediates appear to be compressed into much closer proximity, so that cyclizations of the type **140** → **141** are extraordinarily facile. For example, the primary amide **142** (obtained by cleavage of **11** with KNH$_2$) was observed to spontaneously relactonize simply upon standing in ether solution at room temperature. It was all too clear that our plan for the elaboration of the gibberellin A ring had been undermined by an inopportune exemplification of R. B. Woodward's dictum that *"enforced propinquity often leads to greater intimacy."*[68]

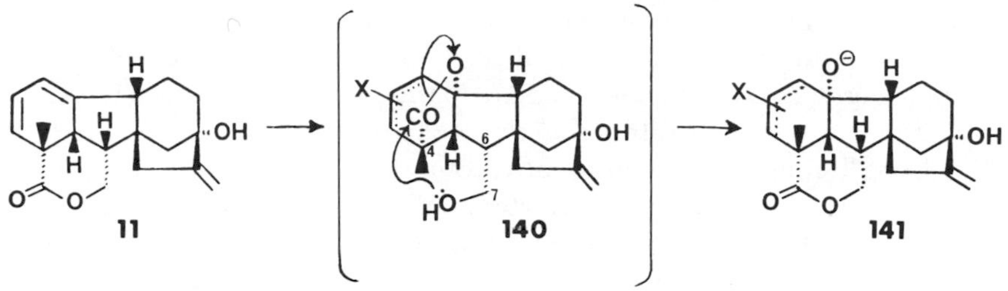

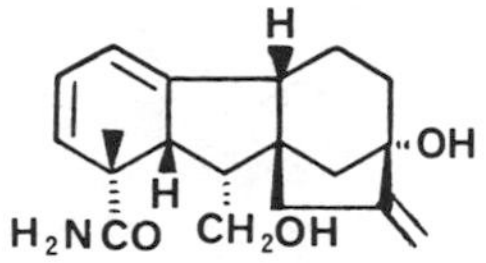

This then was the nadir of our investigation, when the prospects for the completion of the total synthesis of gibberellic acid indeed began to appear grim. Undaunted, however, Professor Corey devised yet another scheme for the elaboration of the AB-ring functionalities. At this point he also presented Gary Keck and myself with his personal copy of Sir Edmund Hillary's classic account of the conquest of Mt. Everest,[69] and assigned to us the study in particular of Chapter 11: "The Summit." Inspired by this account of adversity overcome, we redoubled our efforts and undertook the investigation of the new route to the gibberellin A ring.

XIV.   The Total Synthesis of Gibberellic Acid

The halo ether approach to the elaboration of the gibberellin AB-ring functionality is outlined in the following scheme. Our new strategy was quite simple: reduction of the pentacyclic lactone **11** and subsequent halo ether formation would provide a protected A-ring derivative that would obviously be immune to the lactonization process that had defeated the halo lactone strategy.

In the event, reduction of **11** with lithium borohydride in tetrahydrofuran furnished the desired triol **143**, which was converted to the iodo ether **144** by the action of 1.1 equivalents of iodine and 3 equivalents of sodium bicarbonate in tetrahydrofuran at 0°C. The lability of the allylic iodide moiety in **144** prompted us to convert this intermediate to the more stable bromo ether, and this transformation was easily accomplished by the reaction of **144** with excess anhydrous lithium bromide in tetrahydrofuran at room temperature. Next, the alcohol **145** was efficiently converted to the methyl ester **149** by means of the sequence: PCC oxidation, epimerization of the resulting aldehyde with DBU, Jones oxidation to the corresponding carboxylic acid, and diazomethane esterification. Exposure of **149** to 5 equivalents of aluminum amalgam in wet tetrahydrofuran then afforded the triene **150**, which was oxidized to the Corey–Carney acid in two steps by sequential treatment with pyridinium chlorochromate and then silver oxide. This 10-step sequence provided the Corey–Carney acid in 20–25% overall yield, and formally completed the total synthesis of gibberellic acid since **8** had already been converted to GA_3 in our earlier preliminary studies.

144 : X = I, R^1 = CH$_2$OH, R^2 = H
145 : X = Br, R^1 = CH$_2$OH, R^2 = H
146 : X = Br, R^1 = CHO, R^2 = H
147 : X = Br, R^1 = H, R^2 = CHO
148 : X = Br, R^1 = H, R^2 = CO$_2$H
149 : X = Br, R^1 = H, R^2 = CO$_2$CH$_3$

The inefficiency of this cumbersome final sequence prompted us to continue our investigation in the hope of developing a more direct route for the transformation of **11** to the Corey–Carney acid. These further studies were greatly facilitated by the following method, which we devised for the preparation of the key pentacyclic lactone **11** from the Corey–Carney acid, which, it will be recalled, was itself available in quantity through the degradation of gibberellic acid. An especially appealing feature of this stratagem was that it now exploited to our benefit the unusual reactivity of certain AB-ring derivatives due to the "enforced propinquity" of carbons 7 and 18.

Saponification of the Corey–Carney acid with aqueous potassium hydroxide proceeded smoothly to afford **151** in 95% yield. The conversion of this diacid to the anhydride **152** was then effected in 73% yield by the combined action of excess triethylamine and 1 equivalent of dicyclohexylcarbodiimide. This remarkable transformation involves the contrathermodynamic $6\beta \rightarrow 6\alpha$ epimerization of an activated C-6 carboxyl derivative, driven by the intermolecular capture of the 6α isomer by the proximate C-4 carboxyl group.

8 : R = CH₃
151 : R = H

152

11

Finally, selective reduction at the more accessible carbonyl of the anhydride was achieved by treatment of **152** with 0.5 molar equivalent of lithium borohydride in dimethoxyethane at $-25°C$. The desired pentacyclic lactone **11** was produced in 50% yield from **152** in this manner. It should be mentioned that our resolved, synthetic lactone **11** proved to be identical in every respect with the material thus produced by the degradation of gibberellic acid.

With a bountiful supply of the pentacyclic lactone at our disposal, we were now well equipped to prosecute an extensive investigation of methods for the conversion of this key intermediate to the Corey–Carney acid. The efficiency of our synthesis was thus vastly improved when we discovered that this transformation could in fact be achieved in just two steps, employing the following procedure. The optically active lactone **11** was first saponified by

11

153

8

heating with excess aqueous potassium hydroxide, and the resulting solution was then treated at room temperature with an alkaline solution of 2.07 equivalents of sodium ruthenate.[70] Under these conditions the C-6 hydroxymethyl group is oxidized to the corresponding formyl derivative, which then epimerizes to the more stable β configuration before finally being further oxidized to the diacid **153**. Selective monoesterification of this intermediate was then accomplished by the reaction of **153** with 1.5 equivalents of triethylamine and 1 equivalent of p-toluenesulfonyl chloride in tetrahydrofuran, followed by the quenching of the resulting mixed sulfonic anhydride with excess methanol.

With this last refinement, an efficient route to optically active, synthetic Corey–Carney acid was finally in hand. Six years earlier, at the outset of these studies, we had developed methodology for the transformation of this key intermediate to GA_3. In 35 steps we had thus achieved the first total synthesis of gibberellic acid.

"Durch die Nacht führt unser Weg zum Lichte."

ACKNOWLEDGMENTS

The total synthesis of gibberellic acid was achieved through the combined efforts of an unusually capable and dedicated group of chemists in Professor Corey's laboratory. Drs. S. Chandrasekaran, G. E. Keck, P. Siret, J.-L. Gras, R. L. Carney, T. M. Brennan, S. D. Larsen, B. Gopalan, and G. L. Thompson all participated in the investigations described in this account. Needless to say, the entire project was conceived and directed by E. J. Corey, whose brilliant strategic and tactical leadership was responsible for the ultimate success of this synthetic adventure.

REFERENCES

1. E. J. Corey, R. L. Danheiser, S. Chandrasekaran, P. Siret, G. E. Keck, and J.-L. Gras, *J. Am. Chem. Soc.* **100,** 8031 (1978).
2. E. J. Corey, R. L. Danheiser, S. Chandrasekaran, G. E. Keck, B. Gopalan, S. D. Larsen, P. Siret, and J.-L. Gras, *J. Am. Chem. Soc.* **100,** 8034 (1978).
3. For general discussions of the chemistry and biology of the gibberellins, see references *4–6*, and J. MacMillan and R. J. Pryce, *in* "Phytochemistry" (L. P. Miller, ed.), Vol. III, pp. 283–326. Van Nostrand-Reinhold, Princeton, New Jersey, 1973.
4. H. N. Krishnamoorthy, ed., "Gibberellins and Plant Growth." Wiley, New York, 1975.
5. P. Hedden, *ACS Symp. Ser.* **111,** 19 (1979).
6. R. L. Danheiser, Ph. D. Dissertation, Harvard University, Cambridge, Massachusetts (1978).
7. B. E. Cross, J. F. Grove, J. MacMillan, J. S. Moffatt, T. P. C. Mulholland, J. C. Seaton, and N. Sheppard, *Proc. Chem. Soc., London* p. 302 (1959).
8. F. McCapra, A. I. Scott, G. A. Sim, and D. W. Young, *Proc. Chem. Soc., London* p. 185 (1962).
9. J. A. Hartsuck and W. N. Lipscomb, *J. Am. Chem. Soc.* **85,** 3414 (1963).
10. J. S. Moffatt, *J. Chem. Soc.* p. 3045 (1960).
11. B. E. Cross, *J. Chem. Soc.* p. 4670 (1954).
12. B. E. Cross, J. F. Grove, and A. Morrison, *J. Chem. Soc.* p. 2498 (1961).
13. G. Stork and H. Newman, *J. Am. Chem. Soc.* **81,** 5518 (1959); see also J. W. Cornforth, as quoted by B. E. Cross, *Chem. Ind. (London)* p. 183 (1959).
14. For a discussion of gibberellin syntheses prior to 1978, see reference *6* and E. Fujita and M. Node, *Heterocycles* **7,** 709 (1977).
15. Prior to this date a variety of other approaches to gibberellic acid had been examined in Professor Corey's laboratory. Space does not permit a full discussion of these earlier studies.
16. E. J. Corey, T. M. Brennan, and R. L. Carney, *J. Am. Chem. Soc.* **93,** 7316 (1971).
17. Samples of gibberellic acid were donated for this study by Imperial Chemical Industries Ltd., Merck and Co., Abbott Laboratories, and Chas. Pfizer and Co.
18. P. A. Bartlett and W. S. Johnson, *Tetrahedron Lett.* p. 4459 (1970).
19. E. J. Corey and R. L. Danheiser, *Tetrahedron Lett.* p. 4477 (1973).
20. R. W. Kierstead, R. P. Linstead, and B. C. L. Weedon, *J. Chem. Soc.* p. 3616 (1952).

21. This compound was previously prepared by lengthier, less efficient routes [B. L. Nandi, *J. Indian Chem. Soc.* **11,** 277 (1934); H. Prinzbach and H. D. Martin, *Chimia* **23,** 37 (1969); H. C. Stevens, J. K. Rinehart, J. M. Lavanish, and G. M. Trenta, *J. Org. Chem.* **36,** 2780 (1971); J. D. Roberts, F. O. Johnson, and R. A. Carboni, *J. Am. Chem. Soc.* **76,** 5692 (1954)].

22. F. Straus and W. Voss, *Ber. Dtsch. Chem. Ges.* **59,** 1681 (1926).

23. W. J. Balfour, C. C. Greig, and S. Visaisouk, *J. Org. Chem.* **39,** 725 (1974).

24. For a recent review providing leading references to this and other esterification methods, see E. Haslam, *Tetrahedron* **36,** 2409 (1980).

25. This methylation was the first example of the alkylation of an enolate derived from a 3,6-dihydrobenzoic acid derivative. A general study of the scope of this reaction later appeared [R. K. Boeckman, M. Ramaiah, and J. B. Medwid, *Tetrahedron Lett.* p. 4485 (1977)].

26. E. J. Corey, M. Narisada, T. Hiraoka, and R. A. Ellison, *J. Am. Chem. Soc.* **92,** 396 (1970).

27. E. J. Corey and R. L. Carney, *J. Am. Chem. Soc.* **93,** 7318 (1971).

28. For discussions, see references 6 and 14.

28a. G. Stork, S. Malhotra, H. Thompson, and M. Uchibayashi, *J. Am. Chem. Soc.* **87,** 1148 (1965).

29. For the similar application of trimethylsilylchloride to improve the yields of acyloin reactions, see K. Rühlmann, *Synthesis* p. 236 (1971).

30. E. J. Corey and C. U. Kim, *J. Am. Chem. Soc.* **94,** 7586 (1972).

31. First observed by A. J. Birch and M. Smith, *Proc. Chem. Soc., London* p. 356 (1962).

32. K. B. Sharpless, M. A. Umbreit, M. T. Nieh, and T. C. Flood, *J. Am. Chem. Soc.* **94,** 6538 (1972).

33. T. Mukaiyama, T. Sato, and J. Hanna, *Chem. Lett.* p. 1041 (1973).

34. J. E. McMurry and M. P. Fleming, *J. Am. Chem. Soc.* **96,** 4708 (1974).

35. See also S. Tyrlik and I. Wolochowicz, *Bull. Soc. Chim. Fr.* p. 2147 (1973).

36. For a full account of our investigation of methods for the reductive coupling of ketones and aldehydes, see E. J. Corey, R. L. Danheiser, and S. Chandrasekaran, *J. Org. Chem.* **41,** 260 (1976).

37. For a description of the synthesis of this diketone, see reference 36.

38. For a full discussion of the synthesis of this intermediate, see reference 6.

39. L. L. Shchukovskaya, R. I. Pal'chik, and A. N. Lazerev, *Dokl. Akad. Nauk SSSR* **164,** 357 (1965).

40. R. A. Ruden, *J. Org. Chem.* **39,** 3607 (1974).

41. This step accomplishes desilylation of the initially produced α-trimethylsilylacetate derivative.

42. J. E. McMurry and J. Melton, *J. Am. Chem. Soc.* **93,** 5309 (1971).

43. E. J. Corey and J. G. Smith, *J. Am. Chem. Soc.* **101,** 1038 (1979).

44. E. J. Corey and J. E. Munroe, *J. Am. Chem. Soc.* **104,** 6129 (1982).

45. G. Stork, R. K. Boeckman, D. F. Taber, W. C. Still, and J. Singh, *J. Am. Chem. Soc.* **101,** 7107 (1979). For the first disclosure of the Stork synthesis of the tricyclic intermediate **19,** see R. L. Harlow and S. H. Simonsen, *Cryst. Struct. Commun.* **6,** 689 (1977).

46. See L. W. Butz and A. W. Rytina, *Org. React.* **5,** 136 (1949) and A. S. Onishchenko, "Diene Synthesis." D. Davey & Co., Jerusalem, 1964.

47. R. B. Woodward, F. Sondheimer, D. Taub, K. Heusler, W. M. McLamore, *J. Am. Chem. Soc.* **74,** 4223 (1952).

48. See M. F. Ansell and A. H. Clements, *J. Chem. Soc. C* p. 269 (1971); C. Schmidt, *J. Org. Chem.* **35,** 1324 (1970); F. Bohlmann, W. Mathar, and H. Schwarz, *Chem. Ber.* **110,** 2028, (1977) and references cited therein.

49. In an earlier approach oxidative cleavage of the allyl group was deferred until after the Diels-Alder reaction with 2,4-pentadienol. This route failed when it proved difficult to

satisfactorily differentiate between the allyl and B ring double bonds using a variety of oxidizing agents.

50. R. Pappo, D. S. Allen, R. U. Lemieux, and W. S. Johnson, *J. Org. Chem.* **21,** 478 (1956).

51. R. P. Singh, *Can J. Chem.* **44,** 1994 (1966); P. A. Wehrli and F. Pigott, *Org. Synth.* **52,** 83 (1972).

52. H. M. Van Dort and H. J. Geursen, *Recl. Trav. Chim. Pays-Bas* **86,** 520 (1967); L. H. Vogt, J. G. Wirth, and H. L. Finkbeiner, *J. Org. Chem.* **34,** 273 (1969).

53. R. G. Glushov and O. Y. Magidson, *Med. Prom. SSSR* **16,** 27 (1962).

54. S. Oida and E. Ohki, *Chem. Pharm. Bull.* **17,** 1990 (1969).

55. No conjugate reduction of the enone system occurs under these conditions [K. E. Wilson, R. T. Seidner, and S. Masamune, *J. Chem. Soc., Chem. Commun.* p. 213 (1970)].

56. G. Höfle and W. Steglich, *Synthesis* p. 619 (1972).

57. S. K. Roy and D. M. S. Wheeler, *J. Chem. Soc. p.* 2155 (1963).

58. J. E. McMurry and M. P. Fleming, *J. Org. Chem.* **41,** 896 (1976).

59. For example, see ref. *6* and H. O. House and D. G. Melillo, *J. Org. Chem.* **38,** 1398 (1973).

60. E. J. Corey and C. U. Kim, *Tetrahedron Lett.* p. 287 (1974).

61. Swern (*62*) has reported the oxidation of alcohols to ketones using a DMSO-trifluoroacetic anhydride reagent. In our hands this method generally produced 10–20% of the trifluoroacetate derivative of the alcohol; our modification obviates this side reaction.

62. For a review, see D. Swern, *Synthesis* p. 165 (1981).

63. E. J. Corey, J.-L. Gras, and P. Ulrich, *Tetrahedron Lett.* p. 809 (1976).

64. For examples, see T. Harayama, M. Ohtani, M. Oki, and Y. Inubishi, *Chem. Pharm. Bull.* **21,** 1061 (1973) and A. S. Kende, T. J. Bentley, R. A. Mader, and D. Ridge, *J. Am. Chem. Soc.* **96,** 4332 (1974).

65. V. Van Rheenen, R. C. Kelly, and D. Y. Cha, *Tetrahedron Lett.* p. 1973 (1976).

66. Preliminary experiments employing the propiolyl ester **135** indicated that this compound is a less satisfactory precursor to **138**. Diels-Alder cyclization and methylation afforded **138** in only 25% yield, due to the intervention of the aromatization of the Diels–Alder adduct as a troublesome side reaction.

67. W. H. Pirkle and J. R. Hauske, *J. Org. Chem.* **42,** 2781 (1977).

68. R. B. Woodward, *Pure Appl. Chem.* **17,** 519 (1968).

69. E. P. Hillary, "High Adventure." Dutton, New York, 1955.

70. D. G. Lee, D. T. Hall, and J. H. Cleland, *Can. J. Chem.* **50,** 3741 (1972).

Chapter 3

A PROSTAGLANDIN SYNTHESIS

Josef Fried

> Department of Chemistry
> The University of Chicago
> Chicago, Illinois

I. Introduction

Retracing a road traveled over the past 15 years provides for a nostalgic journey, both in terms of the scientific challenges that had to be faced and in looking back at the collaborators who populated the laboratory and the joys experienced when another bastion was conquered on the way to the solution of the problem.

The problem was a synthesis of a recently discovered new class of biologically highly active substances freely, ubiquitously, and efficiently produced in the mammalian organism, the prostaglandins. The challenge to become involved in the chemistry of this class of compounds for someone who had become somewhat tired of an initially equally challenging subject, the steroids, was inevitable.

The goal in this enterprise was easily defined: a stereospecific synthesis of all the then known prostaglandins in optically active form. A central candidate that would lend itself to extension to the other members of the class was

PGF$_{2\alpha}$ (**1**). Considering in retrospect the wealth of successful approaches that have been employed during the last decade, there was obviously much to choose from.

OH

CO$_2$H

OH   OH

1

We settled at the outset on following a strategy in which a preformed cyclopentane ring with appropriately spaced, sterically defined substituents would serve as the base of operations, to which the two side-chain appendages would be attached by stereochemically well-defined reactions. This would allow for a convergent approach, in which the eight-carbon side chain containing one of the chiral centers could be prepared in optically active form and thus serve to resolve the cyclopentanoid portion of the molecule containing the remaining chiral centers in the appropriate relative stereochemistry.

Early in our planning we were impressed by a reaction that Nagata[1] had reported involving the opening of steroidal epoxides with diethylaluminum cyanide to form *trans*-2-hydroxycyanides. This reaction proceeded at ambient temperature in excellent yield and avoided the base-catalyzed β elimination and other side reactions when potassium cyanide is employed. It occurred to us that equally facile reactions might be observed if the isoelectronic acetylene moiety were to be used in place of cyanide. We were quite naive then in the field of alane chemistry but were struck by the fact that in the epoxide opening, cyanide anion formed the new carbon–carbon bond rather than the much more reactive ethyl anion. It was hoped that this preference of the more stable anion to react would be preserved with the acetylenic alanes as well. Indeed, the reagent prepared from dialkylchloroalane and a variety of lithium acetylides produced *trans*-2-hydroxyacetylenes in quantitative yield with complete stereoselectivity at ambient temperature.[2] Parallel reactions with lithium acetylides and alicyclic epoxides require high temperatures and produce poor yields.

II. 7-Oxaprostaglandins

We decided to exploit this finding to synthesize the first heteroprostaglandin, namely, 7-oxa-PGF$_{1\alpha}$ (**2**),[3] as well as 7-oxaprostaglandins lacking one or more of the hydroxyl groups. Among the latter 7-oxa-13-prostynoic acid (**3**) has gained some prominence as an inhibitor of prostaglandin action *in vitro*.[4] On the chemical side this gave us experience with this type of chemistry and the handling of these largely noncrystalline compounds on a small scale.

2   **3**

The starting material, *cis*-2-cyclopentene-1,4-diol (**4**), which we learned to prepare on a large scale by photooxygenation of cyclopentadiene, followed by reduction of the endoperoxide, lent itself to the preparation of the *cis*- and *trans*-epoxy-*cis*-diols, depending on whether the epoxidation was performed on the free enediol resulting in the all-cis compound or on a protected diol such as its dibenzyl ether. The *cis*-dibenzyl ether (**5**) served as the starting point for the reaction with octynyldiethylalane. The resulting product (**6**)

4   **5**   **6**

required only alkylation to produce the skeletal structure of the 7-oxaprosta-glandins (cf. **2**). All four chiral centers on the cyclopentane ring were thus in place, and there remained the removal of the protecting groups, reduction of the acetylene to the trans olefin, and allylic hydroxylation. During these early studies it was observed that prostaglandins possessing the acetylenic group were biologically active and completely resistant to the action of a ubiquitous enzyme, capable of oxidizing the 15-hydroxyl group to the 15-keto group, the 15-hydroxyprostaglandin dehydrogenase. Since the 15-keto derivatives possess less than 10% of the activity of the 15-S alcohols, the acetylenic side chain provided a welcome chemical modification to increase duration of action.

As stated earlier, it was a major objective of this approach to utilize the hydroxylated eight-carbon side chain in optically active form to effect resolu-tion of the cyclopentanoid moiety and obtain the two diastereomeric 7-oxa-prostaglandins. Commercially available octyne-3-ol was resolved as the α-phenethylamine salt of the hemiphthalate ester, and the S alcohol was converted to the *tert*-butyl ether. In later work the trimethylsilyl ether was also used. The diethylalane prepared from (*S*)-3-*tert*-butyloxyoctyne reacted smoothly in 72% yield with the epoxide (**5**) to form the desired mixture of diastereomers **7a** and **7b**, which could not be separated by TLC. This failure

7a   7b

to achieve separation of diastereomers at the acetylenic stage has been a consistent observation in every case so far examined. This problem has been solved only recently (see below). After de-*tert*-butylation, followed by lithium aluminum hydride reduction, the diastereomeric olefins could be separated with ease and the synthesis completed, as above, resulting in optically active **2**[5].

III. Prostaglandin $F_{2\alpha}$

It seemed then that the alane chemistry was well suited for the introduction of the fully developed side chain with inversion of configuration at an epoxide site. Utilization of the intermediate **5**, conversion of the hydroxyl group created in the alane reaction to a good leaving group, and introduction of the carboxylate side chain (or a precursor of it) by carbanion chemistry would have resulted in the wrong stereochemistry at C-8. Even after inversion of that hydroxyl group, this path did not seem inviting to us. We rather decided to utilize the protected (protecting group P) *trans*-epoxide **8** by treating it with an appropriate alkyl lithium cuprate (trans opening) in order to introduce the substituent S, which was later to be transformed to the carboxylate side

8   9   10   11

chain (**9**). The newly formed hydroxyl group would be trans to the neighboring hydroxyl group at C-11 and could, after tosylation (**10**) and deprotection, be converted to a second epoxide (**11**). This latter epoxide would have the correct orientation for introducing the eight-carbon side chain and for generating the 11-hydroxyl group in the correct stereochemistry. This new epoxide would no longer be symmetrical and would therefore present a regiochemical problem: namely, how could alkynylation be directed to the 12- rather than the 11-position? Although we had some ideas of how to solve this problem, there was no certainty. Yet we felt a solution would be of wider interest and well worth pursuing.[6] We therefore proceeded along those lines. The reaction of the epoxide (**8**: P = ϕCH_2) with allyllithium prepared from tetraallyl lead and butyllithium in hexane and dissolving the precipitate in ether led to partial debenzylation, probably by Wittig rearrangement. The use of the less basic allyllithium cuprate produced **9** (S = allyl) in 95% yield as a crystalline compound, which gave the tosylate **12** in quantitative yield.

It should be mentioned at this stage that several options were available on how to modify the three-carbon side chain, keeping in mind that an aldehyde function was eventually required to employ the Wittig chain elongation according to Corey. We opted for the ethylene acetal **13** because of its ready convertibility to the aldehyde and in the expectation that the acetal oxygens would serve to anchor the alane reagent in such a way as to deliver the acetylenic anion at C-12 rather than at C-11. The crystalline tosylate **12** was therefore ozonized, and the crystalline ozonide was converted to the aldehyde with zinc in acetic acid, followed by acetalization with ethylene glycol to give **13**. The crystalline acetal tosylate was now subjected to hydrogenolysis with palladium-on-carbon to remove the benzyl protecting groups. This caused some problems in that the debenzylation was accompanied by hydrogenolysis of the acetal to form a hydroxyethyl ether. We ascribed this "over-reduction" to prior hydrogenolysis of some of the tosyloxy groups, releasing toluenesulfonic acid into the solution, which would catalyze this unusual side reaction. Buffering the reducing medium with sodium acetate proved an effective remedy and produced the diol **14**, which was converted to the epoxide **15** in 75% yield for the two steps. The stage was now set for the reaction with the alane reagent both in the form of its (S)-*tert*-butoxy and trimethylsiloxy derivatives (**16a** and **16b**). The desired objective, preferred substitution at C-12, was not achieved, the ratio between alkynylation at C-11 and C-12 ranging between 1.0 and 1.6, depending on the nature of the substituent at C-9 (Table I, entries 4 and 5). It became clear, however, from examining the ratios observed with hydrocarbon side chains at C-8, devoid, as they are, of oxygen capable of interacting with the alane reagent, that the acetal oxygens of **15** had indeed facilitated entry of the acetylenic substituent at the sterically more hindered 12-position (entries 1, 2, and 3). In these cases even more unfavorable ratios ranging from 2.7 to 5.0 were obtained. These findings

TABLE I

ISOLATED YIELDS AND RATIOS OF SUBSTITUTION AT C-11 AND C-12
FOR DIFFERENT SUBSTITUENTS (S) AT C-8

		Side chain					
		16b			**16a**		
		Substitution yield (%)		Ratio	Substitution yield (%)		Ratio
	Epoxide (**11**) S =	C-11	C-12	C-11/C-12	C-11	C-12	C-11/C-12
1	$(CH_2)_7OH$				50	10	5.0
2	$(CH_2)_7OTMS$, 9-OTMS	45	15	3.0	51	19	2.7
3	$CH_2CH=CH_2$, 9-OTMS	47	13	3.6			
4	CH_2CH (O, O epoxide)	22	23	1.0	26	21	1.2
5	CH_2CH (O, O epoxide), 9-OTMS	43	27	1.6	32	22	1.5
6	$(CH_2)_2OH$	7	35	0.2	0	60	0
7	$(CH_2)_2OTMS$	36	14	2.6			
8	$(CH_2)_3OH$				10	50	0.2
9	$CH_2CHOHCH_2OH$	4	20	0.2			
10	CH_2OH	50	15	3.3			

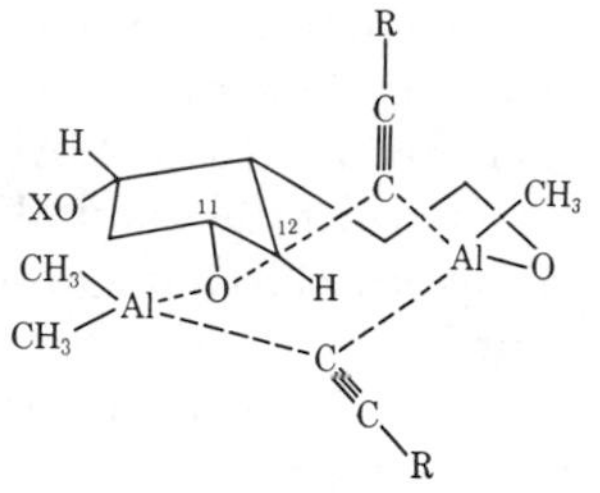

FIG. 1. Model of the transition state of the reaction between **16** and **17**.

suggested the hydroxyethyl side chain (entry 6) as the substituent of choice, since it offered an opportunity for covalent attachment of the alane reagent. Indeed, substitution now occurred exclusively at C-12, providing a solution to the problem of regiospecificity.[7] A transition state, pictured in Fig. 1, in which the reacting centers are held in a seven-membered ring, accounts for these results. Lengthening the side chain by one carbon to permit formation of an eight-membered ring (entry 8) still gives satisfactory regioselectivity, whereas shortening it by one carbon (entry 10) is thoroughly detrimental to substitution at C-12. As expected, silylation of the hydroxyethyl group (entry 7) abolishes the favorable effect of the primary hydroxyl group, leading to preferred substitution at C-11.

With this solution of the problem of regioselectivity in hand, only a minor change in the previously outlined reaction sequence was necessary. The crystalline ozonide prepared from **12** was reduced with borohydride, the benzyl groups were removed with 10% Pd–C in moist ethyl acetate, and the epoxide **17** was formed with 1 equivalent of KOH in methanol. The four steps were performed in an overall yield of 82%. As indicated in Table I, the introduction of the acetylenic side chain was preferably performed with the 3-*tert*-butoxy-alane (**16a**), leading in a 60% yield exclusively to the 12-substituted product, which consists, of course, of a mixture of diastereomeric *tert*-butyl ether triols **18a** and **18b**. To achieve a satisfactory alkynylation reaction it was necessary to convert the free hydroxyl groups to their anions, which was accomplished, rather uneconomically, by employing an excess of the alane reagent. (The

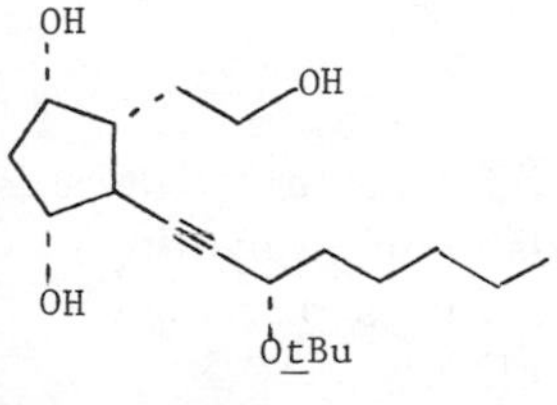

18a 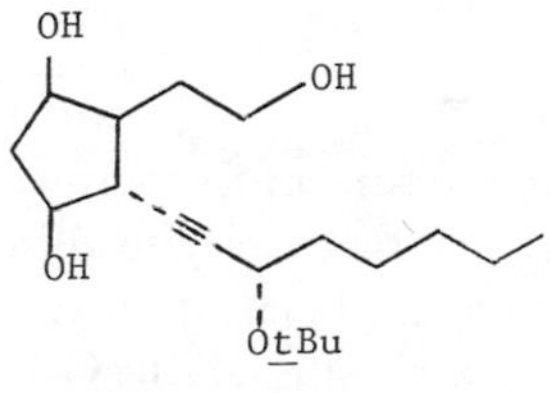   **18b** 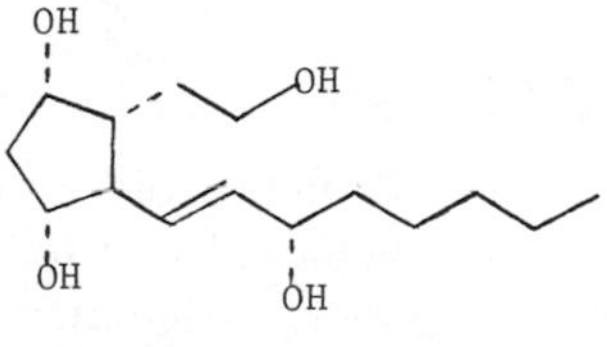   **19**

unused side chain was always recovered by vacuum distillation). The *tert*-butyl ether triols were debutylated with trifluoroacetic acid in 87% yield to the tetrols, and the latter were reduced with LiAlH$_4$ to the corresponding trans olefins **19**. (Only the isomer with the natural configuration is shown.) The latter were readily separated by chromatography, and the synthesis was continued with the more polar isomer. It is worth commenting at this point that the acetylenic diastereomeric tetrols could not be separated by chromatography, nor could any of their esters or ethers be isolated individually.

To complete the synthesis, there remained the extension of the two-carbon side chain and the introduction of the 5,6-cis double bond. Obviously, the most efficient process would be a selective oxidation of the primary hydroxyl group to the aldehyde followed by a Wittig reaction. Although we eventually succeeded in following such a plan, our first approach consisted of the conventional maneuvering of protecting groups: tritylation of the primary hydroxyl group (55%), followed by acetylation, detritylation, and oxidation with CrO$_3$·2py in CH$_2$Cl$_2$ (85% from the monotrityl ether). Deacetylation and Wittig olefination gave optically active PGF$_{2\alpha}$ (**1**) in 55% yield. Our dissatisfaction with the above protection–deprotection sequence led us to try a reaction that had impressed me for a long time and that I had been anxious to use to special advantage: the catalytic dehydrogenation of primary alcohols with platinum and oxygen. The selectivity of this reaction can, under certain circumstances, approach that of an enzymatic reaction. In the case of the prostaglandins, the geometry turned out to be unusually favorable for selectivity. When the tetrol **19** was treated with Pt and oxygen under carefully controlled conditions,[8] only the primary site was affected. Unfortunately, it was not possible to stop the reaction at the aldehyde stage, and mixtures of the hemiacetal **20** and the lactone **21** were produced. It was therefore more

                    20   **21**

expedient to prolong the reaction time until complete conversion to the lactone was achieved and to reduce the lactone with diisobutylaluminum hydride. The dehydrogenation reaction has been applied in excellent yield to many different substrates, including a prostaglandin C-1 alcohol.

One day, shortly after these results had been reported in the literature, John Sih, trying to duplicate the synthesis that C. H. Lin had worked out with so much skill, confessed with unbridled emotion that the regiospecificity of the alkynylation reaction with the epoxydiol **17** could no longer be reproduced. This seemed to be a terrible blow, but believing in a cause-and-effect universe, we felt that there had to be a rational explanation for this failure. We therefore examined every possible experimental facet of the reaction and the possible changes that might have been made at the time the disaster had struck. It soon became evident that the only potentially significant change was the use of a cylinder of dimethylchloroalane in toluene since the glass-bottled material previously in use had been consumed. (The use of dimethyl- in place of diethylchloroalane had been instituted long before, when the former became more readily available.) The bottle was sealed with the customary rubber septum, and since that bottle had been used over a long period, the septum could have been badly worn. As a result, oxygen could have entered and converted some of the dimethylchloroalane (DMCA) to methoxymethyl-chloroalane (MMCA). This hypothesis was quickly checked by contaminating tank-derived DMCA with MMCA by either air oxidation or addition of methanol, with the result that complete regiospecificity was restored when the "contaminated" reagent was used. Extensive experimentation indicated that a fixed ratio of DMCA · MMCA was not essential and that the arbitrary decision of converting DMCA to MMCA by addition of 1 mol of methanol achieved the desirable standardization of reaction conditions.

Coincident with the above described loss in the regioselectivity there was a marked increase in the rate of the reaction when the tank-derived alane was used. Again, with the use of MMCA the lower rates returned. The change in the regioselectivity of the reaction may well be related to the rate change, the slower rate permitting bond formation between the hydroxyl group of **17** and the alane reagent so as to assure the transition state shown in Fig. 1. Alternatively, the rate of exchange between the hydroxyethyl group of the substrate and the methoxyl group of MMCA may be faster than the protonation of a methyl group of DMCA with loss of methane.

IV.   13,14-Dehydroprostaglandins

It has been mentioned before that prostaglandins possessing an acetylenic group in place of the trans double bond possess unusual biological properties. Their ready availability by our synthetic route made such compounds attractive targets of synthesis. The first examples of this class were 13,14-dehydro-PGF$_{2\alpha}$ (**22**) and its diastereomer (**23**).[9] No new chemistry would be

22   **23**

required if our original strategy could be employed, which, to repeat, involved resolution of the racemic cyclopentane nucleus by combination with the optically active alkyne. To our great disappointment none of the chromatographic methods tried proved suitable for the separation of the acetylenic diastereomers. Several attempts in the laboratories of one of the major manufacturers of HPLC equipment ended in failure, even in the face of our offer to purchase their instrument if they could separate such mixtures. Acylation with a variety of acids was equally unproductive. Similar reports came from other laboratories. Only a decade later, when we took up the problem again, was a general method developed for the separation of such mixtures of acetylenic prostaglandins or their intermediates.

In the meantime, the functionalized cyclopentane nucleus had to be resolved.[6,9] This was conveniently performed on the intermediate **24** via the α-phenethylurethane **25** prepared with (S)-α-phenethylamine isocyanate. The rotation of the latter is negative in chloroform but positive in benzene, which became a source of considerable confusion because of failure of some of the suppliers to indicate the absolute configuration. Alkaline hydrolysis, or better, $LiAlH_4$ reduction, led to (+)-**24**, possessing the absolute configuration shown. The latter was determined by conversion of (+)-**24** to the epoxy ketone (−)-**26**, whose Cotton effect was related to that of the known (S)-2-methylcyclopentanone.

24   **25**   **26**

The synthesis of **22** and **23** in crystalline form from (+)-**24** and its antipode, respectively, in gram amounts followed in a straightforward manner the outlined synthesis of $PGF_{2\alpha}$. The intermediates **18a** and **18b** turned out to be the most satisfactory ones for the catalytic oxidation.

Our finally successful attack on the problem of isomer separation was sparked by the necessity of preparing the fluorinated 13,14-dehydroprostaglandin **27** in pure, optically active form.[10] From biological tests on a mixture of diastereomers it was concluded that this compound had to possess extraordinary activity. The resolved cyclopentane nucleus was now available but the required fluorinated side chain (**28**) resisted all attempts at resolution. The racemic material was prepared easily enough by Grignard reaction of 2-fluorohexanal with Mg acetylide, which resulted mainly in the desired erythro isomer. When the above synthesis was completed, using racemic **28** and (+)-**24**, the resulting mixture of diastereomers proved inseparable, consistent with previous observations.

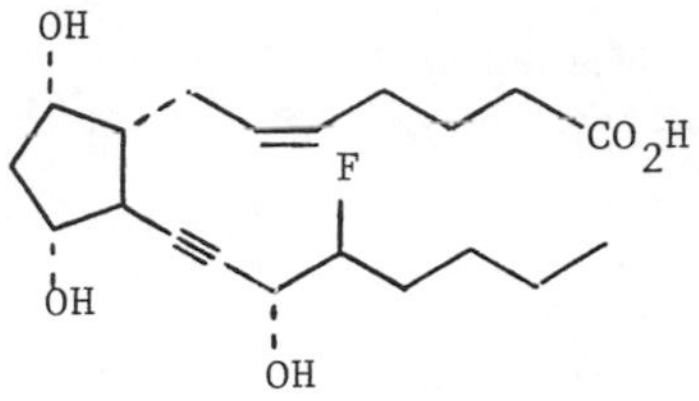

27

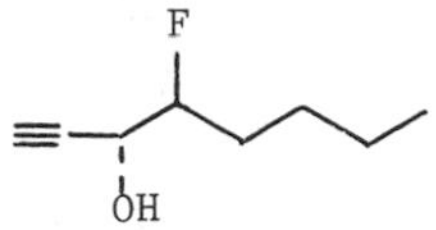

28

A general solution to the problem was finally achieved by changing the cylindric geometry of the acetylenic bond to a structure of lower symmetry by reaction with dicobalt octacarbonyl. The resulting dicobalt hexacarbonyl complexes, which possess extraordinary stability, could be readily separated by chromatography.[11] The complex **29** derived from such a mixture with its 15,16-epimeric diastereomer could then be reconverted to the parent acetylene by oxidation with ceric ammonium nitrate, furnishing the pure isomer **27** in optically active form. For practical reasons it was preferable to effect separation at the stage of the fluorinated lactones **30**, which showed

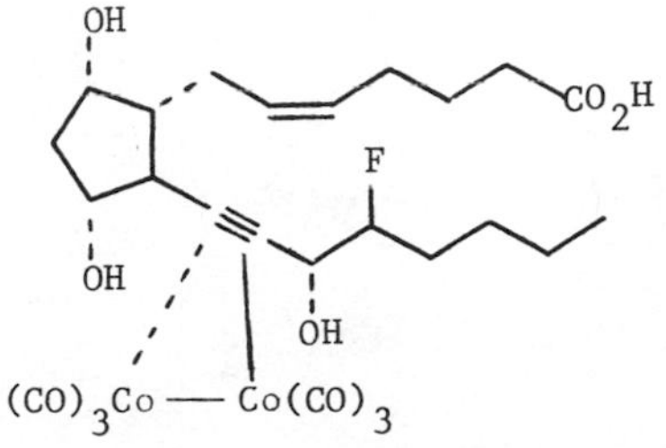

29

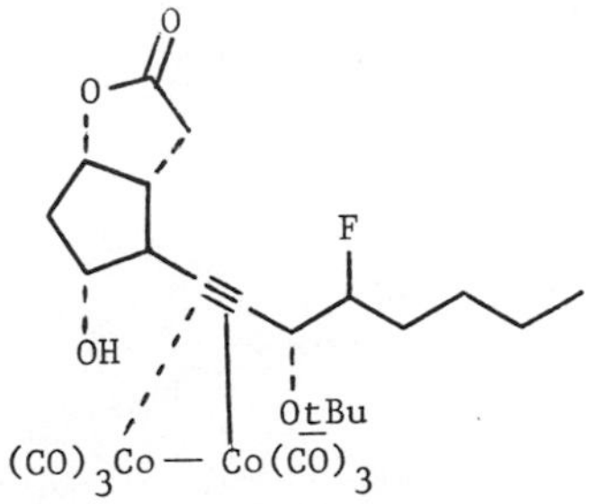

30

a very satisfactory difference in R_f values. In all cases tried, including that of the mixture of **22** and **23**, the method proved successful.

It should be noted that several of the compounds described in this chapter, as well as related compounds, have been prepared in optically active form by the above methodology on a several-hundred-milligram to several-gram scale.

REFERENCES

1. W. Nagata, M. Yoshioka, and T. Okamura, *Tetrahedron Lett.* pp. 847–852 (1966).
2. J. Fried, C. H. Lin, and S. H. Ford, *Tetrahedron Lett.* pp. 1379–1381 (1969).
3. J. Fried, S. Heim, S. J. Etheredge, P. Sunder-Plassmann, T. S. Santhanakrishnan, J. Himizu, and C. H. Lin, *Chem Commun.* pp. 634–635 (1968).
4. J. Fried, T. S. Santhanakrishnan, J. Himizu, C. H. Lin, S. H. Ford, B. Rubin, and E. O. Grigas, *Nature (London)* **223,** 208–210 (1969).
5. J. Fried, M. M. Mehra, and W. L. Kao, *J. Am. Chem. Soc.* **93,** 5594–5595 (1971).
6. J. Fried, C. H. Lin, J. C. Sih, P. Dalven, and G. F. Cooper, *J. Am. Chem. Soc.* **94,** 4342–4343 (1972).
7. J. Fried, J. C. Sih, C. H. Lin, and P. Dalven, *J. Am. Chem. Soc.* **94,** 4343–4345 (1972).
8. J. Fried and J. C. Sih, *Tetrahedron Lett.* pp. 3899–3902 (1973).
9. J. Fried and C. H. Lin, *J. Med. Chem.* **16,** 429–430 (1973).
10. J. Fried, M. S. Lee, Y. Yoshikawa and D. C. Mammato in Biochemical Aspects of Prostaglandins and Thromboxanes. Eds. N. Kharasch and J. Fried, Academic Press, 1977, p. 215.
11. C. O-Yang and J. Fried, *Tetrahedron Lett.* pp. 2533–2536 (1983).

Chapter 4

SYNTHESIS OF INDOLE ALKALOIDS

Philip Magnus

Department of Chemistry
Indiana University
Bloomington, Indiana

I. Introduction

The title of this book, *Strategies and Tactics in Organic Synthesis*, should evoke, at least in the student of organic synthesis, a question. What is meant by strategy and tactics as applied to organic synthesis? These two nouns are often used freely, without too much attention to exactly what differentiates one from the other, or how one influences the other. Now that we are in the era of computerized organic synthesis, this differentiation becomes even more important, since a computer, at least presently, appears to be tactically competent, but strategically naive. The most interesting and particularly appropriate difference is to be found in the subtle analogies of organic synthesis to the game of chess. A competent chess player implements a plan, usually based upon the control of key areas of the board, and attempts to execute this plan by positioning pieces on places that will have a direct, or more deviously, indirect outcome upon the control of the disputed area. Once

83

the key area of the chess board is in control, the process of completing the conquest can be carried out in two ways depending on the degree of territory controlled. The conquest can be completed in an elegant, efficient way by an imaginative combination of moves that offers no tangible resistance other than the massive depletion of forces, or by a more prolonged struggle that ends in the slow strangulation of all resistance. There are many fascinating plans, called strategies, and these are implemented by a direct response to the immediate circumstances, called tactics. Tactics are the methods used to carry out the strategy. In chess a poor strategy usually leads to defeat. Only in exceptional situations can superb tactics or methodology save the day, and lead to success. If excellent strategy is employed, poor tactics can frequently survive because the outcome is inevitable, although difficult. The most satisfying situation is a carefully planned strategy and a direct way of applying it.

In the practice of organic synthesis the above analogies are frequently true, and can perhaps be summarized by saying that a flexible strategy allows the maximum latitude in the choice of methods used to carry out a synthesis. Also new methods can enable completely different strategies to be developed. This is particularly true in the area of organometallic chemistry, where the recent surge of innovative methodology is leading to different types of strategies that could not be previously considered, since there was no way to implement them.

One of the most desirable features of a particular synthetic design is that it be applicable to more than one compound within any class of structurally similar types, whether they are natural products or not. This has been thematic to many of the syntheses of alkaloids, especially indole alkaloids, and has frequently been inspired by biosynthetic theories or models.

Within the auspices of my group, for the last 5 years or so, we have studied the uses of some aspects of organosilicon chemistry in synthesis. This area has become so popular recently that it seems as though every other paper concerned with synthesis involves organosilicon chemistry. Being convinced that the law of diminishing returns applies equally to synthesis as to economics, it seemed appropriate to expand the areas of research being conducted in my laboratories. Fortunately, through a series of chemical events that I'll describe next, a very opportune situation arose that has enabled my group to develop a new strategy for the synthesis of indole alkaloids.[1]

II. Early Thoughts

One of the potentially most useful ways in which a trimethylsilyl group can be incorporated into a synthetic plan is to view it as a "masked carbanion,"

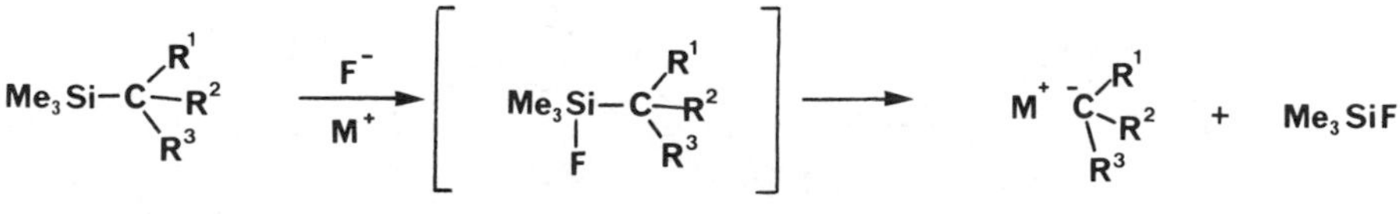

SCHEME 1

waiting to be exposed by treatment with fluoride ion (Scheme 1). An obvious advantage of this way of making carbanions is the simple access to counterions (M^+), such as Cs^+, that are not at all readily available by any other methodology. At the time we became interested in the application of this method of generating carbanionic species, a great deal of work had been published, and was currently being published, describing natural product syntheses, in particular in the ring A aromatic (estrogenic) steroid area, that capitalized on the intermediacy of highly reactive quinodimethanes or *o*-xylylenes (**1**). This type of reactive intermediate can be made by the types of reactions shown in Scheme 2, which although diverse, share the common handicap of requiring temperatures of 180°C or more, or of the inaccessibility of precursors with the necessary regiochemical control for intramolecular trapping. Utilizing the simple ideas expressed in Scheme 1, it seemed that a benzyltrimethylsilane, molded into a 1,4-Grob-type fragmentation reaction, would serve as a precursor to **1** under exceptionally mild conditions. The driving force of such a reaction is supplied by the extremely strong Si—F bond (~ 140 kcal mol^{-1}). Rather than pursue simplified model systems, it was just as straightforward to test this idea directly on an estrogenic steroid precursor. Treatment of **2** (prepared as outlined) with CsF in diglyme at 20°C cleanly gave 11α-hydroxyestrone *O*-methyl ether (**4**), thus validating the above ideas (Scheme 3).[2] An underlying incentive for developing this way of making quinodimethanes **3** was not just to have a mild system compatible with thermally sensitive

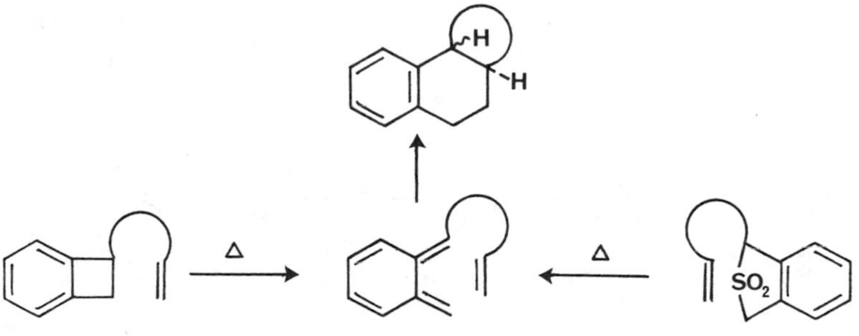

1

SCHEME 2

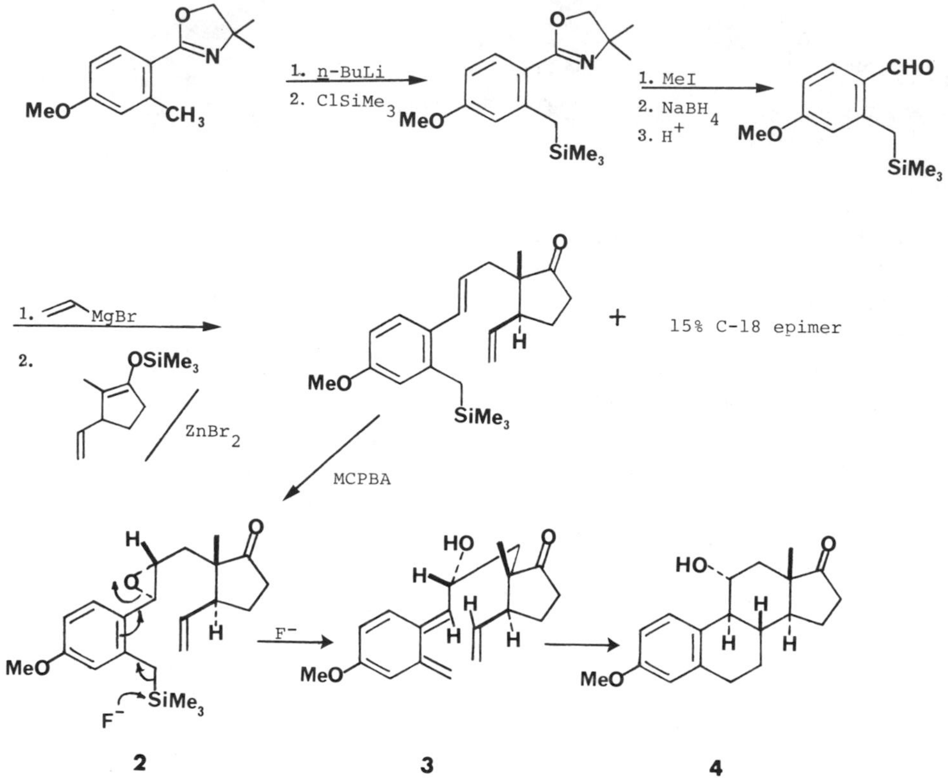

SCHEME 3

functional groups, but the benzyltrimethylsilane strategy should be readily adaptable to heteroaromatic systems. The benzocyclobutene method of making quinodimethanes, when extended to heteroaromatic compounds, such as pyridine, furan, thiophene, pyrrole, and indole would require the synthesis of a four-membered ring fused to these heterocycles: a formidable task in itself, and when further combined with the regiochemical problems

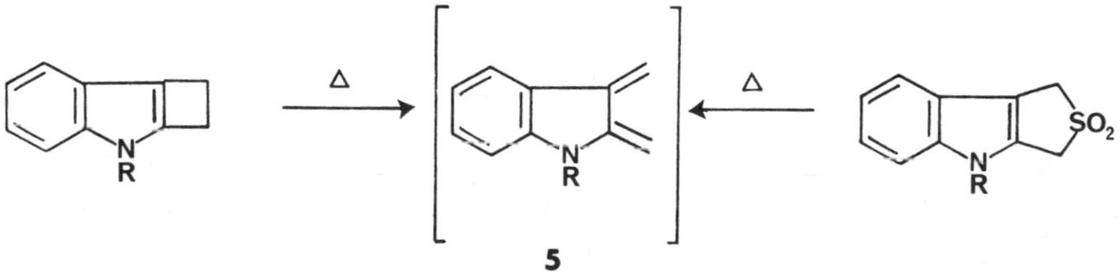

SCHEME 4

of either intermolecular or intramolecular cycloaddition trapping, it is hardly surprising that there have been no examples of heterocyclic quinodimethanes used for the synthesis of natural products.[3] The same constrictions apply to the utilization of sulfone precursors. This is illustrated in Scheme 4 for an indole-2,3-quinodimethane intermediate (**5**).

III.   Initial Results: First Phase

Obviously, the benzyltrimethylsilane strategy, at this stage in our research, seemed an ideal way of extending monobenzenoid quinodimethanes to hetero-aromatic systems. Of the variety of heteroaromatic possibilities, it did not take too much time to decide that the indole-2,3-quinodimethane system (Scheme 4) immediately suggested some intriguing ways of making indole alkaloids, particularly those of the *Aspidosperma* species. And before describing our experimental adventures, it is well to reveal that we did at least have some preconceived thoughts as to how this broad strategy might work out.

If the expression in Scheme 4 can be realized, using a 2-trimethylsilylmethyl group to generate an indole-2,3-quinodimethane intermediate (Scheme 5),

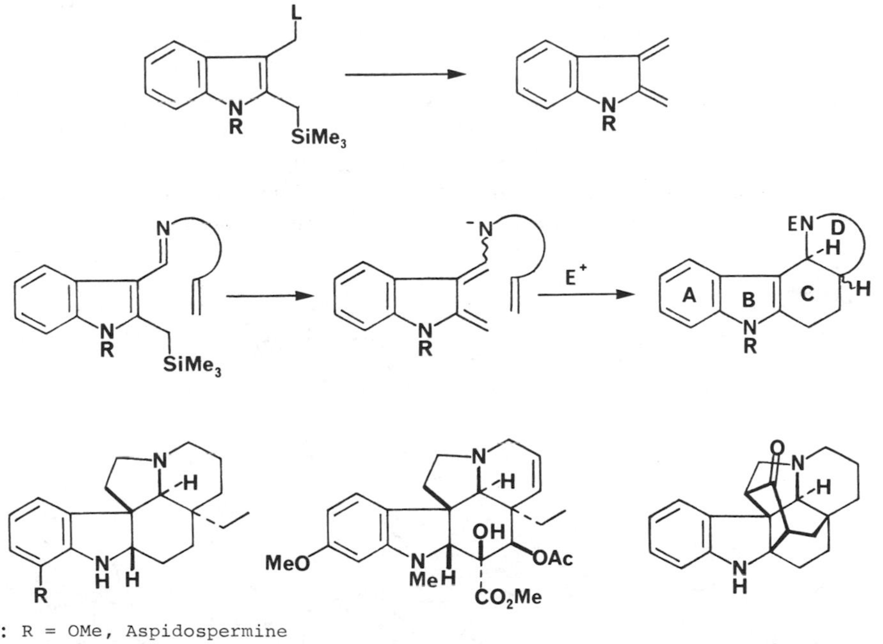

6: R = OMe, Aspidospermine

     R = H, Aspidospermidine

7 (Vindoline)

8 (Kopsanone)

SCHEME 5

then we could extend this to an intramolecular version and rapidly assemble the basic tetracyclic framework of the *Aspidosperma* alkaloids. Exemplary structures of alkaloids of the *Aspidosperma* group are aspidospermidine (**6**) and more highly functionalized versions of this, such as vindoline (**7**). A somewhat more ambitious objective that might be realized, if a number of problems could be solved, is the synthesis of the highly condensed members of the *aspidosperma*, such as the kopsanes (**8**) and eventually some of the fascinating dimeric indole alkaloids.

One of the most important outcomes of the intramolecular trapping of an indole-2,3-quinodimethane is the relative stereochemistry of the newly formed ring fusion C–D. The intermolecular trapping of a monobenzenoid *o*-quinodimethane has been generally assumed to have the geometry indicated in **9**. At first sight, this seems reasonable since the products formed, in particular those that have been used in estrone syntheses (Scheme 3), have the newly formed ring junction (B–C for steroids) predominantly trans-fused. The steric interaction between the exocyclic methylene group and the X group, usually CH_2 or NR^1, is assumed to destabilize the configuration **9a** in favor of **9**. On the other hand, the steric repulsion between the X group and a substituent R could be particularly unfavorable and destabilize the configuration **9**. It is particularly important that the indole-2,3-quinodimethane intermediate give predominantly a cis-fused C–D product since it will be necessary later to form a bond from C-11—C-12, and this can only readily be achieved if the C–D rings are cis-fused. Also, all the naturally occurring *Aspidosperma* alkaloids have the C–D rings cis-fused. It is amusing to remember that it is possible to epimerize at C-19, if the compound has a C=N double bond, by a retro-Mannich–Mannich sequence. This was discovered during the Stork and Ban synthesis of aspidospermine (**6**) (Scheme 6).[4] But before

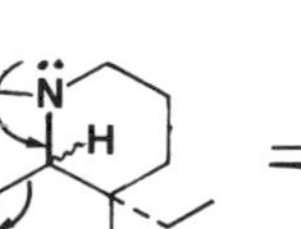

9: X = CH_2 or NR_2^1 **9a**

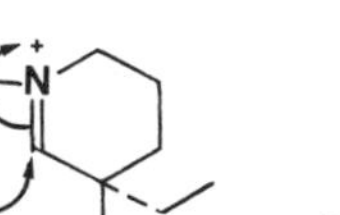

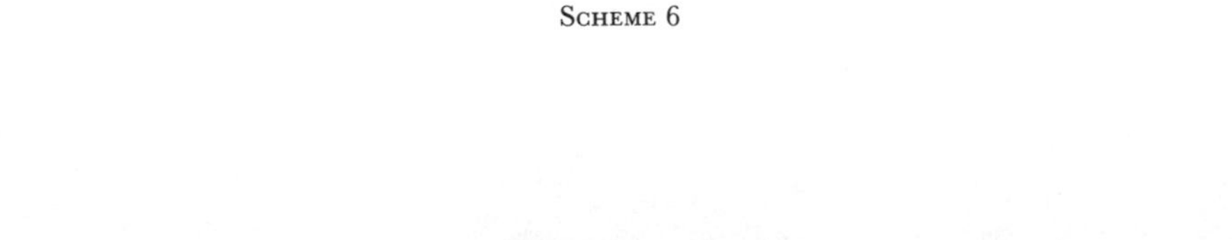

SCHEME 6

undue concern about stereochemical problems became too conjectural, it was of prime importance to know if the indole-2,3-quinodimethane strategy (Scheme 5) had any experimental reality.

It was decided to use a 4-methoxybenzenesulfonyl group to protect the indolic N atom for two main reasons: first, to remove any vinylogous amide character present in indole-3-carboxaldehyde derivatives and second, to direct lithiation into the 2-position of the indole nucleus. The easily made indole derivative **10** was treated with *n*-butyllithium, followed by trimethyl-silyl chloride, in an attempt to prepare the required 2-methyltrimethylsilyl system, **11**, gave mainly the aldehyde precursor to **10**. The ^{1}H-NMR data on the crude product suggested that the desired product was present, but extensive desilylation occurred on chromatography.

10: R = $SO_2C_6H_4OMe$-p throughout **11**

The extremely facile removal of the trimethylsilyl group suggested that the protodesilylation is proceeding via the desired indole-2,3-quinodimethane intermediate. Also, the desilylation observation indicated that the 2-methyl group of **10** must be considerably more acidic than one might have expected, although there are no quantitative data available in the literature. Maybe the trimethylsilyl group was not necessary. This was immediately seen as an exciting prospect because it increased the simplicity of the idea of making indole-2,3-quinodimethane intermediates and was easily tested in the laboratory.

Condensation of N^1-(4-methoxybenzenesulfonyl)-2-methylindole-3-car-boxaldehyde with 4-pentenylamine gave the imine **12**. When **12** was heated in acetic anhydride at 140°C for 4 h, a clean transformation took place to give a single compound (**13**) in 64% yield. While the spectral data were consistent with the gross structure being **13**, no information about the stereochem-

12   **13**

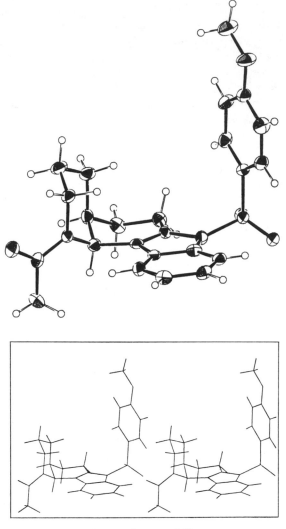

FIG. 1. Compound **13**.

istry of the newly formed ring C–D junction could be deduced with any
certainty. Indeed, amide resonance caused the data to be somewhat con-
fusing. As a result, a single crystal X-ray structure determination was undertaken, and the result is shown in Fig. 1. The C–D ring junction is gratifyingly
cis. While it is possible that the formation of **13** is the result of thermodynamic
equilibration of the initially (kinetically) formed trans-fused product **14**
through a reversible fragmentation that is somewhat typical of gramine derivatives, we have accumulated substantial evidence that this is not the case.

14   **14a**

15   **16**   **17**

18 exo-(E)   **19** exo-(Z)

For example, the acetylenic system **15** readily cyclizes at 130°C in the presence of methyl chloroformate and diisopropylethylamine to give the 1,4-dihydro-carbazole **16**, without irreversibly undergoing 1,4-elimination to the aromatized carbazole **17**. Consequently, it appears that **13** is the kinetic product and therefore arises from a transition state that we have designated exo-(Z) (**19**) rather than exo-(E) (**18**), which would lead to trans-fused products. Presumably, the steric repulsion between the 4-position of the indole and the imine–enamine N-substituent is sufficient to destabilize the normally assumed configuration (see earlier discussion of quinodimethane geometry for simple monobenzenoid systems) of the diene **18** and make **19** preferable. We have carried out some 30 different examples of this type of reaction to date and have not detected trans stereoisomers.

This very promising result immediately suggested a number of extensions, but first it was important to establish two facts. First, how are we going to make the C-11—C-12 bond and carry the two-carbon, so-called tryptamine bridge C-10—C-11 in this strategy. Secondly, does the indole-2,3-quinodimethane cyclization work when the alkene intramolecular trap carries an ethyl group, which is necessary if the basic *Aspidosperma* alkaloids, such as **6**, are to be made in highly convergent sequence (Scheme 7).

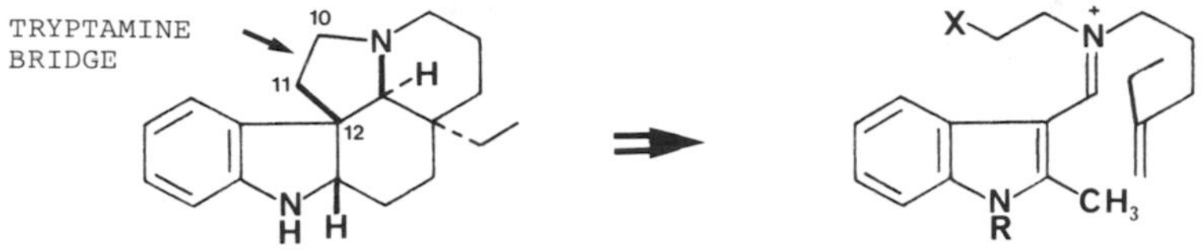

SCHEME 7

IV. Flexible Strategy–Flexible Tactics: Second Phase

The first problem we decided to tackle was the formation of the C-11—C-12 bond, which, of course, includes within this problem the two-carbon tryptamine bridge C-10—C-11. Our first inclinations proved to be somewhat frustrating and difficult but, by their very nature, uncovered some interesting chemistry and, not as an incidental, eventually solved the above problems.

Since **13** was readily available, we attempted to hydrolyze the N-acetyl group to form the secondary amine **20** (E = H). Perhaps not too surprisingly, the amide was not hydrolyzed without complete destruction of the rest of the molecule. Immediately this negative result provided a direct incentive to look at other electrophiles to trigger the conversion of the imine **12** into tetracyclic indole derivatives, where the functional group attached to the secondary amine can be removed under suitably mild conditions without destroying the rest of the molecule. Treatment of **12** with a range of chloroformates in chlorobenzene, in the presence of diisopropylethylamine, at 138–140°C gave the octahydropyridocarbazoles **20** in the yields shown. The reason for the complete failure of benzyl chloroformate is its decomposition below the temperature required to make **20**.

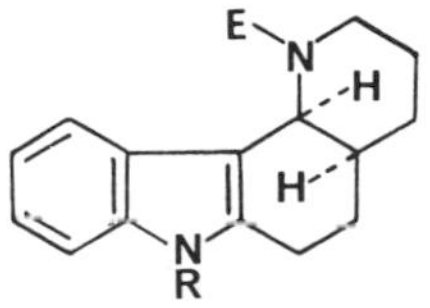

For E =	$CO_2CH_2CH_2Cl$ (92%)
	$CO_2CH_2CCl_3$ (79%)
	CO_2Me (88%)
	CO_2Ph (68%)
	CO_2Et (43%)
	CO_2CH_2Ph (0%)

20

The simplest candidate among the series of carbamates **20** is the N-carbo-methoxy system (E = CO_2Me). Removal of the N-p-methoxybenzenesulfonyl group with aqueous potassium hydroxide was straightforward. It must be remembered that sulfonamides are notoriously difficult to remove, and the reason for the comparative ease of removal of this one is because the pK_a of indole is approximately 16, and therefore it is still a reasonable leaving group. Treatment of the desulfonylated carbamate **21** with trimethylsilyl iodide, generated *in situ*, gave the air-sensitive diamine **22** (66%). As in chess, our first analogy, if the order of moves is reversed the result is different. Treatment of **20** (E = CO_2Me) with trimethylsilyl iodide for many hours in acetonitrile, heated at reflux, gave no reaction. Evidently, the removal of the indole pro-tecting group provides a mechanism to activate the N-carbomethoxy group. A very plausible way to do this is shown below, and is in keeping with other experimental results we have accumulated, and, of course, typical of gramine chemistry.

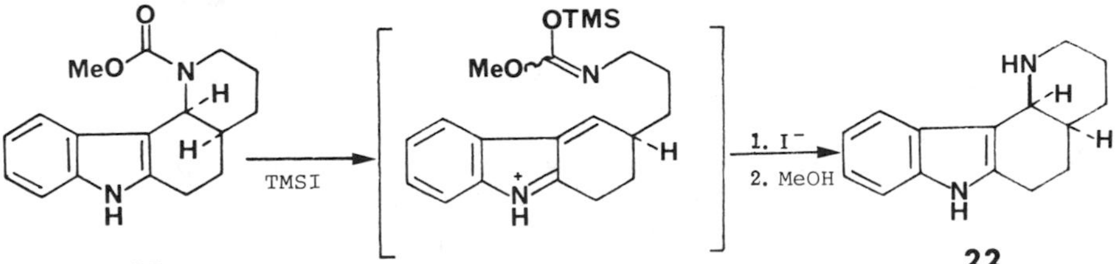

It is somewhat ironic that this order of deprotection of the two nitrogen atoms proved to be very crucial to the overall strategy, for in the overall scheme of things it might not appear too significant. Where it becomes problematic is in combination with the two-carbon tryptamine bridge. Cer-tainly, in principle, Scheme 8 would appear to present the perfect combination of what we know so far but, unfortunately, molecular life is not so simple. If the acetylating agent **23** is a two-carbon unit with the inherent capability of dealing, in a mechanistic sense, with the forming of the hindered (quater-nary) C-11—C-12 bond, then it seems inevitable that the protons adjacent to the carbonyl and X groups are quite acidic. This means that the two-carbon C-10—C11 progenitor can readily form a ketene and remove the imine as an unwanted β-lactam; and true to course, this is exactly what happens. Consequently, we must proceed to a carbamate derivative (**20**), remove it, and replace it by a group such as **23**. While this introduces unwanted steps, it eventually offers certain advantages that emerge later but, to be honest, only after the fact. It is good to be wise before the fact but better to be wiser after it. Combined with these problems, if the diamine **22** is converted into a range of reasonable precursors to pentacyclic systems, such as **24**, these

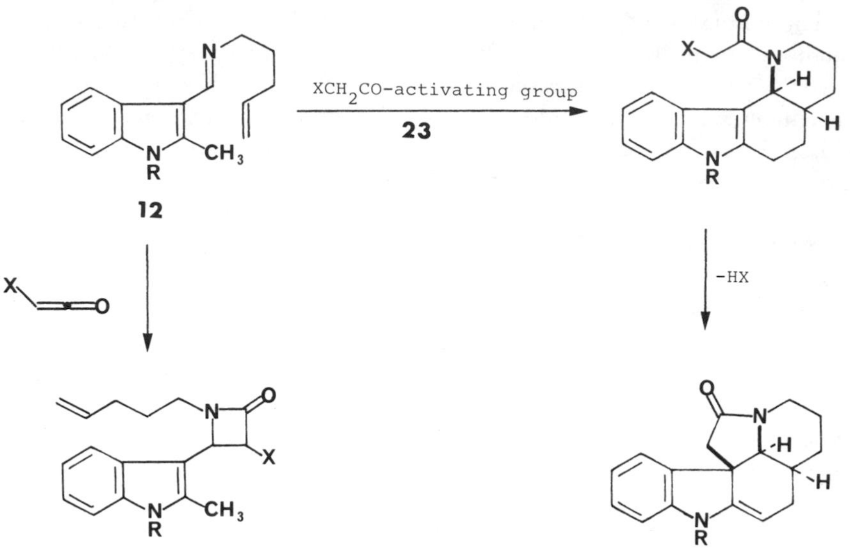

SCHEME 8

do not close the C-11—C-12 bond to give pentacyclic compounds, at least, not under a wide range of plausible conditions. This is shown in Scheme 9.

Of particular significance in this sequence was that we subsequently found that if the indole nitrogen was inductively deactivated, the intramolecular Pummerer-type reaction depicted in Scheme 9 was successful. This negative

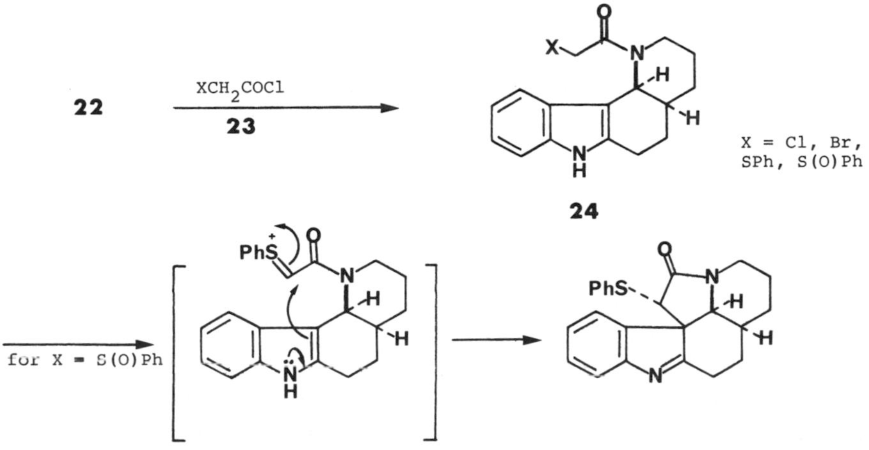

SCHEME 9

observation immediately posed the problem of how to construct a tetracyclic indole system (**20**) having a group that can be removed from the secondary amine nitrogen atom without disturbing the p-methoxyphenylsulfonyl group attached to the indole nitrogen atom. Of the series of carbamates **20**, those that possess the potential to be removed by reductive means appeared to offer the most promise. The 2-chloroethylcarbamate **20** ($E = CO_2CH_2CH_2Cl$) was treated with a number of strong reducing agents ($CrCl_2$, Bu_3SnH) including $Zn-H_2O-THF-B_{12}$(aquo). In no case did we observe any of the required secondary amine **25**. Obviously, an extremely potent reducing agent was required. As one travels down the periodic table, the electropositive nature of the elements increases, and in the lanthanide series the outermost electrons, being far from the nucleus, are easily removed. Consequently, we looked at the possibility of using a low-valent lanthanide to reduce the 2-chloroethylcarbamate and were immediately attracted to the fascinating work of Kagan, involving samarium diiodide. Samarium diiodide is easily made by adding 1,2-diiodoethane to a suspension of samarium metal in THF. Ethylene gas is evolved and a deep turquoise solution of SmI_2 is formed. The exclusion of oxygen is imperative. When the 2-chloroethylcarbamate **20** ($E = CO_2CH_2CH_2Cl$) was added to the SmI_2 solution at 70°C, a clean conversion to the secondary amine **25** (70%) took place.[5] The overall yield

20 SmI_2 ⟶   **25**

of **25** from **12** was 64%. While this sequence is chemically more interesting, an alternative way of making the secondary amine **25** was to treat **20** ($E = CO_2CH_2CCl_3$) with $Zn-AcOH-H_2O$ to give **25** (86%) in an overall yield of 69%. At this point the stage was set to investigate the formation of the C-11—C-12 bond. It is well to mention briefly that other investigators have looked at the formation of the C-11—.C-12 bond, and their results are summarized in Scheme 10. All of these reactions involve, in the transition state, a pseudo-pentacoordinate such as **26**. It is also possible that an aziridinium species **27** is involved. It should be noted that the formation of the C-11—C-12 bond involves making a quaternary carbon atom, a usually difficult problem. Changing the hybridization at C-11 from sp^3 to sp^2 reduces the steric requirements since only trigonally hybridized intermediates are involved (Scheme 11). Scheme 11 predicts explicit stereochemistry for the group X at C-11.

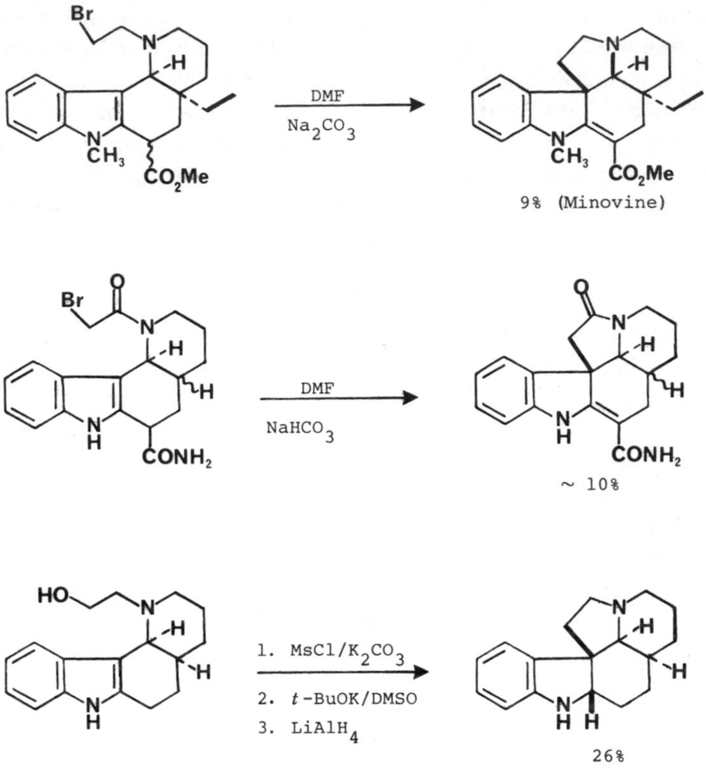

9% (Minovine)

~ 10%

1. MsCl/K$_2$CO$_3$

2. t-BuOK/DMSO

3. LiAlH$_4$

26%

Scheme 10

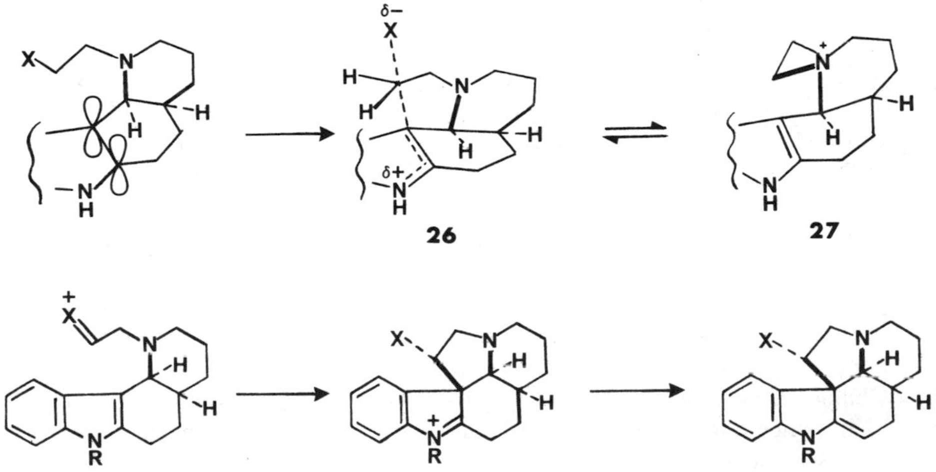

26   27

Scheme 11

Alignment of the 2,3 π-bond of the indole in a trans-coplanar fashion to the C=X⁺ species leads to the stereochemistry for X shown in Scheme 11. A further advantage, if such a scheme were to work, is the functional handle that a suitable X group would impart to the C-11 position. In the longer term prospects, this could prove to be extremely valuable if this strategy is to be applied to the more complicated indole alkaloids. Consequently, we first converted the amine **25** to the phenylthioacetyl derivative **28** and oxidized it to the sulfoxide **29**. Exposure of the above sulfoxide to trifluoroacetic anhydride in dichloromethane at 0°C for 10 min, then addition of chlorobenzene,

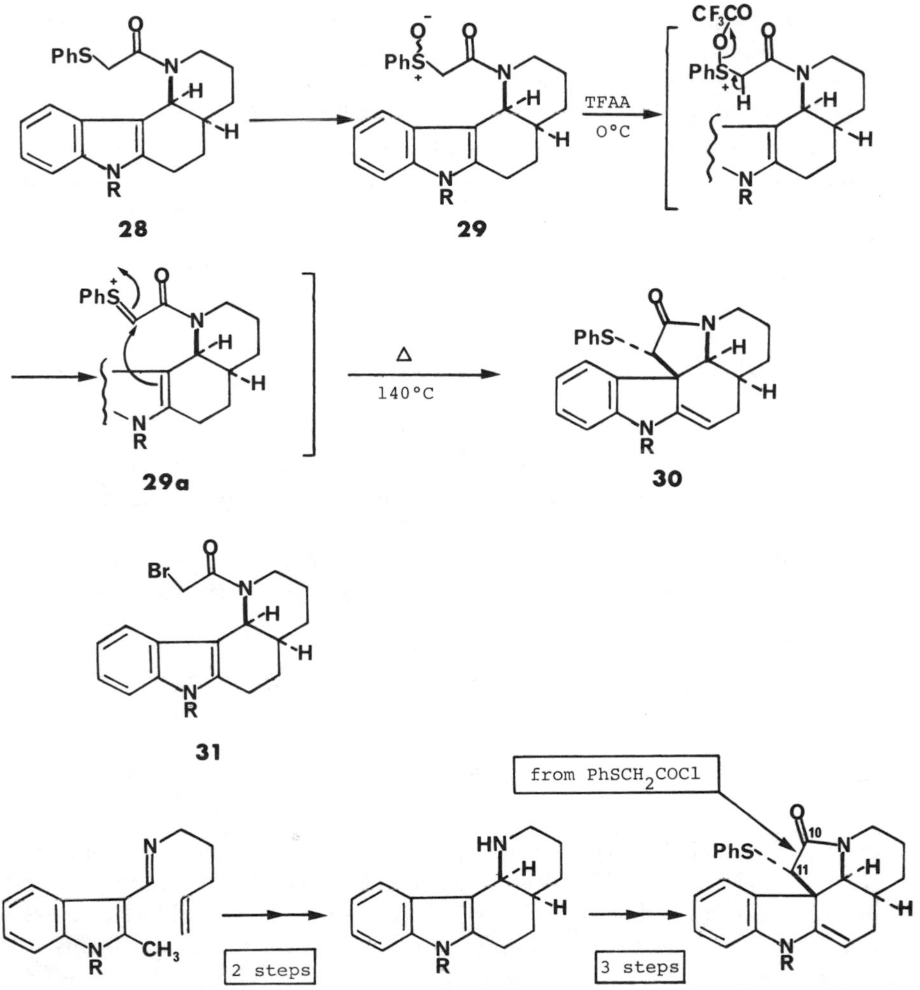

SCHEME 12

followed by rapid heating to 140°C, gave the desired pentacyclic system **30** (55%). These precise experimental conditions were rationalized by the observation that the sulfoxide **29** was consumed at 0°C, but none of the product **30** was formed. Not until the mixture was heated to 140°C was **30** formed. Presumably the intermediate sulfonium ion **29a** does not have sufficient activation energy at 0°C to close to **30** and requires quite considerable thermal activation. Also, the indole-2,3 π-bond is at best only weakly nucleophilic because the N-sulfonyl group strongly withdraws the nitrogen lone pair, although only inductively and not resonancewise; the indole nitrogen atom is pyramidal (see Figs. 1, etc.). Interestingly, and by way of contrast, the α-bromoacetamide **31** could not be induced to form the C-11—C-12 bond under a variety of basic conditions. The intramolecular Pummerer reaction method for completing the pentacyclic skeleton **30** proved in a number of crucial cases to be ideal. Frequently yields as high as 90% were obtained. The problem of the C-11—C-12 bond formation having been solved, we could proceed directly to the synthesis of a simpler member of the *Aspidosperma* alkaloids, aspidospermidine (**6**: R = H), and see if the indole–quinodimethane methodology is successful when an ethyl group is present at the C–D ring fusion. Our progress at this point is summarized in Scheme 12.

V.   Synthesis of (±)-Aspidospermidine: Third Phase

The first requirement of this important third phase was a convenient way to make 4-ethyl-4-pentenylamine (**32**). A superbly simple solution (a quick combination, using chess parlance) was available in the organometallic literature.[6] The readily available triol **33** (propanal–formaldehyde) was converted into the trichloride **34**, and then treated with magnesium metal, followed by carbon dioxide, to give the acid **35**. Straightforward methods

33: X = OH   **35**   **32**
34: X = Cl

converted the acid **35** into the amine **32**. Treatment of the imine **36** (made by condensation of **32** with N-4-methoxyphenylsulfonyl-2-methyl-3-formylindole in the presence of molecular sieves) with a range of alkyl chloroformates gave the tetracyclic carbamates **37**. While the cis-relative stereochemistry was known in the desethyl series, it was by no means a foregone conclusion that the ethyl series has the same stereochemistry. A suitable derivative for single-crystal X-ray crystallography was found to be the free indole **38**, and Fig. 2 shows that the relative stereochemistry of the newly formed ring junction is indeed cis.

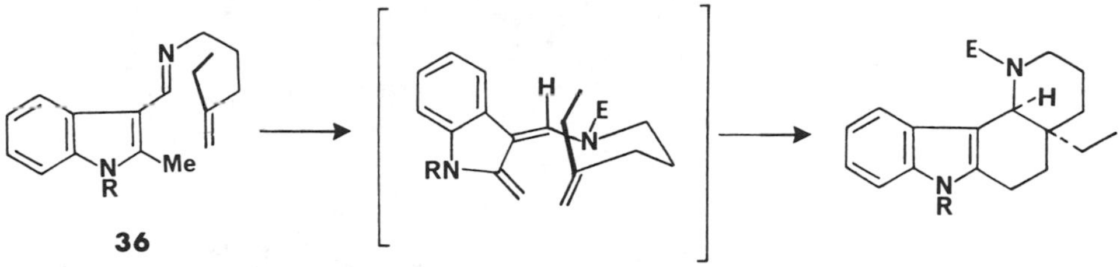

37 : E = CO$_2$Me (54%)

E = CO$_2$CH$_2$CH$_2$Cl (70%)

E = CO$_2$CH$_2$CCl$_3$ (46%)

E = H

E = COCH$_2$S(O)Ph

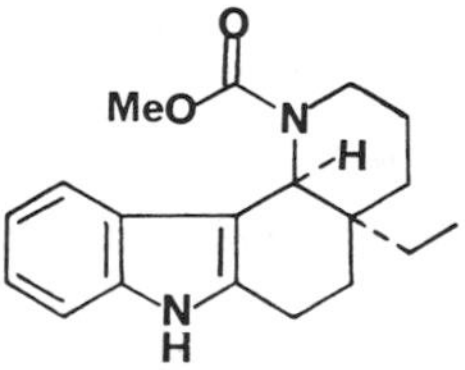

38

The most convenient way to complete the sequence was to remove the trichloroethylcarbamate group in **37** (E $=$ CO$_2$CH$_2$CCl$_3$) with Zn–AcOH to give **37** (E $=$ H) and convert it to the sulfoxide **37** [E $=$ COCH$_2$S(O)Ph]. Treatment of the sulfoxide **37** [E $=$ COCH$_2$S(O)Ph] with trifluoroacetic anhydride, as for **30**, gave the pentacyclic system **39** (91%). Standard desulfurization (Raney nickel) gave **40**, which was reduced with LiAlH$_4$ to give ($\pm$)-aspidospermidine (**6**: R $=$ H). For this particular sequence, the

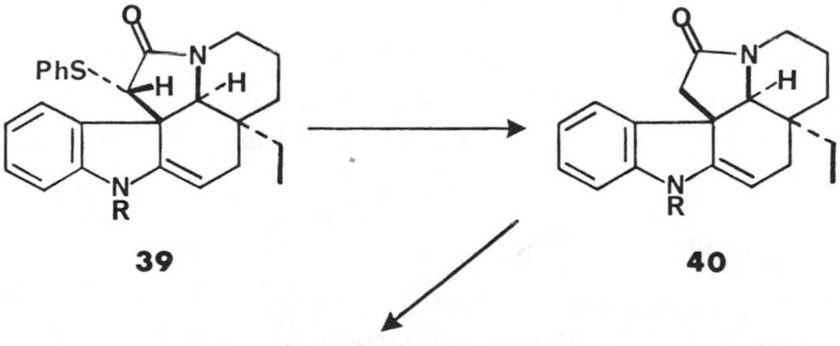

39

40

6: R = H

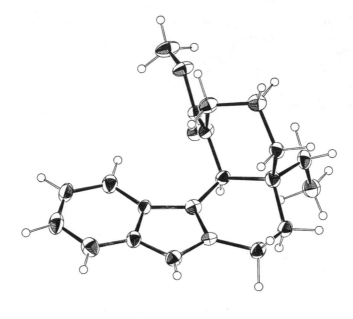

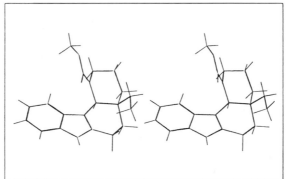

Fig. 2. Compound **38**.

overall yield from the *N*-sulfonylated indole is 6.3% through eight steps. While this is a relatively short synthesis, the deficiencies outlined in Scheme 8 still apply and suggested our next extension of this type of strategy.

VI.   Reverse Roles: Fourth Phase

If we make the C-10—C-11 two-carbon tryptamine bridge part of the imine and the triggering electrophile contain the alkene component to

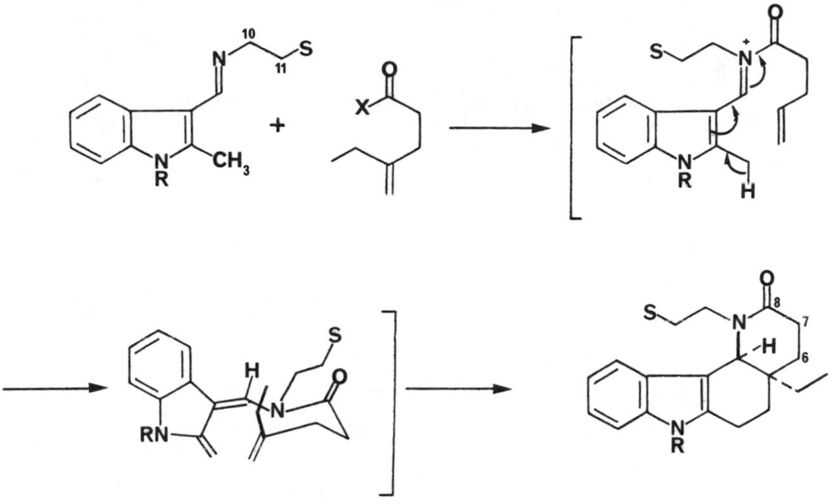

SCHEME 13

intramolecularly trap the indole-2,3-quinodimethane, then it is possible to assemble all the atoms of aspidospermidine in a single step (Scheme 13). This reversal of roles not only removes two steps (deprotection and reacylation of the piperidine nitrogen atom) but places the amide carbonyl at C-8 rather than at C-10. This functionality offers the potential to introduce a double bond at C-6—C-7 and an ethyl group at C-7. Both transformations lead to other interesting indole alkaloids.

The basic ideas represented in Scheme 13 work, but the complications that arose decreased its basic efficiency but compensated by providing many valuable discussions that resulted in many useful experiments. We treated a wide range of imines **41** with the mixed carbonic anhydride derived from ethyl chloroformate and 4-pentenoic acid in chlorobenzene, initially at 20°C, then heated to 135–140°C, and obtained the required tetracyclic amides **42** along with two other products, **43** and **44**. Rather than describe the many different types of conditions that have been used in these cyclizations and the large number of different S groups (only three are shown here) used, I will concentrate in detail on the system where $S = CH_2CH_2SPh$ for two important reasons. First, it is this system that readily enabled the C-11—C-12 bond to be made, and secondly the mass balance for the conversion of the imine **41** ($S = CH_2CH_2SPh$) to **42/43/44** ($S = CH_2CH_2SPh$) is 83% (the remaining 17% consists of intractable material and a small amount of hydrolysis of the imine).

Obviously, if the formation of the ethoxy adduct **43** and the β-lactam **44** could be decreased, or completely removed and replaced by **42**, then we

41: S = CH$_2$CH$_2$SPh

S = CH$_2$CH$_2$SePh

S = CH$_2$CH(OMe)$_2$

42: S = CH$_2$CH$_2$SPh (60%)

S = CH$_2$CH$_2$SePh (22%)

S = CH$_2$CH(OMe)$_2$ (56%)

43: S = CH$_2$CH$_2$SPh (13%)

S = CH$_2$CH$_2$SePh (15%)

S = CH$_2$CH(OMe)$_2$ (19%)

44: S = CH$_2$CH$_2$SPh (10%)

S = CH$_2$CH$_2$SePh (5%)

could expect the yield of **42** to increase from 60 to 83%. Scheme 14 depicts a reasonable working hypothesis upon which to plan corrective experiments.

The probable intermediate acyliminium ion **45** can lose a proton to give the tetracycle **42** (via an indole-2,3-quinodimethane intermediate). It can be trapped by the liberated ethoxide–ethanol to give the adduct **43**. The mixed carbonic acid anhydride can fragment to a ketene, which in a [2 + 2] cycloaddition to the imine bond results in a classical way to make the β-lactam **44**. Base catalysis by addition of Et$_3$N or iPr$_2$EtN, with the intention of assisting the proton-loss pathway, removed the adduct **43**, but the β-lactam **44** became a major by-product (>30%). Acid catalysis (TsOH) caused complete destruction to uninteresting products. With the simple solutions out of the way, more detailed rationalization was required. It might have been reasonably expected that the ethoxy adduct **43** would reversibly eliminate ethanol on heating to give back the intermediate acyliminium ion **45**, which can irreversibly lose a proton to give the tetracyclic amide **42**. Unfortunately, this was not the case. Prolonged heating of **43** (180°C) gave the starting imine **41** and ethyl 4-pentenoate. Conducting the reaction of **41** with the mixed anhydride at 180°C instead of 135°C did not give the ethoxy adduct **43**, but the β-lactam **44** became the major by-product. The required tetracyclic amide **42** was formed in the usual 60% yield.

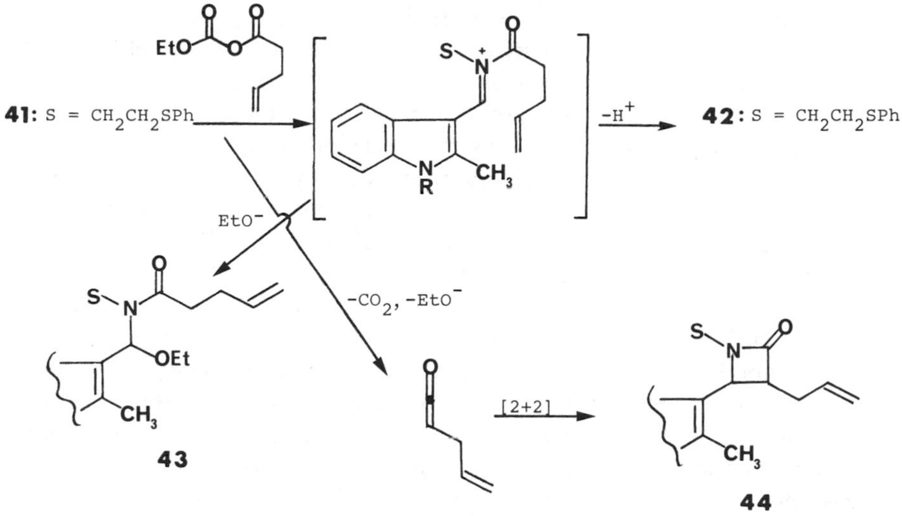

SCHEME 14

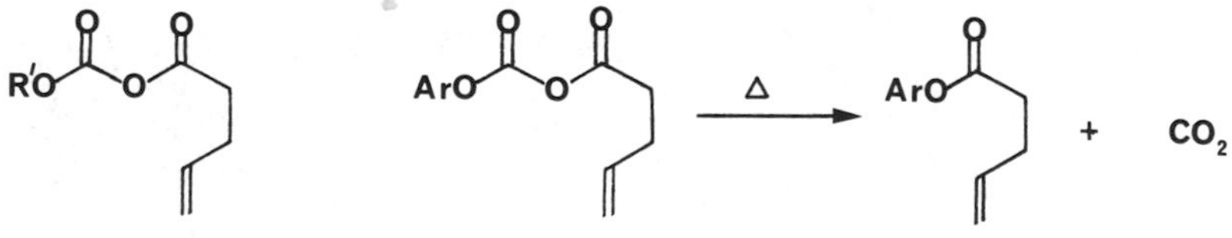

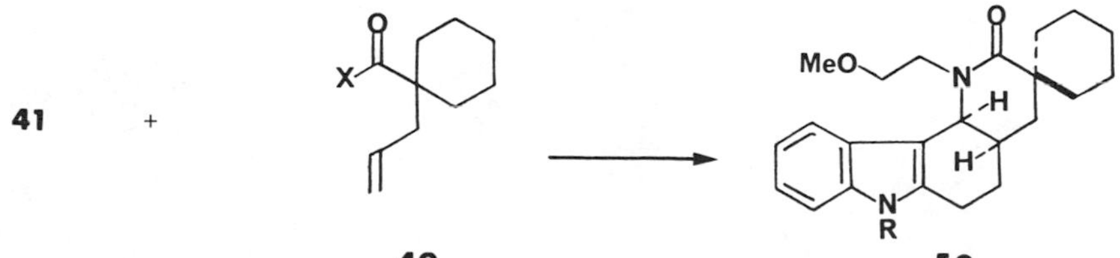

Efforts to reduce the amount of ethoxy adduct **43** by using different mixed carbonic anhydrides did not work. For the sequence **46**: R′ = Me, Et, *n*-Bu, all gave the alkoxy adducts in approximately the same yield, 13–20%. We could not use *tert*-butyl or isopropyl mixed carbonic anhydrides because they decomposed at the temperature (135°C) needed to conduct the cyclization. The aryloxy carbonic anhydride **47** rearranged to the ester **48** at approximately 100°C. Other acylating species, such as acid chlorides, acylimidazoles, acylimidazolium salts, or acylsulfonates, gave absolutely none of the tetracycle **42**. It appears that very reactive acylating agents lead predominantly to β-lactams via the ketene pathway. If the ketene formation is blocked by geminal alkylation, then the cyclization works. The conversion of **49** to **50** illustrates this.

According to our working hypothesis (Scheme 14) the acyliminium ion is trapped in a bimolecular process by the nucleophilic alkoxide fragment. If we designed an alkoxide fragment that would destroy itself in an intramolecular fashion, this should overwhelmingly compete with any bimolecular trapping. Two interesting candidates that we tried were **46** (R′ = CH=CH$_2$) and **46** (R′ = CH$_2$CH$_2$Cl). In the first case, the mixed carbonic anhydride is easily made from vinyl chloroformate. Although no alkoxy adduct **43** was formed by using this system, the yield of **42** was the same. In the second case, the 2-chloroethyl system should destroy itself by forming either ethylene oxide and carbon dioxide or ethylene carbonate. Sadly, the alkoxy adduct **43** was formed in the usual yield. This quite unexpected result required us to modify our working hypothesis. The most plausible explanation is that the adducts are formed concertedly with the acylation step via a pseudo intramolecular addition of the alkoxide to the developing iminium ion (Scheme 15, Mode A).

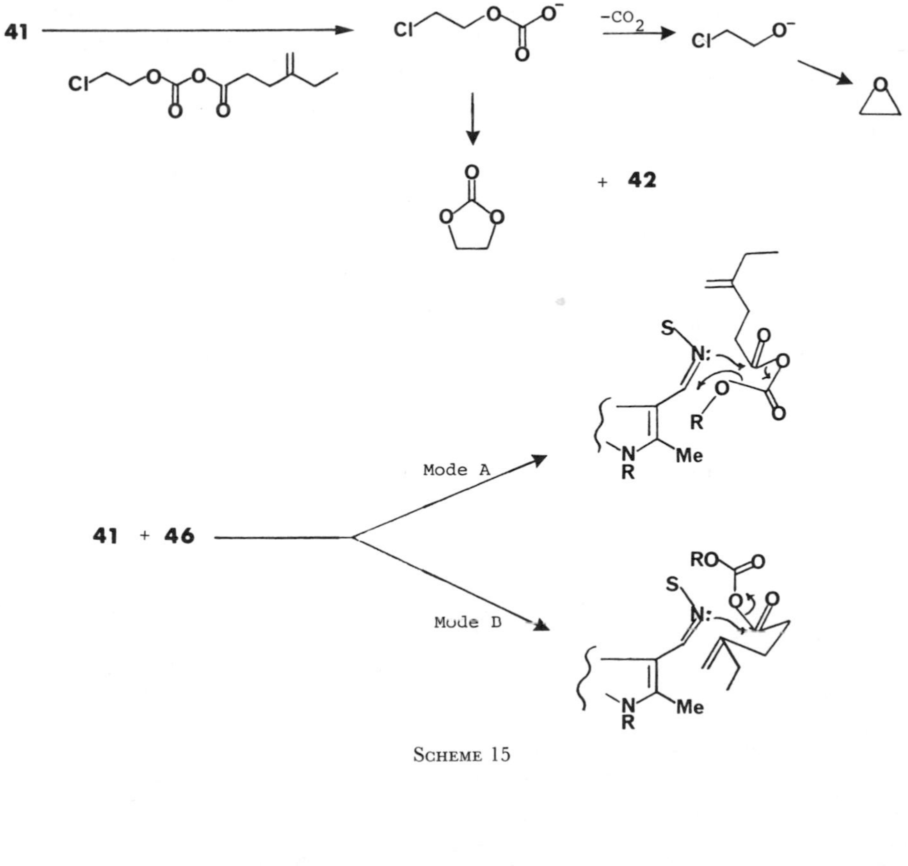

SCHEME 15

The imine **41** can interact with the mixed anhydride in two orientations for the nitrogen lone pair of the imine to have a low energy trajectory (maximum overlap) with the π-orbital of the ester carbonyl group. Mode A situates the alkoxy group directly over the π-system of the developing acyliminium ion in a six-membered chair-like transition state and leads to the observed adducts **43**, whereas turning the mixed carbonic acid anhydride around, as in Mode B, allows the acylation step but releases the carbonate group into the surrounding environment.

All of these complications are exacerbated in the ethyl series. Treatment of the imine **41** with the mixed ethyl carbonic anhydride from 4-ethylpentenoic acid at 135°C in chlorobenzene gave the tetracyclic amide **51** (33%, after chromatography and recrystallization) and the ethoxy adduct **52** (21%) along with small amounts of the β-lactam **53**. To establish unambiguously

the relative stereochemistry of **51**, a single crystal X-ray crystallographic determination was carried out. Figure 3 shows the result. While the tactic of the intramolecular indole-2,3-quinodimethane cyclization only works in 33% yield, in this strategy, in a single step, all the carbon atoms of aspidospermidine are assembled. Furthermore, the relative stereochemistry of the C–D ring fusion, a key stereochemical feature, is correct.[7]

The conformation of **51**, in the crystalline state (Fig. 3) is very unusual. The phenylthioethyl group attached to the amide nitrogen atom is held over the indole ring and parallel to the 4-methoxyphenylsulfonyl group. This conformation situates the C-11 carbon atom directly over the C-12 position, approximating to the positions these atoms must be in when the C-11—C-12 bond is made in the Pummerer reaction step. Of course, this applies only to the crystalline state but is nonetheless provocative and unusual.

Before moving on to the conversion of the tetracyclic amide **51** into (±)-aspidospermidine, a few final comments about the indole-2,3-quinodimethane cyclization are in order. The very best yield is over 90%, and the worst workable yield is 33%. In general, the ethyl series cyclizations proceed in approximately 20% lower yield than the desethyl series. Although we have tried

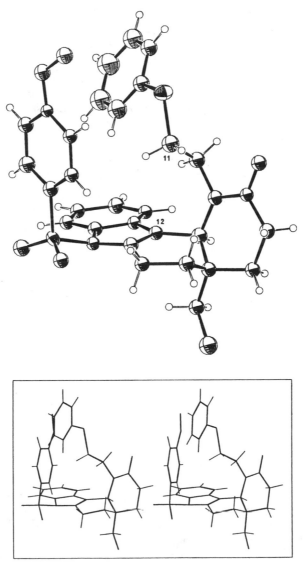

Fɪɢ. 3. Compound **51**.

many ways to improve the yields of the poorer cyclizations and have yet to make any real improvements, I believe that it is possible because we know what the other products are and should be able to do something about it. At least, it presents a formidable challenge to an enthusiastic student.

VII. Completion of the "Model Work"

The tetracyclic amide **51**, having been made readily available, was converted into $(\pm)$-aspidospermidine in a straightforward sequence. Oxidation of **51** (MCPBA) gave **54**, which was treated with TFAA–CH$_2$Cl$_2$ at 0°C and then rapidly warmed in chlorobenzene to give the pentacyclic amide **55** (81%). The stereochemistry of the thiophenyl substituent at C-11 was

confirmed by X-ray (Fig. 4) and substantiates the mechanistic picture explained earlier (Scheme 11). It is interesting to note that the newly formed C-11—C-12 bond is exceptionally long, 1.563 Å. Desulfurization of **55** (Raney nickel W-2) gave **56** (81%), which was reduced with LiAlH$_4$–THF to give $(\pm)$-aspidospermidine (**6**: R = H). The second synthesis of aspidospermidine is summarized in Scheme 16.

The two syntheses of aspidospermidine serve as a somewhat complicated set of model studies whose results should enable us to make plausible plans

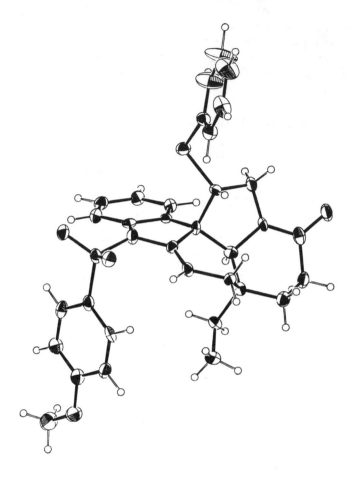

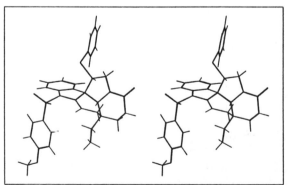

FIG. 4. Compound **55**.

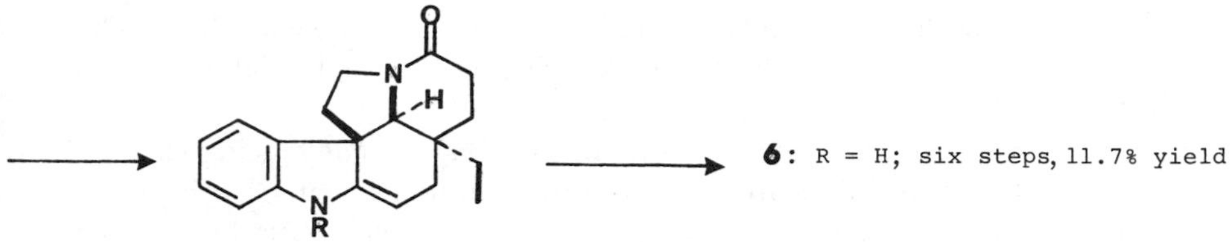

6: R = H; six steps, 11.7% yield

SCHEME 16

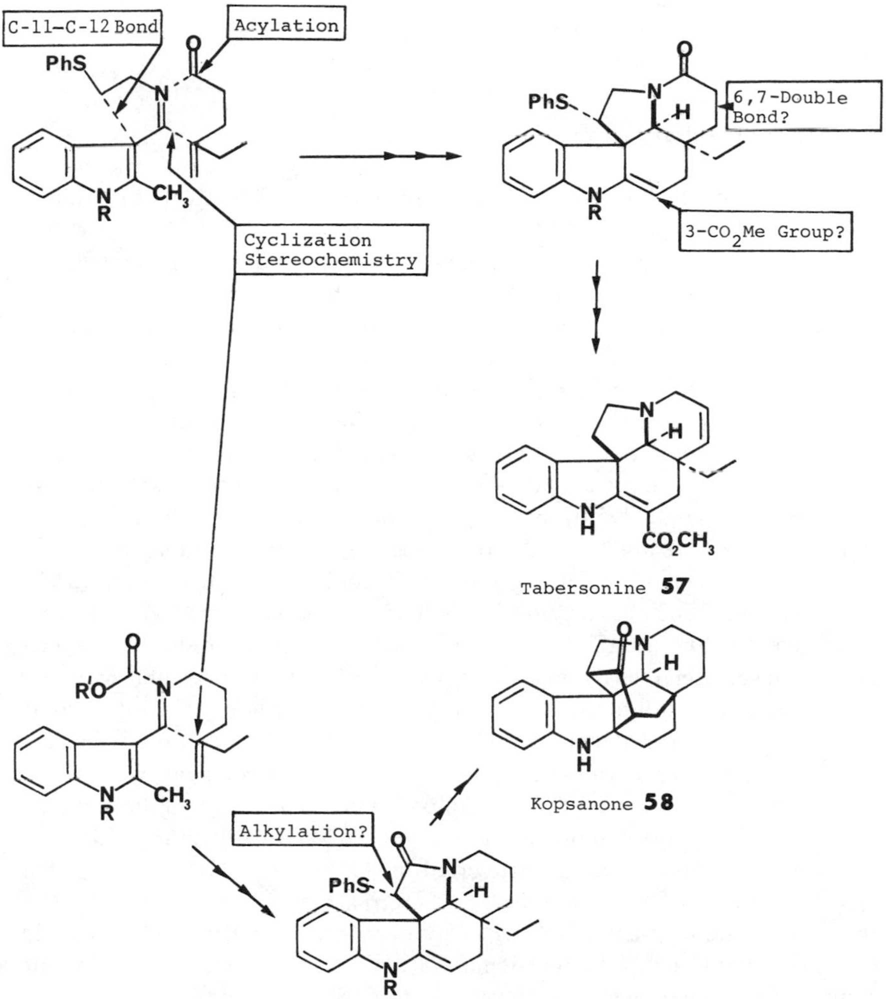

C-11—C-12 Bond

Acylation

PhS

6,7-Double Bond?

CH₃

R

3-CO₂Me Group?

Cyclization Stereochemistry

Tabersonine 57

Kopsanone 58

R'O

CH₃

R

Alkylation?

PhS

R

SCHEME 17

for synthesizing the more condensed and functionalized indole alkaloids. The summary in Scheme 17 illustrates the key findings and potential extensions of these results.

If the strategy developed so far is to be of real value in advancing alkaloid synthesis, then it should be able to make inroads into problems that have not been previously solved. Consequently, of the future problems posed in Scheme 17, the 6,7 double bond and 3-CO_2Me group lead to solutions for alkaloids such as tabersonine (**57**), which has already been synthesized, whereas alkylation at C-11 offers the intriguing challenge of synthesizing the heptacyclic indole alkaloid kopsanone (**58**).

VIII. New Territory

The kopsane alkaloids are a group of heptacyclic alkaloids that were first reported as long ago as 1890. It was not until the early 1960s that the extraordinary complex cage structure of the kopsane alkaloids was elucidated. For some years it was thought that they had a strychnine-type structure. This is somewhat ironic since their biological properties more closely parallel the *Aspidosperma* alkaloids. Eventually, in the early 1960s, their structures were elucidated by mass spectrometry and confirmed by single crystal X-ray crystallography. I have drawn out some of the main structures in a more visual three-dimensional representation that is intended to show the highly fused nature of these attractive structures. The compounds kopsanone (**58**), 10,22-dioxokopsanone (**59**), and kopsine (**60**) are shown as the N^1-H substances, but N^1-Me and N^1-CO_2Me analogs have also been isolated.[8]

Another very closely related group of alkaloids are the fruticosanes (**61**). A complex sequence of degradative reactions have converted kopsine (**60**) to fruticosine (**61**), although, strangely enough, this relationship or correlation has never been referred to as such.[9] The details of this particular sequence are not really relevant to this work, at this stage, but we hope they will become important when and if kopsine (**60**) is synthesized. I have mentioned the fruticosane structure because it is part of the retrosynthetic analysis.

Our first thoughts in designing a viable plan for the synthesis of the kopsanes was to decide upon the most suitable target. 10,22-Dioxokopsane (**59**) is an easy choice because it can be converted into kopsanone (**58**) (LiAlH$_4$, followed by DMSO–DCC oxidation), and on treatment with MeOH–MeONa the nonenolisable β-ketoamide is cleaved to give the hexacyclic alkaloid pleiocarpinilam (**62**). This particular alkaloid has the potential to be one component of the dimeric system pleiomutine (**63**) (Scheme 18).

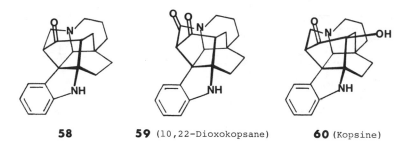

58   **59** (10,22–Dioxokopsane)   **60** (Kopsine)

61 (Fruticosine)

Obviously, alkylation at C-11 (Scheme 17) is a very plausible way to introduce the necessary three-carbon bridge. This viewpoint leads to the retrosynthetic analysis **64 → 65/66**. The allyl group attached at C-11 on the
concave face of the homoannular diene **64** can, in principle, cyclize in two
possible orientations. One leads to the kopsane ring structure **65**, thus forming
a combination of five-, six-, and seven-membered rings; whereas the other
forms three six-membered rings, namely, the fruticosane structure **66**. The
synthesis of **64** presents two substantial problems: first, the construction of the
homoannular diene and secondly the allylation at C-11 from the concave
face of the homoannular diene. A carbonyl group at C-10 (amide) could, in
principle, enable epimerization at C-11 and, under conditions of kinetic protonation, force the C-11 allyl group into the concave vicinity of the diene.

59 $\xrightarrow[\text{MeONa}]{\text{MeOH}}$

62   **63** (Pleiomutine)

SCHEME 18

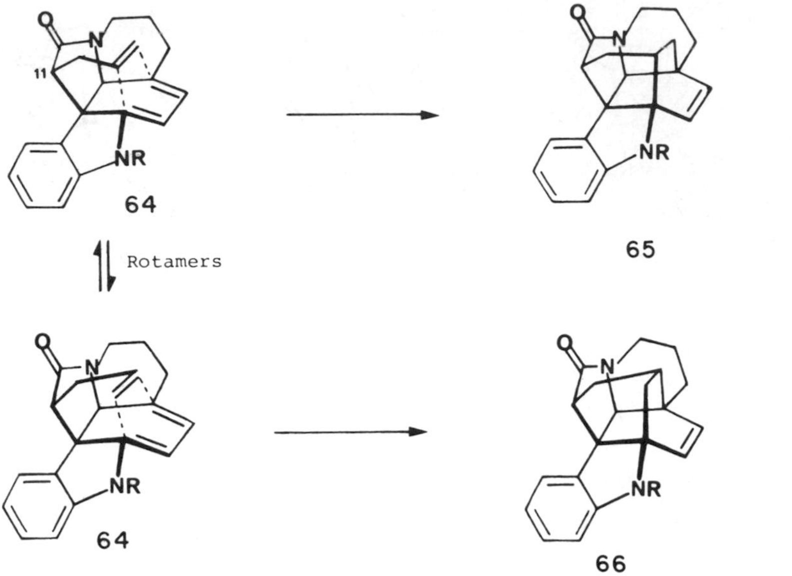

64

Rotamers

65

64

66

The first objective was to make the homoannular diene **67** and to examine its allylation. I shall return to a more detailed analysis of the allylation at C-11 later. It seemed very appropriate to examine the intramolecular trapping of an indole-2,3-quinodimethane intermediate with an acetylenic dienophile (Scheme 19). Treatment of **41** with the appropriate acetylenic carboxylic acid mixed anhydride at 120°C in chlorobenzene gave the dihydrocarbazole **69**. Higher temperatures resulted in large amounts of aromatiza-

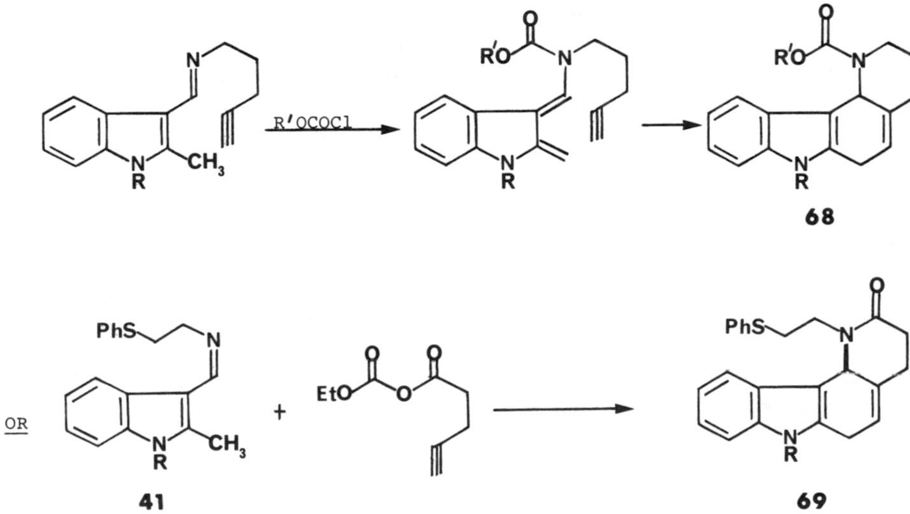

68

OR

41

69

Scheme 19

tion from 1,4-elimination. Oxidation of **69** ($MCPBA–CH_2Cl_2$) gave the sulfoxide **70**, ready for the intramolecular Pummerer reaction to form the C-11—C-12 bond. Treatment of **70** with trifluoroacetic anhydride, even at low temperatures, immediately resulted in 1,4-elimination to the carbazole **71**. We were unable to detect or induce any reactions that resulted in the homoannular diene **67**. These results demonstrate that the order of events

70   TFAA   **71**

Normal Pummerer

67

would benefit from being reversed. The C-11—C-12 bond must be formed before the formation of the homoannular diene; and in this way we reasoned that the aromatization reaction would be blocked. This approach needs a masked double bond (C-4—C-5) and a simple method of revealing it at the right moment (in chess parlance, concealed check). Mate has not yet been inflicted, but the opponent is on the defensive.

The first requirement is a simple primary amine that contains a functionalized dienophile, that at a later moment in the plan can be transformed to the C-4—C-5 double bond. The (E)-vinylchloroamine **72** is available by standard homologation procedures from (E)-1,3-dichloropropene. Condensation of **72** with our standard building block N-[(4-methoxyphenyl)sulfonyl]-2-methyl-3-formylindole (**73**) gave the imine **74** ($>98\%$). Of the various procedures that we have tried and described to date, the most practical was treatment of **74** with Cl_3CCH_2OCOCl (2.0 equivalents) and $iPr_2EtN–PhCl$ at 120°C to give the tetracyclic carbamate **75** in 50% yield. The *secondary* chloro substituent is equatorial and contrasts with the (Z)-vinylchloro isomer, which gave the corresponding axial chloro isomer of **75** in very low yield ($<5\%$). Removal of the 2,2,2-trichloroethyl group and concomitant decarboxylation by treatment with $Zn–AcOH–THF–H_2O$ at 20°C gave the secondary amine **76** (91%). The secondary equatorial chloride remains intact during the deprotection step. The two-carbon unit C-10—C-11 was attached to the piperidine nitrogen atom in the usual way and was prepared for the Pummerer reaction by oxidation with MCPBA to give the diasteromeric sulfoxides **77** (91%).

The stage is set for a key reaction. It is imperative that the C-11—C-12 bond is made before the elimination of HCl, because the previous results show that if the elimination of HCl precedes the C-11—C-12 bond formation, then only aromatization results.

When the sulfoxide **77** was treated with trifluoroacetic anhydride at 0°C, then heated at 130°C, a remarkably clean transformation took place, and the homoannular diene **67** crystallized directly from the reaction mixture in 72% yield. This very desirable result confirms the hypothesis that the C-11—C-12 bond is made first, and, as a result, the chlorine atom (equatorial) is now allylic and readily eliminated to give **67**. Now the stage is set to examine the

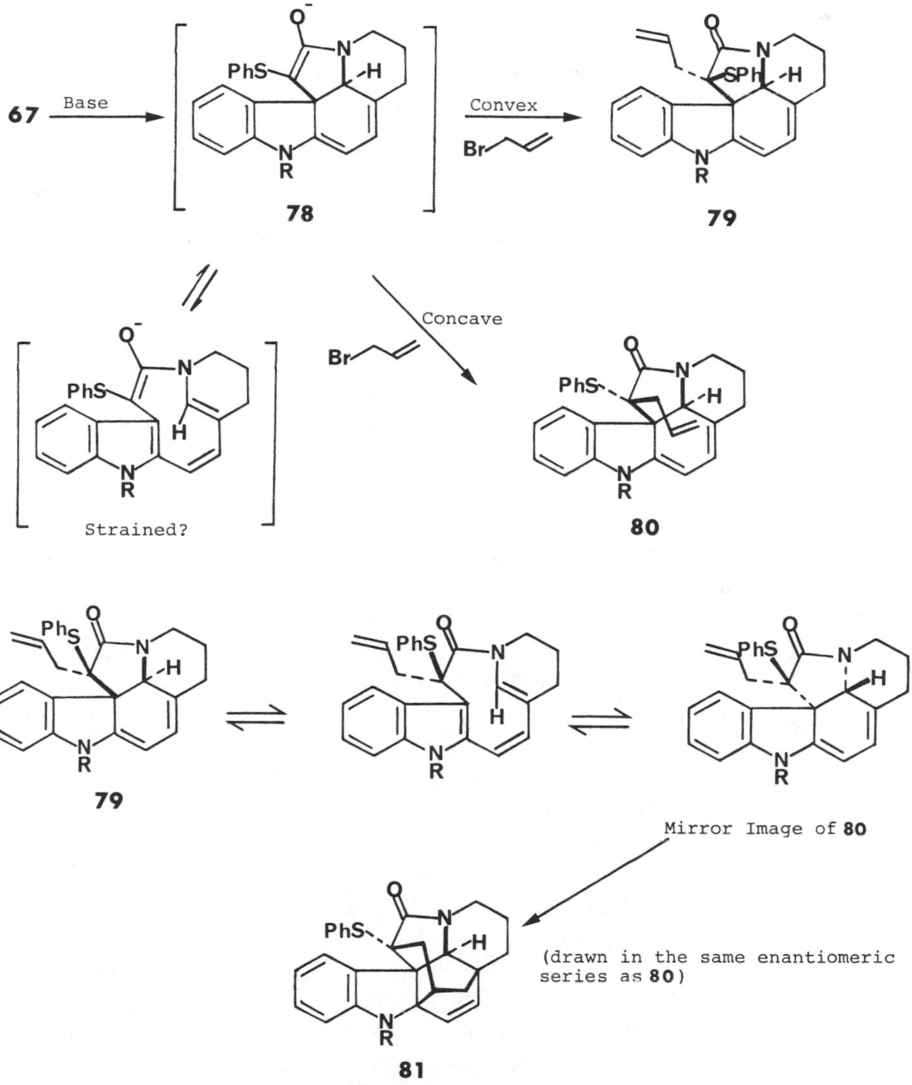

SCHEME 20

allylation at C-11. It is, of course, essential that the allyl group is on the concave face of the homoannular diene (**64 → 65/66**). We considered in some detail the stereochemical possibilities involved in this key transformation.

First, it is not absolutely clear which face of a planar enolate species, such as **78**, is more accessible toward an allylating agent. Consequently, it was predicted that an epimeric mixture at C-11 would result: **79** and **80**. The undesired product **79** has the inherent capability to be converted to the mirror image of **80** by a sequence of cycloreversion–recyclization transformations. This sequence (Scheme 20) inverts the configuration at both C-12 and C-19, which has the overall effect of turning the allyl group from the convex to the concave face of the diene. Once the dienophile is situated directly over the concave face of the diene, it should irreversibly cyclize to the basic kopsane structure **81/65**. The alternative fructicosane structure **66** is far less likely because it is much more strained. One of the purposes of this analysis was to provide a pathway to the correct product (**80**) even if the stereochemical outcome of the allylation initially gave **79**.

Another much more intriguing aspect of the allylation at C-11 is the possibility that it could take place with retention of configuration. It could well be that the carbanion at C-11 is not a delocalized species, such as **78**, since delocalization of negative charge into the amide carbonyl group destroys the amide resonance. The inductive effects of both the SPh and $CONR_2$ groups should be sufficient to stabilize the C-11 carbanion as the pyramidal species **82**.

Treatment of **67** with $KN(SiMe_3)_2$–THF at 0°C, followed by allyl bromide, gave the key allyl "intermediate" **80** in 91% yield after chromatography and crystallization. The melting point of **80** is 158–160°C followed by conversion to **81**, mp 234–235°C. The structures of both **80** and **81** were confirmed by single crystal X-ray crystallography (Figs. 5 and 6). The exceptionally mild conditions required to convert **80** to **81** reflects the fact that **80** sits in a small potential energy well that requires only a small energy nudge (very small ΔS) to give **81** (Scheme 21).

The most practical way to make **81** is to heat a solution of **80** in toluene at 100°C. In this way **81** was isolated in 86% yield. Currently we are examining the mechanism(s) that convert **67** to **80**, then into **81**.

Since the synthesis of kopsanone (**58**) and 10,22-dioxokopsane (**59**), in principle, ought to require the treatment of **82/78** with acrolein or acryloyl derivatives. This was tried and to date has not been successful. As a consequence, we were forced to consider ways to convert **81** into **58** and **59**.

An intriguing possibility that could be viewed as extremely high risk, yet nevertheless a possibility, is to transfer the oxidation state at C-11 [S(O)Ph] to C-22. It was, of course, immediately evident that to accomplish such a process could require the intervention of an **anti-Bredt** compound. This risky speculation is outlined in Scheme 22.

Reduction of the isolated double bond in **81** with dimide, generated *in situ* by treatment of TsNHNH$_2$ with NaOAc–EtOH, gave **83** (**95%**). Oxidation of the SPh group at C-11 with *m*-chloroperbenzoic acid at $-70°$C gave a

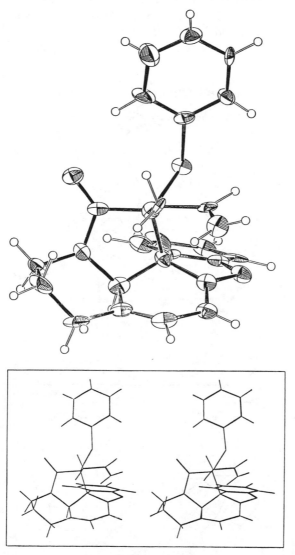

Fig. 5. Compound **80**.

mixture of the two diastereomeric sulfoxides **84** (72%) and **85** (15%). The assignments of relative configuration at sulfur is based upon the subsequent thermolysis reaction. Only one of the sulfoxides **84/85** can orient the sulfur–oxygen bond in a syn-coplanar fashion to the β-hydrogen atom to undergo syn elimination. The other enantiomer must force the S-phenyl group into the indoline ring in order to achieve the same conformation for syn elimination of phenylsulfenic acid. A priori we had no way of knowing whether the major or minor sulfoxide would lead to the torsionally strained α,β-unsaturated amide **86** or, indeed, if the syn elimination would take place at all. When in a

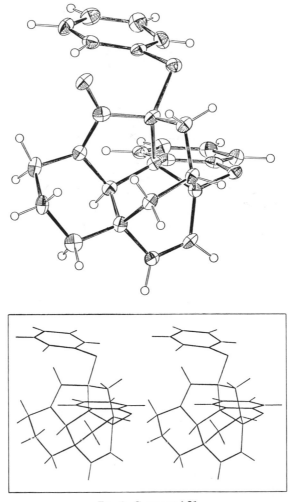

FIG. 6. Compound **81**.

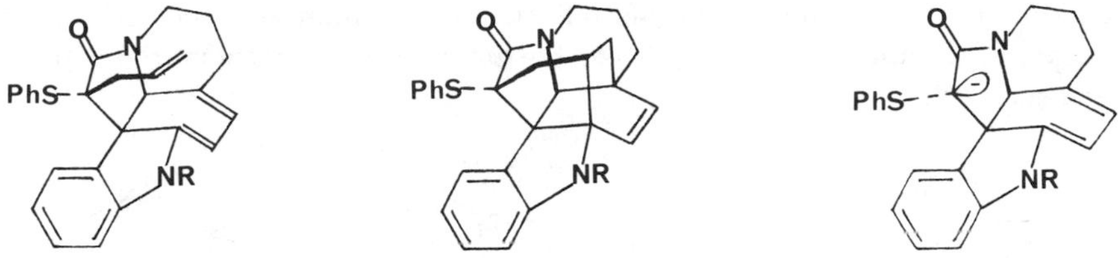

80   **81**   **82**

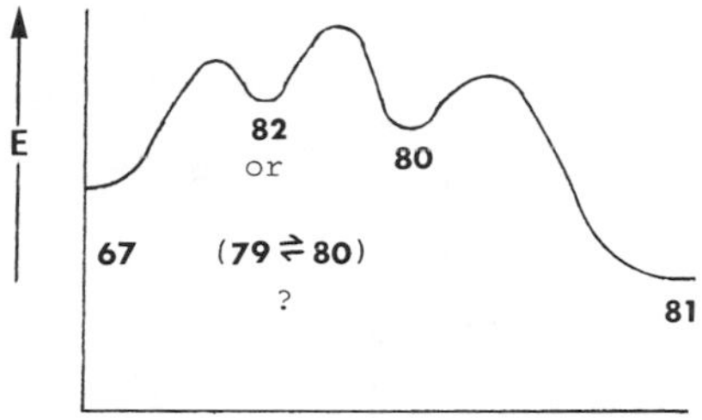

SCHEME 21

tight corner in a chess game, it is best to make the most forthright move and see what happens.

The mixture of sulfoxides **84/85** was heated in toluene in a sealed tube at 230°C. The major sulfoxide **84** disappeared and was replaced by a single new sulfoxide **87** (95%), and the minor sulfoxide **85** remained unchanged. The minor sulfoxide was subsequently recycled by reduction to **83** and reoxidation to **84** and **85**. Assignment of the configuration at the sulfoxide sulfur atom in **87** is based on the cis addition of phenylsulfenic acid to the torsionally strained α,β-unsaturated amide **86**. This unusual sequence transfers the PhS(O) group from C-11 to C-22, in the correct position to be converted to a carbonyl group. The extremely high temperature (230°C) required for the syn elimination of phenylsulfenic acid should be contrasted with the usual

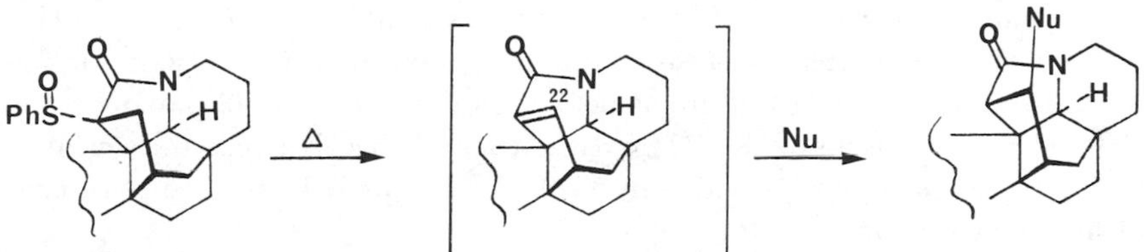

84/85

SCHEME 22

conditions (120°C) and, of course, reflects the extremely strained anti-Bredt nature of the unsaturated amide **86**. Fortunately, the most forthright move was successful.

When the sulfoxide **87** was treated with TFAA–CH$_2$Cl$_2$ at 0°C, then warmed to 130°C, the β-ketoamide **88** was formed (70%). One remaining problem needs to be solved: how to remove the *p*-methoxyphenylsulfonyl group from the indoline nitrogen atom without destroying the rest of the molecule. Sulfonamides are notoriously difficult to remove, and there was no appropriate model system that could have provided any testbed for this crucial final reaction. There are two ways, both reductive, that can straightforwardly lead to kopsanone (**58**) and 10,22-dioxokopsane (**59**). Reduction of **88** with LiAlH$_4$–THF gave an epimeric mixture of kopsanols (**89**), which were oxidized using a modification of the Moffatt oxidation (DMSO–MDCC) to give racemic kopsanone (**58**). When **88** was treated with Li–NH$_3$–THF, it was cleanly converted to **90** (as an epimeric mixture at C-22). Reoxidation of **90**, using the same procedure as before, gave racemic 10,22-dioxokopsane (**59**). The comparisons of IR, ^{1}H-NMR, MS, and TLC between the synthetic and authentic samples, provided by Professor Manfred Hesse (Zurich), confirmed their identity.[10]

The overall indole-2,3-quinodimethane strategy has vindicated itself well in providing the early part of the first synthesis of the kopsanes. The interesting reactions that arose during this synthesis are currently being studied to elucidate their mechanisms. In particular, the mechanism of the extraordinary alkylation of the homoannular diene **67** to give **80** is being examined in the chiral series made by resolution of the amine **76**.

Our current plans are to synthesize a dimeric indole alkaloid, such as pleiomutine (**63**), in which the bond that attaches the two alkaloids together is made before both components are completely constructed.

In summary, what started as an extension of our organosilicon research has developed into a separate program that itself is rapidly expanding into different areas of alkaloid synthesis. The ability of any strategy to cope with complicated problems, and simplify them, is the best measure of its value.

ACKNOWLEDGMENTS

It is a great pleasure to thank Dr. Timothy Gallagher who single-handedly carried out all the work described here. Dr. Christopher Exon has made substantial contributions to other aspects of the alkaloid work. Dr. John Huffman is thanked for the X-ray structures. The National Institutes of Health is thanked for financial support.

REFERENCES

1. T. Gallagher and P. Magnus, *Tetrahedron Symp.* p. 3889 (1981).
2. S. Djuric, T. Sarkar, and P. Magnus, *J. Am. Chem. Soc.* **102,** 6885 (1980).
3. C. Kaneko and T. Naito, *Heterocycles* **19,** 2183 (1982).
4. G. Stork and J. E. Dolfini, *J. Am. Chem. Soc.* **85,** 2872 (1963); Y. Ban, Y. Sato, I. Inoue, M. Nagai, T. Oishi, M. Terashima, O. Yonemitsu, and Y. Komaoka, *Tetrahedron Lett.* p. 2261 (1965).
5. T. P. Ananthanarayan, T. Gallagher, and P. Magnus, *J. Chem. Soc., Chem. Commun.* p. 709 (1982).
6. E. L. McCaffery and S. W. Shalaby, *J. Organomet. Chem.* **8,** 17 (1967).
7. T. Gallagher, P. Magnus, and J. Huffman, *J. Am. Chem. Soc.* **104,** 1140 (1982); C. Exon, T. Gallagher, and P. Magnus, *ibid.* **105,** 4739 (1983); T. Gallagher and P. Magnus, *ibid.* **105,** 4750 (1983).
8. M. Greshoff, *Ber. Dtsch. Chem. Ges.* **23,** 3537 (1890); B. W. Bycroft, D. Schumann, M. B. Patel, and H. Schmid, *Helv. Chim. Acta* **47,** 1147 (1964); D. Schumann, B. W. Bycroft, and M. Schmid, *Experientia* **20,** 202 (1964); C. Kump, J. J. Dugan, and H. Schmid, *Helv. Chim. Acta* **49,** 1237 (1966); A. R. Battersby and H. Gregory, *J. Chem. Soc.* p. 22 (1963); H. Achenbach and K. Biemann, *J. Am. Chem. Soc.* **87,** 4944 (1965); C. Djerassi, H. Budzikiewicz, R. J. Owellen, J. M. Wilson, W. G. Kump, D. J. LeCount, A. R. Battersby, and H. Schmid, *Helv. Chim. Acta* **46,** 742 (1963); B. M. Craven, B. Gilbert, and L. A. Paes Leme, *Chem. Commun.* p. 955 (1968).
9. R. H. F. Manske, ed., "The Alkaloids," Vol. 11. Academic Press, New York, 1968.
10. T. Gallagher and P. Magnus, *J. Am. Chem. Soc.* **105,** 2086 (1983).

Chapter 5

SYNTHESIS OF TYLONOLIDE, THE AGLYCONE OF TYLOSIN

William P. Jackson
Linda D.-L. Lu Chang
Barbara Imperiali
William Choy
Hiromi Tobita
Satoru Masamune

> *Department of Chemistry*
> *Massachusetts Institute of Technology*
> *Cambridge, Massachusetts*

I. Introduction

The macrolides constitute a structurally unique and medicinally important class of natural products. These compounds exhibit a variety of biological activities, and indeed many of them are widely used as antibiotics.[1]

A subclass of this large family consists of polyoxomacrolide antibiotics, which have the following structural features:[1] a 12-, 14-, or 16-membered lactone incorporating a plethora of chiral centers in various oxidation states arrayed around the ring, and one or more sugars attached to the ring through the glycoside linkage. These sugars, one of which is usually an amino sugar, have been shown to be necessary for the macrolide to have antibiotic activity.

Tylosin (**1**)[2] is a representative polyoxomacrolide antibiotic. It was first isolated from the fermentation broths of *Streptomyces fradiae* in the 1950s and has been marketed for many years by Eli Lilly Company as an animal food-stuff additive (Tylar®).[3] The natural product can be produced in huge quantities by fermentation, therefore the challenge of synthesis was not for the purpose of preparing analogs or stereoisomers for biological testing, but rather to establish a general approach to the synthesis of polyoxomacrolides. There are seven chiral centers embedded in the macro ring, which constitutes the aglycone **2** of the antibiotic. Thus our aim was to show that we could develop a general method or methods for the stereoselective introduction of each chiral center of any macrolide regardless of substitution patterns and oxidation levels.[4]

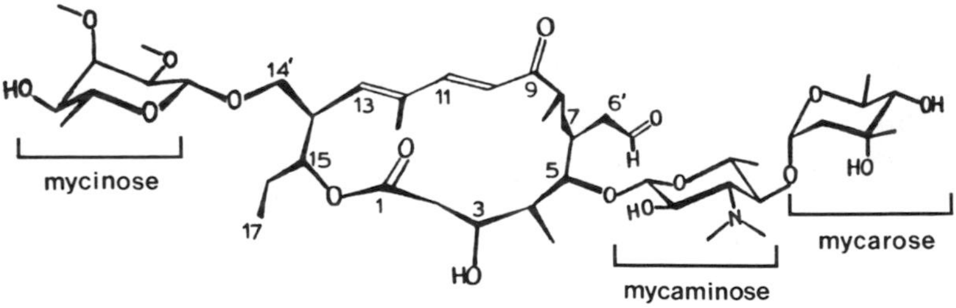

1

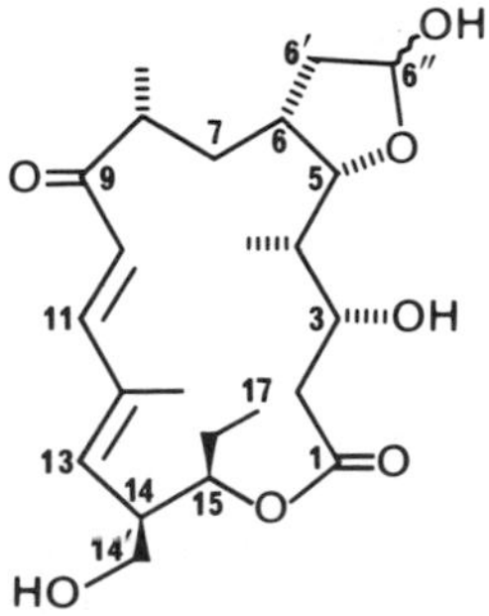

2

This task was by no means easy when the concept of *acyclic* stereocontrol or chiral induction only began to emerge as our initial studies were undertaken and when no macrolide, other than methymycin[5] which is simpler in structure than **1**, had yet been synthesized.

II. General Strategy

Considering the magnitude of the task before us, it was first necessary to set forth a basic synthetic strategy for the macrolide. The best practical way to construct a large molecule such as **1** is by a convergent synthesis, i.e., the molecule is broken down (hypothetically) into smaller fragments which can be joined together after the synthesis of each fragment. The advantages of this approach over the linear one are evident and have been demonstrated on numerous occasions in the past.

In our scheme, which we hoped would be applicable for the synthesis of all macrolides, the macrocycle is made in a final stage by forming the lactone between an "activated" carboxylic acid and a suitably disposed alcohol group. This open chain "seco-acid" needed for lactonization is in turn prepared by the coupling of a right-hand fragment (having a suitably protected acid) with a left-hand fragment (that contains a hydroxyl group). This overall strategy is now called the seco-acid approach (see Scheme 1).[1] Once the aglycone has been prepared, the synthesis of an antibiotic will be completed with the attachment of an appropriate sugar or sugars to the macrocycle. However, as glycosidation is another challenging problem in its own right and our objective was to construct the macrocycle, we set the aglycone tylonolide (**2**) as our prime target.

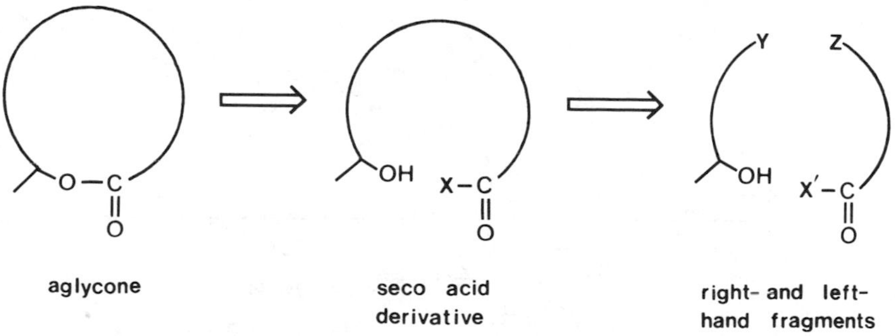

aglycone   seco acid   right- and left-
                  derivative   hand fragments

Scheme 1

III. Preliminary Work

It is always advantageous to have some knowledge about the properties of the target molecule prior to the initiation of its synthesis. Thus our first task was to prepare the aglycone **2** by removal of the sugars from the natural product. It was shown that mild acid hydrolysis cleaved the glycoside linkage between mycaminose and mycarose (see **1**) to form demycarosyltylosin ("desmycosin"). Stronger acid treatment (H_2SO_4, pH 1.5, reflux) resulted in removal of two sugars, mycarose and mycinose, to afford O-mycamino-syltylonolide (**3**). The glycoside linkage of an amino sugar (e.g., mycaminose in **1** and **3**) is much more resistant to hydrolysis, and attempts to remove this last amino sugar under more drastic conditions resulted in decomposition of the aglycone. Thus, only a small amount of tylonolide hemiacetal (**2**) was produced in the saponification of the O-mycaminosyl linkage (21 mg from 10 g of desmycosin), an amount obviously not sufficient for further studies. The problem was finally overcome when the amine was converted to its N-oxide; elimination of the nitrogenous group from the sugar can take place either via a Cope or Polonovski reaction. The resulting "neutral" sugar should behave normally and hydrolysis of the glycoside linkage is much easier. In practice, O-mycaminosyltylonolide (**3**) was converted to its tetrakis(trifluoroacetate), which was then treated with m-chloroperoxybenzoic acid to afford the N-oxide, and finally trifluoroacetylated. Hydrolysis of the glycoside and trifluoroacetate groups took place when the resulting compound was heated at reflux in a buffered (sodium acetate) aqueous tetrahydrofuran solution. Chromatography of the hydrolysate afforded the aglycone **2** in approximately 50% yield (Scheme 2).[6] With the intact aglycone now readily available, further degradative experiments could be performed. The intermediates produced in these experiments would serve two purposes. First, they would confirm the identities of those produced by total synthesis. Second, be-

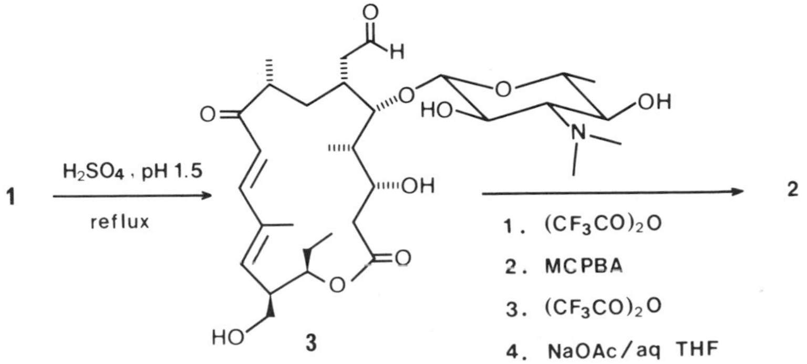

SCHEME 2

cause the projected lactonization of the seco-acid would occur late in the synthesis, degradation would provide another source of the scarce acid, so that the feasibility of the seco-acid approach to **2** could be explored at an early stage. Protection of the hemiacetal (with trimethyl orthoformate) and C-14′ primary hydroxyl group (with dihydropyran) of **2**, followed by reduction of the C-9 ketone, produced a pair of epimeric allylic alcohols (**4**).

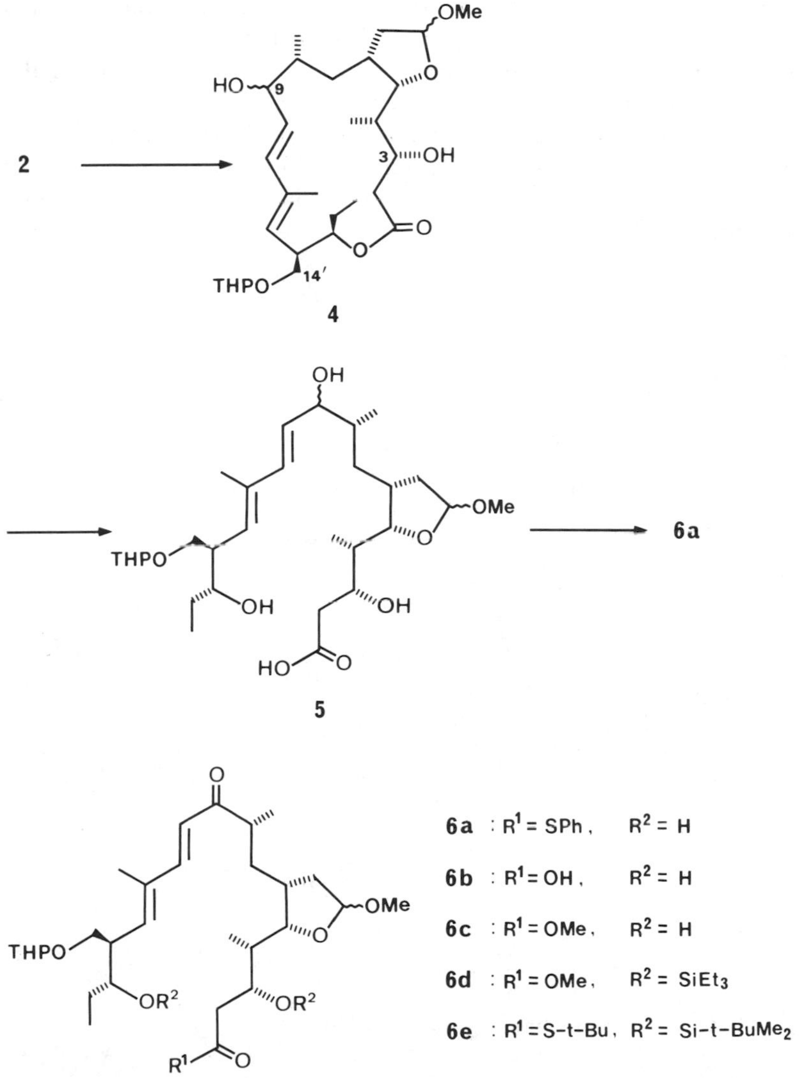
2
OMe
HO
9
3
OH
O
THPO
14′
4
OH
OMe
6a
THPO
OH
OH
HO
O
5
O
OMe
THPO
OR²
OR²
R¹
6a : R¹ = SPh, R² = H
6b : R¹ = OH, R² = H
6c : R¹ = OMe, R² = H
6d : R¹ = OMe, R² = SiEt3
6e : R¹ = S-t-Bu, R² = Si-t-BuMe2

SCHEME 3

Hydrolysis of the lactone linkage in **4** proceeded smoothly under rather mild conditions (~ 1 N NaOH, 60°C, 2 h) because of the presence of the free C-3 hydroxyl group (Scheme 3). In fact, protection of the C-3 hydroxyl group prior to lactone opening necessitated more drastic hydrolysis conditions, which caused a net dehydration to give the α,β-unsaturated acids. (It was also noted that no reverse aldol reaction took place with **4**). The resulting seco-acid (**5**) would not undergo recyclization without the activation of the carboxylic acid, and it was therefore converted via the acid imidazolide to the benzenethiol ester. Allylic oxidation (with MnO_2) afforded the unsaturated ketone **6a**, which underwent ring closure on treatment with mercury(II) methane-sulfonate under buffered conditions to afford, after removal of the protecting groups, a 17% yield of tylonolide **2**. Thus a basic tenet of our seco-acid approach to tylonolide was rendered viable.[6] Although the yield obviously falls short of today's standards, the success of this lactonization itself was quite an achievement at that time.

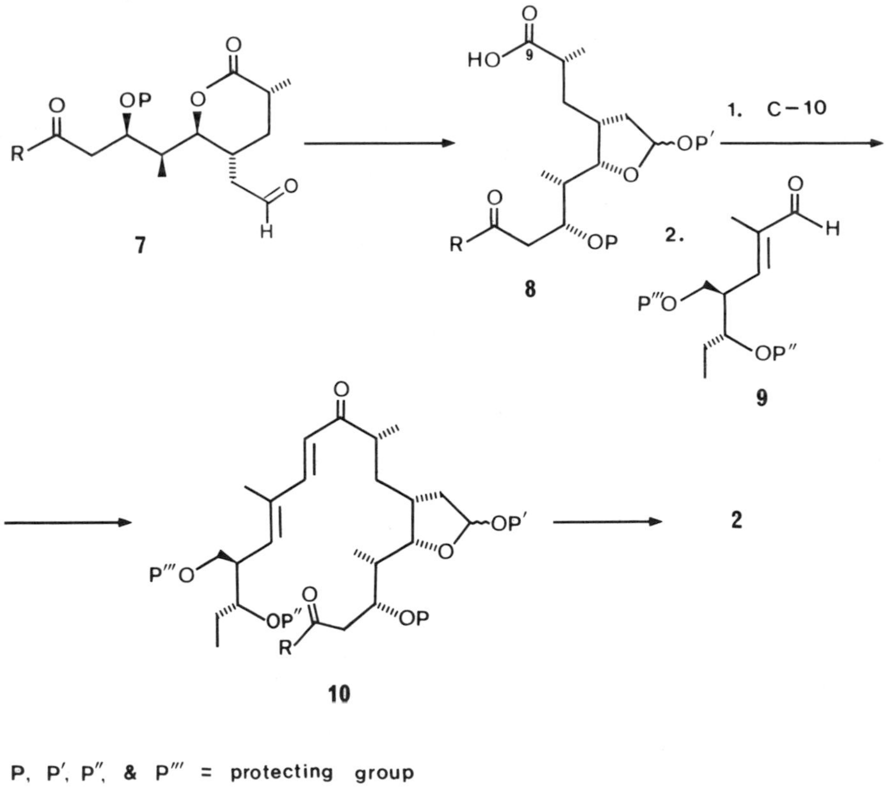

P, P′, P″, & P‴ = protecting group

SCHEME 4

Our synthetic approach to tylonolide is shown in Scheme 4. It was hoped that we could prepare a δ-lactone, such as **7**, which could be rearranged to a protected form of the right-hand fragment **8**. Incorporation of a one-carbon unit (C-10) followed by attachment of the left-hand fragment **9** would then afford the seco-acid derivative **10**, which after suitable activation could be cyclized to tylonolide, as demonstrated above. The rearrangement of **7** to **8** was not as speculative as it may seem, as our work in the leucomycin and carbomycin series had already shown that similar rearrangements were quite facile.[7]

At this point further degradation of the above degradation products to **8**, or even **7**, appeared to be worth pursuing, as compounds **7** and **8** could be compared with those obtained at an early stage in the synthesis. Consequently, the carboxylic acid (**6b**, see above), prepared this time by oxidation (with DDQ) of the alcohol **5**, was methylated (see **6c**) and protected to afford **6d**. Lemieux–Rudloff oxidation then afforded a half acid–ester (**11a**), which contained C-1 to C-9 of tylonolide. All attempts to isolate the left-hand fragment failed. With **11a** in hand, its rearrangement to the δ-lactone corresponding to **7** was attempted; **11a** was converted to the 2-methylpropane-2-thiol ester **11b** via the mixed anhydride (EtOCOCl, Et₃N, TlS*t*Bu).[8] Acid hydrolysis of **11b**, which was intended to liberate the original five-membered hemiacetal, however, produced the seven-membered lactol as shown in **12a**. Reduction of **12a** proved difficult; treatment with sodium borohydride in isopropanol at 10°C removed the thiol functionality from **12a**, whereas reaction in methanol at 0°C left the starting material untouched. Finally, reduction with zinc borohydride in ether at room temperature afforded a 60% yield of the triol **13**. Treatment with mercury(II) trifluoro-acetate afforded the δ-lactone **14**, but only in disappointingly low yield (13%). In order to avoid the complications which might have resulted from the use of mercury(II), the sequence was repeated, using the benzyloxymethyl ester **11c** from which the carboxylic acid could be liberated by catalytic hydrogenolysis. Reduction of **12b** with sodium borohydride in isopropanol cleanly and directly afforded **14** after decomposition of a tight borate complex using methanol containing acetic acid. No catalytic hydrogenolysis was needed in the last step, and the δ-lactone was found to form spontaneously from the hydroxy ester. Thus, after all this, the sequence starting with the methyl ester **11d** was followed to provide **14** in excellent overall yield. Note that in this final step the C-5 hydroxyl group did not form the γ-lactone with the C-1 ester group, and thus a similar sequence could be executed with the C-1 thiol ester as well (see below).

The production of the seven-membered lactols **12a** and **12b** highlighted the fact that it would be necessary to protect the C-3 hydroxyl group during

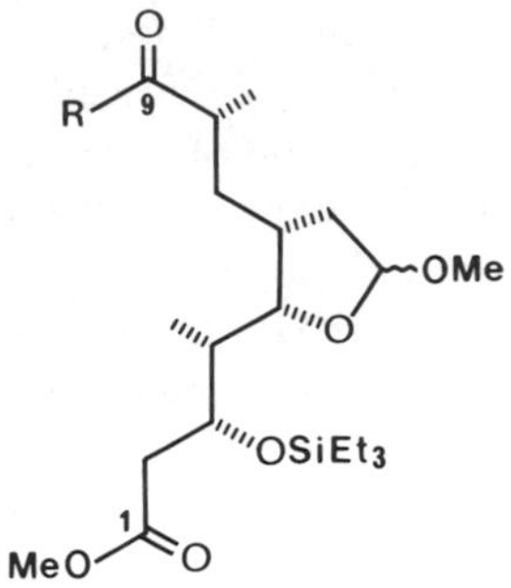

11a : R = OH
11b : R = S–t–Bu
11c : R = CH$_2$OCH$_2$Ph
11d : R = OMe

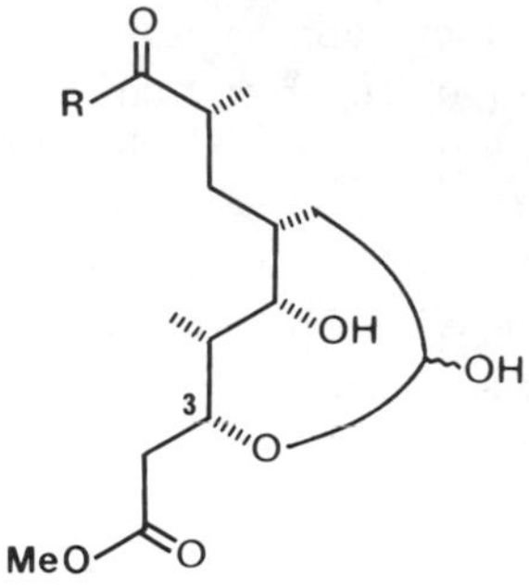

12a : R = S–t–Bu
12b : R = CH$_2$OCH$_2$Ph

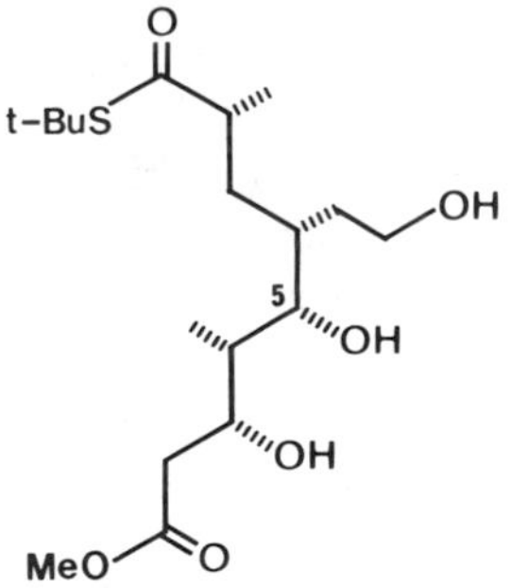

13

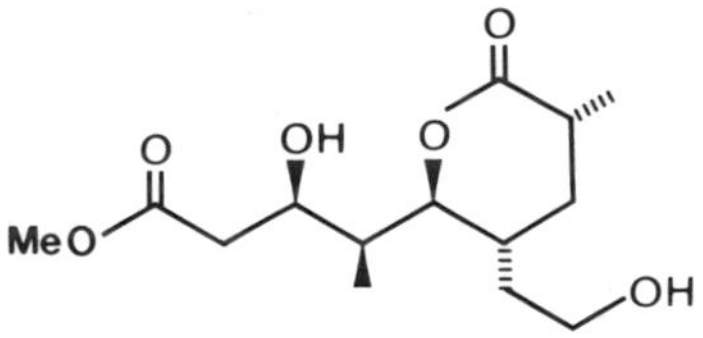

14

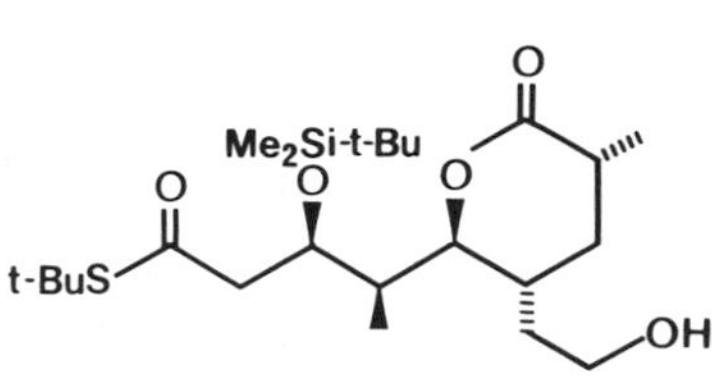

15

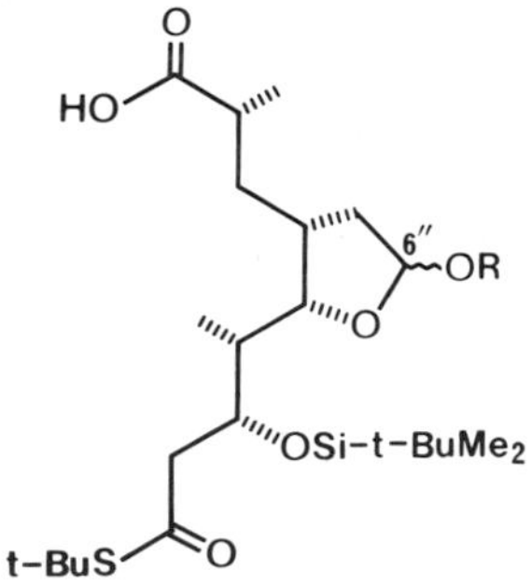

16 : R = H
17 : R = Me

the synthesis. This protection was also important because it would be extremely "tricky" to oxidize selectively the primary hydroxyl group to the aldehyde in the presence of the unprotected secondary alcohol. Thus, **15** became our new degradation target. The 2-methylpropane-2-thiol ester was chosen because it was thought that this would be useful for the final cyclization of the macrolide seco-acid and also because it is somewhat more stable than the benzenethiol ester (see above). Treatment of the mixed anhydride from **6b** with thallium 2-methyl-2-propanethiolate and 2-methyl-2-propanethiol,[8] followed by silylation, gave **6e**, which would be an intermediate in the total synthesis. Lemieux–Rudloff oxidation of **6e**, followed by hydrolysis, gave **16**. Borohydride reduction with the methanol–acetic acid workup mentioned above gave a 40% yield of **15**. To complete the preliminary studies, **15** was oxidized to the corresponding aldehyde with pyridinium chlorochromate and then treated with 70% aqueous acetic acid at 70°C for 1 h. Although the reaction was not brought to completion in this initial study, a 50% yield of **16** was obtained. Protection afforded **17**, the desired right-hand fragment. Thus, the final stages of the planned total synthesis was almost secured so that we were ready to initiate the synthesis of both fragments. Compare this preliminary work with our earlier methymycin synthesis, which was executed all the way through at the risk of failure in the final lactonization stage.[5]

IV. Synthesis

A. The "Kim's Lactone" Approach

In our synthesis of methymycin[5] we had used the Prelog–Djerassi lactonic acid (**18**) (itself a degradation product of methymycin) to construct the right-hand portion of this antibiotic. The key intermediate **18** was prepared from ("Kim's") bicyclic lactone **19a**,[9] which provided all necessary "footholds" to control the stereochemistry at all four centers present in **18**. Our immediate target molecule (**15**) in the synthesis of tylonolide contained these four centers with the same relative and absolute stereochemistry. Thus, in a similar manner, we might be able to prepare a "homo"-P.D.-lactonic acid such as **20**, which would be a precursor for **14**. A scheme for the conversion of **19a** to **22** through homologation of **19a** was readily envisioned.

In order to homologate **19a**, the ester had first to be converted to the corresponding acid in the presence of the lactone. This necessitated the use of a nonhydrolytic method. Although lithium propanethiolate[10] worked well on a simple model, cleavage of the lactone occurred preferentially, and no desired acid was produced. However, reaction of **19a** with *neat* trimethylsilyl iodide[11] at 100°C proceeded without complication to provide the product **19b** in

90% yield. Treatment with diazomethane afforded only **19a**, thus indicating that no epimerization had taken place. The diazo ketone **21**, obtained via the acid chloride, was subjected to the photo-induced rearrangement to afford the homologated acid **22a** in 90% overall yield. Reduction of the carboxylic group through the corresponding mixed anhydride (with NaBH₄ in wet THF) required extremely dilute conditions (35 ml THF/mmol) in order to obtain maximum yield (80%). Oxidation (with PCC–NaOAc) of the resulting primary alcohol (**22b**) to the aldehyde **22c**, followed by protec-

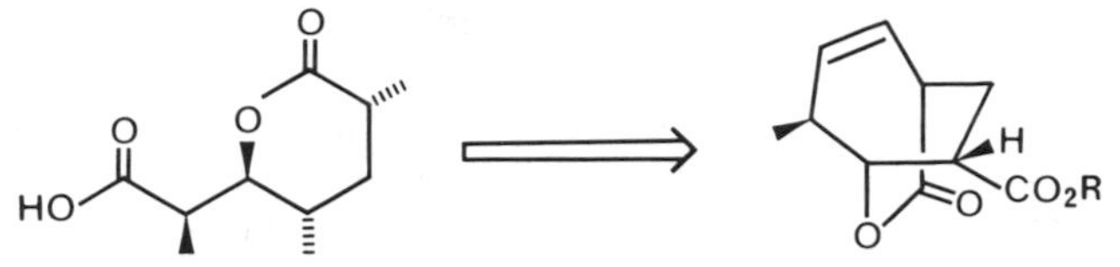

18

19a : R = Me
19b : R = H

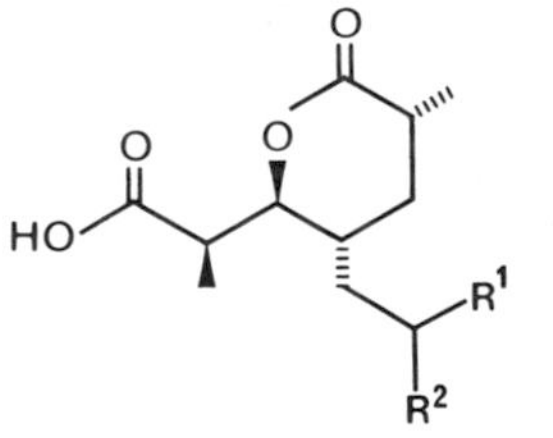

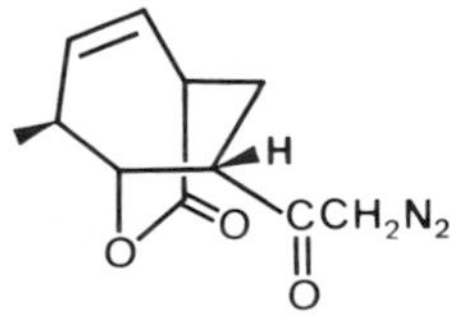

20a : R¹ R² = O   O
20b : R¹ = H , R² = OSi–t–BuMe₂
20c : R¹ = H , R² = OSi–t–BuPh₂

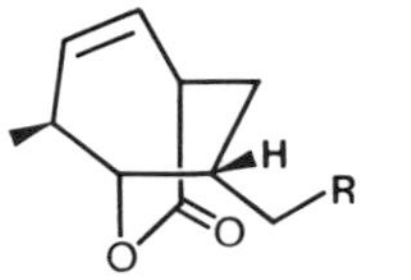

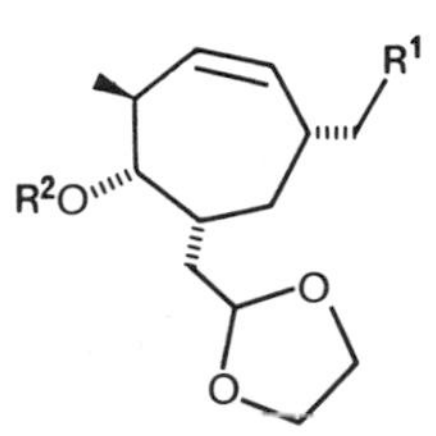

21

22a : R = CO₂H
22b : R = CH₂OH
22c : R = CHO

23a : R¹ = OH , R² = H
23b : R¹ = OTs , R² = SiMe₃
23c : R¹ = H ,   R² = SiMe₃

tion (with ethylene glycol) and reduction (with LAH), afforded the diol **23a**. Although selective monotosylation of the primary hydroxy group of **23a** could be achieved easily, this resulting monotosylate was extremely labile to intramolecular ether formation with the secondary hydroxyl group. Consequently, the secondary alcohol was protected *in situ* to afford **23b**. Reduction (with LAH) and resilylation afforded **23c**.

The next step in the synthesis was a Lemieux–Rudloff oxidation of the double bond. Model studies indicated that the reaction solvent had to be oxygen free in order for the protecting group to stay intact. With this in mind, the reaction was carried out under "normal" conditions, which were successfully applied in the methymycin synthesis (0.1 equivalent $KMnO_4$, 10 equivalents $NaIO_4$, 1 equivalent K_2CO_3, pH 7.6) but no discernible product was obtained. Apparently, oxidation of the acetal group was taking place at a competitive rate. After exhaustive studies, the optimum conditions (1.9 equivalents $KMnO_4$, 31 equivalents $NaIO_4$, and 33.3 equivalents K_2CO_3) were found to provide **20a** in 55% yield. This compound, however, rapidly decomposed even at $-78°C$, presumably due to auto-deprotection catalyzed by the carboxylic acid. This could be overcome by conversion of the acid to the corresponding alcohol via the borane–ammonia complex reduction of its mixed anhydride. A maximum yield of 10% of the lactonic alcohol was obtained. Other reducing agents rapidly attacked the (rather reactive) δ-lactone in the molecule.

In view of the capricious nature and modest yields of the Lemieux–Rudloff oxidation coupled with the instability of the acid **20a**, it was decided that we should use a protected alcohol instead of a protected aldehyde. The protecting group could be removed at an appropriate time and the alcohol then oxidized to the aldehyde. We selected silyl protecting groups, which should be stable throughout the sequence of reactions including Lemieux–Rudloff oxidation. Compound **24a** was prepared by protection of the corresponding alcohol **22b**. Reduction of the lactone with lithium aluminum hydride had to be carried out at -30 to $-20°C$ in order to avoid desilylation. Monotosylation and silylation, followed by reduction and resilylation, proceeded smoothly to afford **25a**. Lemieux–Rudloff oxidation, however, again caused problems and, at best, a 20% yield of **20b** was obtained.

It was shown by model studies that the *tert*-butyldiphenylsilyl protecting group survived the oxidation conditions very well. Yet the Lemieux–Rudloff oxidation of **25b** again proved frustrating. Under the previously "successful" conditions the major product was **26**, the partially oxidized substrate, and once again conditions had to be optimized. Finally, under a set of carefully controlled conditions the desired product **20c** was obtained in 75% yield together with 20% of **26**. Rosenmund reduction provided the corresponding aldehyde **27** without complication. With **27** in hand we were in a position to

24a : R = Si-t-BuMe$_2$
24b : R = Si-t-BuPh$_2$

25a : R = Si-t-BuMe$_2$
25b : R = Si-t-BuPh$_2$

26

study the introduction of a two-carbon unit (C-1 and C-2) to prepare racemic 14. By this time our research had become heavily involved in the boron-mediated aldol reaction, and this was the method by which we chose to introduce this unit. Since 27 was extremely valuable, the Prelog–Djerassi lactonic aldehyde 28 was first used as a model compound. Reaction of 28 with enolates 29a, 29b, and 29c in dichloromethane provided, at best, a 1.3:1 mixture of desired (3,4-syn)[4] and undesired (3,4-anti)[4] products, although at this time we could not assign the stereochemistry of the newly created center. The reaction of 27 with 29a, 29b, and 29c was then carried out in order to determine the best conditions. With 29a and with ether as solvent, a 2:1 mixture of desired and undesired aldol products was obtained in 80% yield. HPLC provided pure samples of both diasteromers. The 250-MHz ^{1}H-NMR spectrum of the major isomer, compared with a derivative of 14 prepared by silylation of the primary hydroxyl group of 14, convinced us that this isomer indeed had the correct stereochemistry as shown in 30. Thus we had *managed* to prepare the precursor to the right-hand fragment of tylonolide, albeit in not a very elegant fashion. Obviously, we were not satisfied with the overall outcome of the efforts described, in particular, in view of the new chemistry that had emerged from the highly stereoselective and efficient synthesis of 6-deoxyerythronolide B just completed at that time.[12] We reevaluated the methodology for the construction of chiral centers and decided to take an entirely different approach. The first "Kim's Lactone" approach, thus abandoned at this stage, however, was not a total loss because racemic

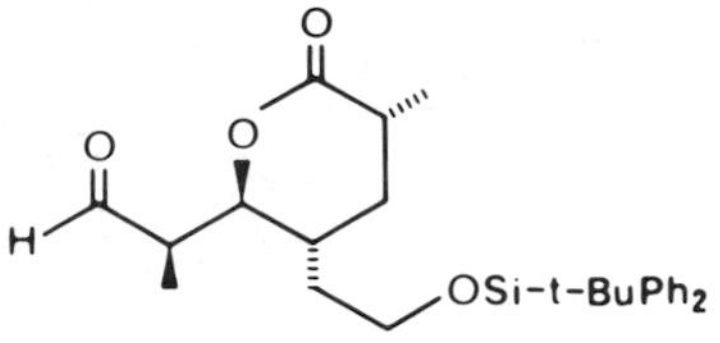

27

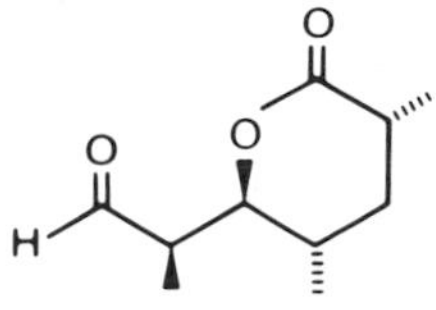

28

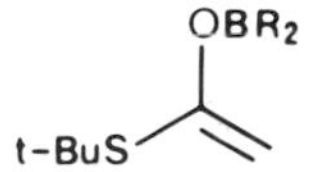

29a : BR$_2$ = 9-BBN
29b : R = n-Bu
29c : R = c-C$_5$H$_9$

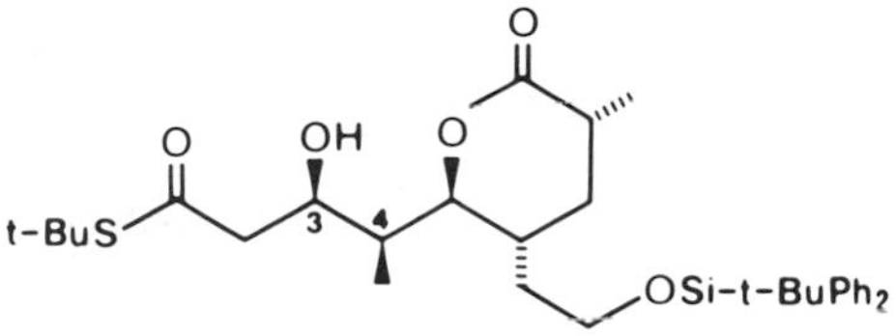

30

20c described above served an important role in characterizing key synthetic intermediates that will appear in the next section.

B. The Aldol Approach

The advent of the boron-mediated aldol reaction coupled with the discovery of the chiral reagents **31a** and **31b** had brought about a new method capable of creating a carbon–carbon bond with high stereoselection.[13] A full account of this reaction with emphasis on a diastereofacial selectivity is already available elsewhere.[4] This development meant that the preparation of tylonolide (**2**) was now considerably simplified.

The main feature of this aldol reaction may be briefly summarized as follows: Reagents **31** react with dialkylborinic triflate in the presence of a base to form a chiral $Z(O)$-enolate, and the resulting chiral enolate (**a**, **b**, or **c**) undergoes an aldol reaction with an achiral aldehyde to afford a mixture of diastereomeric products having 2,3-syn stereochemistry (Scheme 5). The degree of chirality transfer (the ratio of diastereomeric products) is very high (20–100:1) but varies with the size of the ligand on boron [(cyclopentyl)$_2$ > n-Bu$_2$ > 9-BBN] and also with the degree of substitution on the aldehyde.[4] This aldol approach was used in an efficient synthesis of 6-deoxyerythronolide B[12] (as mentioned earlier), which proceeded via the Prelog–Djerassi lactonic acid **18**. Therefore, it appeared to be an easy task to adapt this scheme to prepare **15**.

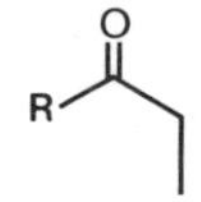

31a : R = 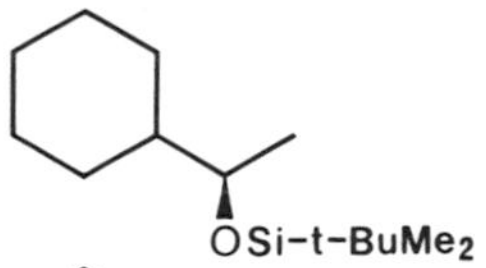 = R_S

31b : R = 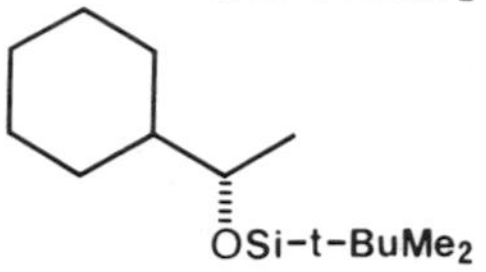 = R_R

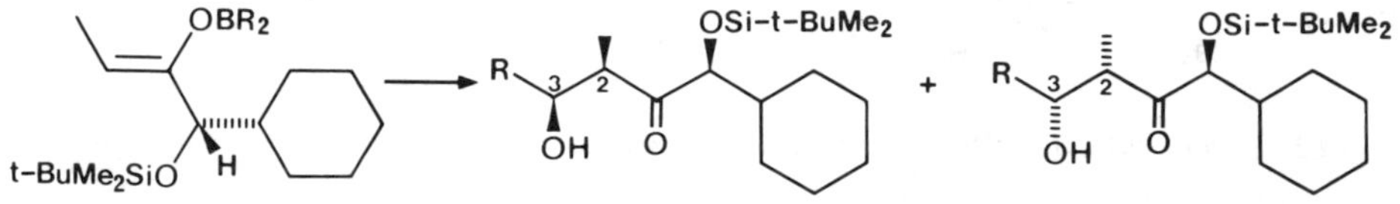

a : BR_2 = 9-BBN   major   minor

b : R = n-Bu

c : R = c-C_5H_9

Aldehyde	Boron enolate	Ratio of major to minor	Major β-hydroxy acid*
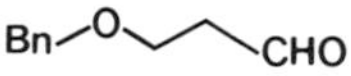 Bn∼O∼∼CHO	a	16 : 1	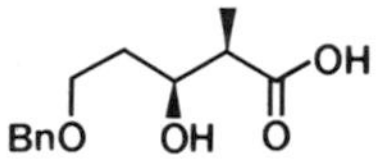
	b	28 : 1	
Bn = benzyl	c	100 : 1	
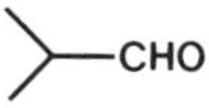 ⟩—CHO	a	100 : 1	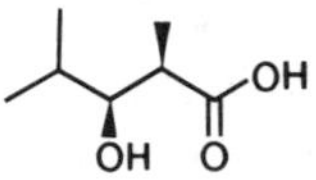
	b	100 : 1	
	c	No reaction	

SCHEME 5. *, Major β-hydroxy acid results after oxidative cleavage of the α-hydroxyketone moiety.

1. *Preparation of the Right-Hand Fragment*

Compound **32**, chosen as our immediate target molecule in the aldol approach, was to be prepared according to the retrosynthetic scheme shown in Scheme 6.

The scheme shows two possibilities: **32a**, $n = 2$ or **32b**, $n = 3$. Although the latter compound (**32b**) would have to be degraded by one carbon at some point in the synthetic scheme, it was thought that this would be the better synthetic intermediate. We knew from previous experience that the presence of an additional oxygen functionality located close to the aldehydic functionaality could disrupt the tight transition state in the boron-mediated aldol reaction, thereby leading to low diastereomeric ratios. Thus, with **33b** as the aldehyde component, we could hope to minimize interference of the oxygen functionality by increasing the distance from the reacting center.

Our initial task was the preparation of the acid **34**, which could be readily converted to **33b**. We hoped to obtain this in optically active form using the method developed independently by two groups led by Sonnett and Evans.[14-14b] The chiral amide **35** was prepared from **36** as shown in Scheme

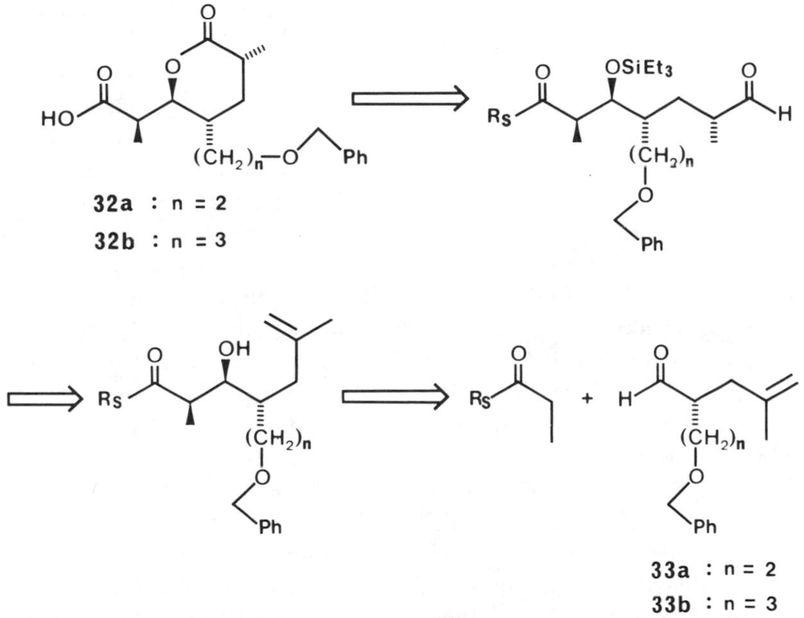

32a : n = 2
32b : n = 3

33a : n = 2
33b : n = 3

SCHEME 6

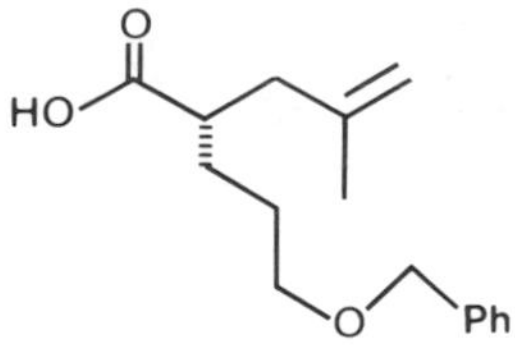

34

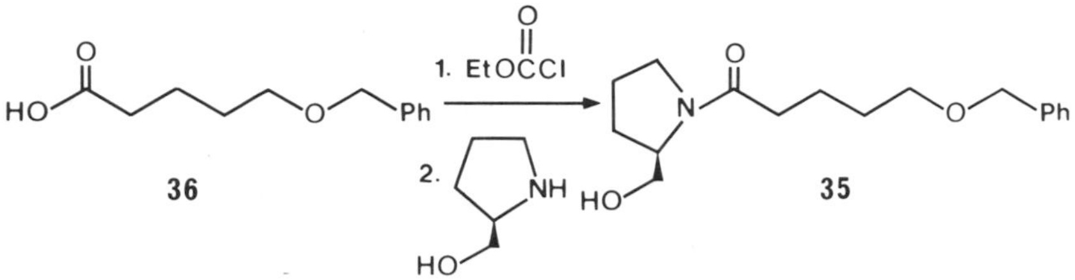

SCHEME 7

7, but alkylation of its dianion failed. Although unsuccessful, in a way this failure was not particularly disappointing. If the reaction had worked, it would have required expensive D-prolinol in order to obtain the correct chirality in **34**. Reversal of the alkylation sequence on amide **37**, obtainable from inexpensive L-prolinol, should provide the correct stereochemistry at C-2 (see Scheme 8). In fact, the ratio of two diastereomeric alkylated products (**38**) turned out to range between 4:1 and 5:1. Because the products were chromatographically inseparable, some resolution of the acid corresponding to **38** would be required. Removal of the chiral auxiliary proved extremely troublesome; treatment with 1 N HCl at 100°C afforded the undesired lactone **39** in high yield. Treatment with HCl in methanol at low temperature afforded **39** together with 8% of the methyl ester **40**. Attempts to reduce the amide **38** with diisobutylaluminum hydride or lithium trimethoxyaluminum hydride gave an approximate 1:1:1 mixture of amino alcohol **41**, starting material, and aldehyde **42**. Finally, protection of **38** as the methoxymethyl ether, followed by reduction with lithium triethoxyaluminum hydride, and then lithium aluminum hydride, gave the alcohol **43** in 30% overall yield (at best). Due to the difficulties in the preparation of **43** and the need for large quantities of starting material, it was decided that resolution would be preferable. We chose **33a** rather than **33b** as our target because of the ready availability of acid **44**. Aldehyde **33a** was prepared according to Scheme 9. All attempts to resolve **45** or **46** failed, and therefore we decided to carry out the aldol reaction of racemic **33a** with a chiral enolate in the hope that the resulting diastereomeric products could be separated by chromatography.

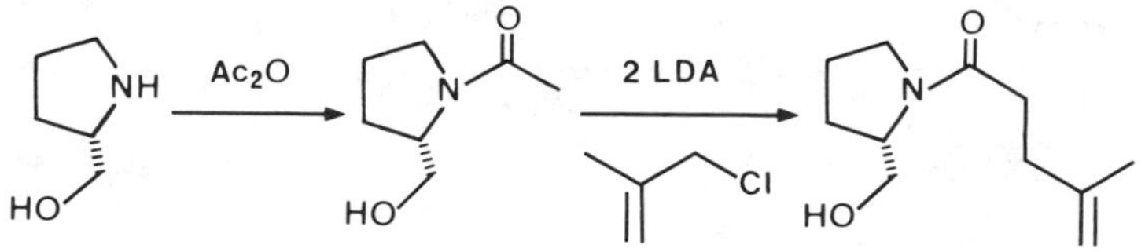

SCHEME 8

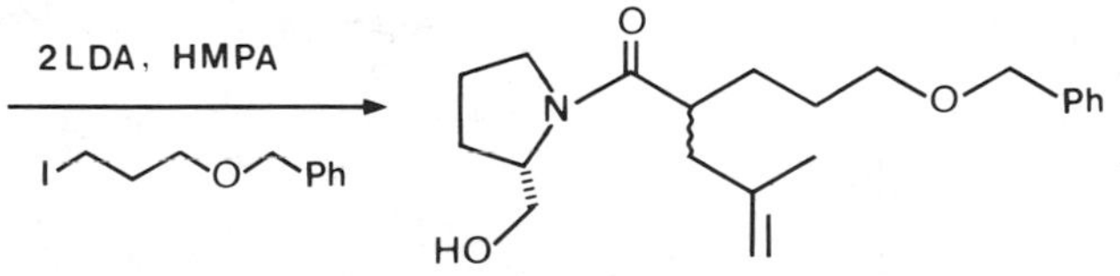

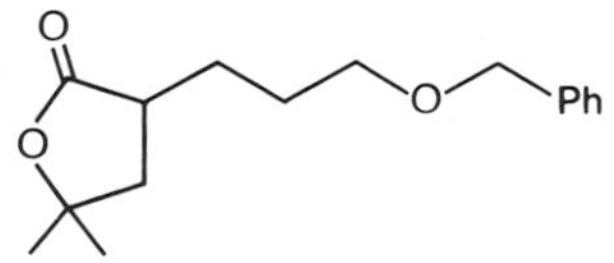

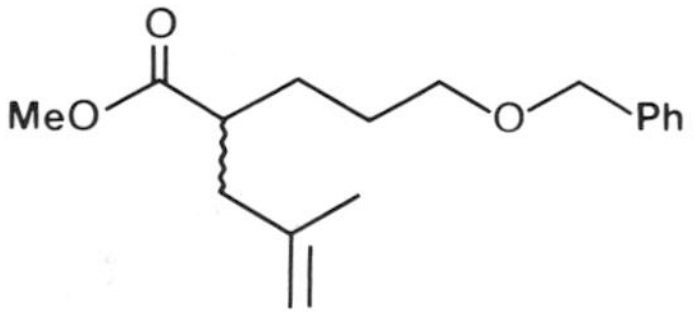

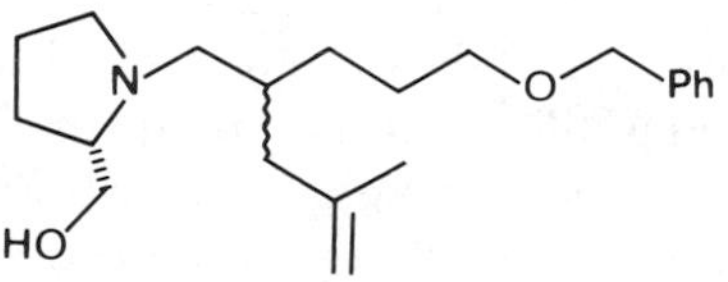

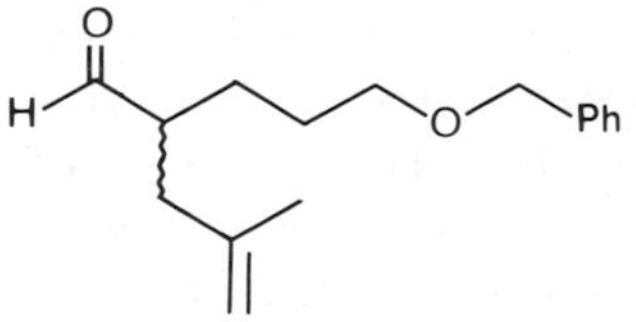

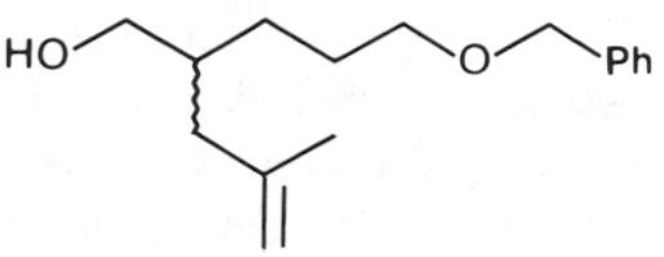

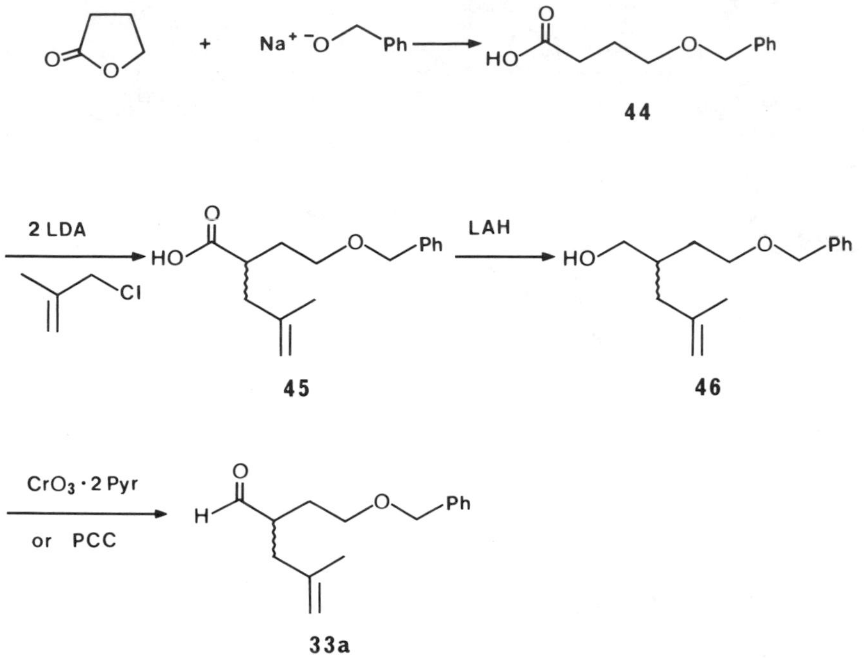

SCHEME 9

Thus, reaction of the chiral reagent **31b** with 9-borobicyclo[3.3.1]non-9-yl triflate and diisopropylethylamine afforded the corresponding $Z(O)$-boron enolate, which reacted with excess **33a** to afford all four possible products. It appeared that our concern about the disruption of the transition state because of the ethereal oxygen in the aldehyde was justified. The four products **47–50**, two major and two minor, were separated chromatographically, but it was not possible to tell which was the desired product. At this stage we could only deduce that each pair of aldol products (**47** and **49**) or (**48** and **50**) had been derived from the respective enantiomers of aldehyde **33a**, because the total yield of each pair had to add up to a maximum of 50%. It was discovered that the major products could be obtained in 32 and 42% yield when an excess of chiral agent **31a** was used. We hoped that the desired isomer **50** was the higher-yielding product, so the ensuing series of reactions (see the next paragraph) was initially carried out using that compound, which was *later* found to be **49**.* Therefore, **50** was actually the 32%-yield product. The problems concerning the rather inferior diastereoselection observed in this aldol reaction were finally overcome by using di-n-butylborinic triflate to

---

* For the sake of brevity, the description of the sequence starting with **49** is abbreviated.

generate the enolate from **31b**; the added bulk of the *n*-butyl ligands prevented the *O*-benzyl oxygen from interfering in the transition state resulting in the *exclusive* formation of **49** and **50**. The use of bulkier reagents normally slows down the rate of the reaction. Thus, the reaction time was prolonged from 2 h at 0°C to 18 h at room temperature. In this fashion, the desired isomer **50** was obtained in 42% yield (which corresponded to 84% based on the correct enantiomer of **33a**). The overall process of converting **44** to **50** was now efficient and highly stereoselective.

47

48

49

50 : R = H

51 : R = SiEt₃

After protecting the hydroxyl group to give the triethylsilyl ether **51**, we then had to hydroborate the double bond to introduce a terminal hydroxyl group (see **52**). Because a basic hydrogen peroxide workup commonly used after hydroboration might epimerize the methyl group α to the carbonyl group, we used instead a (somewhat unusual) peracid workup. Readily available 9-borabicyclononanc (9-BBN) was chosen for the initial studies. Reaction of the protected olefin with 9-BBN, followed by *m*-chloroperoxybenzoic acid workup, afforded **52** as a 2:1 mixture of epimers. The mixture was inseparable by thin-layer chromatography, and it was impossible to tell by NMR which isomer was the desired product. The newly created primary hydroxyl group was oxidized with Collins' reagent to the corresponding aldehyde **53**. Treatment with 70% aqueous acetic acid resulted in removal of the triethylsilyl protecting group and subsequent spontaneous lactol formation (see **54**). The mixture was further oxidized to the lactones. NMR spectral comparison of the major lactone with the racemic lactonic acid **20c**, previously prepared, strongly indicated that it had stereostructure **55** (Scheme 10). The sequence was definitely more efficient than Kim's lactone approach in terms of both the number of steps involved and enantiomeric purity. However, we were still losing a considerable amount of material at the foregoing hydroboration stage. This prompted an investigation to epimerize the undesired lactol, because **54** (with the C-8, β-methyl) was supposed to be more stable than its epimer. Acid treatment, however, effected epimerization only to the extent of 10% and also led to some desilylation of the chiral auxiliary. (Note: R_s carried the *tert*-butyldimethylsilyloxy group.)

Because attempts at epimerization failed, we went back to investigate the effect of bulkier boranes on the isomeric ratio of **52**. We also raised the question of the possible use of a chiral borane to improve the ratio. These two ideas were combined by using the bulky chiral reagents (+)- and (−)-bis(isopinocampheyl)borane [(IPC)$_2$BH], which are readily available in high optical purity.[15] The (+)-enantiomer appeared to be the correct reagent for this present purpose, after the examination of several models proposed earlier.[16] The use of (+)-(IPC)$_2$BH, followed by *m*-chloroperoxybenzoic acid workup, gave a single product. However, the NMR spectrum indicated that this corresponded to the minor isomer from the 9-BBN reaction. To our further amazement, reaction in a similar fashion with (−)-(IPC)$_2$BH also afforded

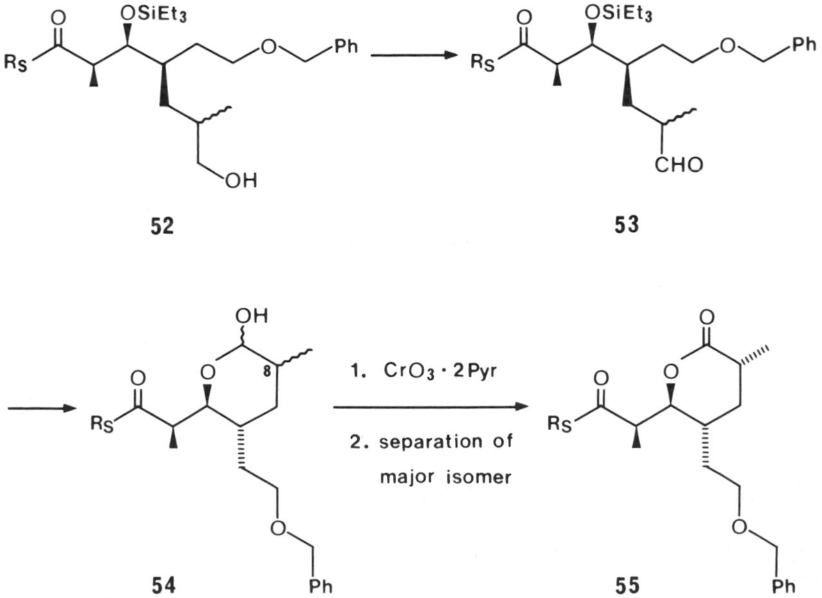

SCHEME 10

another *single* product (**52a**), which was the desired isomer! The remarkably high stereoselectivity observed with both the (+)- and (−)-reagents (>50:1) was not expected for the reaction with a simple methallyl compound.[17] Compound **51** behaves in a unique manner, and apparently the chirality existing in **51** hardly influences the overall steric course of the reaction.[4,18] Further work is still under way to elucidate the mechanistic course of this unusual stereoselectivity. With all of these steps worked out, it proved possible to prepare **55** in high overall yield and with excellent stereoselection (Scheme 11).

An alternative, more straightforward route to **55** proceeded via Fetizon oxidation[19] of the diol (obtained by hydroboration of the free alcohol **50** or by deprotection of **52a**). The reaction did work but required an excessive amount of the reagent and was extremely capricious, depending on the unpredictable quality of the reagent and the reaction temperature. Therefore, this shorter sequence was judged to be of little practical use. The identity of **55** was established at this stage by its conversion to the correct enantiomer of the methyl ester of **20c** through **56** and **57** (Scheme 12). Spectra of the enantiomer were identical with those of the racemate **20c** of established relative stereochemistry.

The next task was the preparation of aldehyde **58**. The method that had proved most effective with the Prelog–Djerassi lactonic acid was Rosenmund reduction of the corresponding acid chloride. Indeed, conversion of acid **57** to the acid chloride, followed by reduction, did give **58**, but large amounts of debenzylated material were also produced. Another alternative was reduction of ketone **56**, followed by treatment with sodium meta-periodate, to yield the aldehyde directly. The reduction of **56** was carried out with borane–ammonia complex, since our previous work in the leucomycin and carbomycin series[7] had shown that the δ-lactone was labile to stronger reducing agents. Cleavage of the reduction product then afforded **58** cleanly (Scheme 13).

Introduction of the final two-carbon unit, C-1 and C-2, into **58** was first attempted, using boron enolate **29a** as already described for aldehyde **27**.

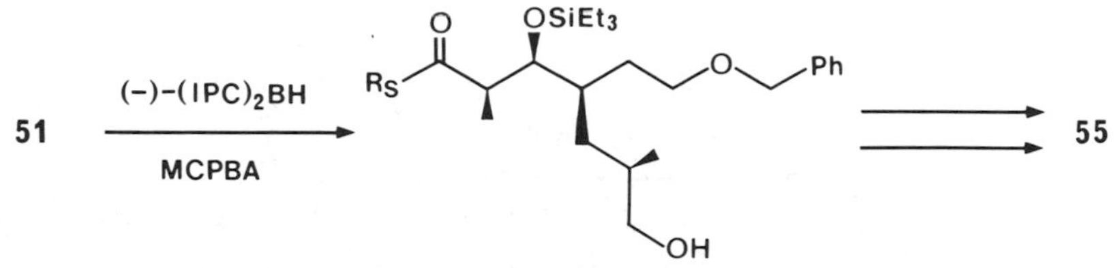

SCHEME 11

In this case a 1:1 mixture of isomers **59** was obtained, clearly indicating a definite need for a chiral enolate reagent with high diastereofacial selectivity analogous to **31**. Indeed, extensive efforts were made to achieve this purpose.

Reagent **60** with X = H had been prepared in connection with other work and found to exhibit virtually no stereoselectivity in the aldol reaction with *achiral* aldehydes, a rather surprising result when compared with the results obtained with the corresponding ethyl ketone **31**. Therefore, other reagents (**60** with X = SR, SeR, SiR_3, SnR_3, or halogens) were examined. The results are briefly summarized below. The sulfur reagents (X = SMe, SPh) exhibited the high selectivity, but removal of the thiol group from the aldol

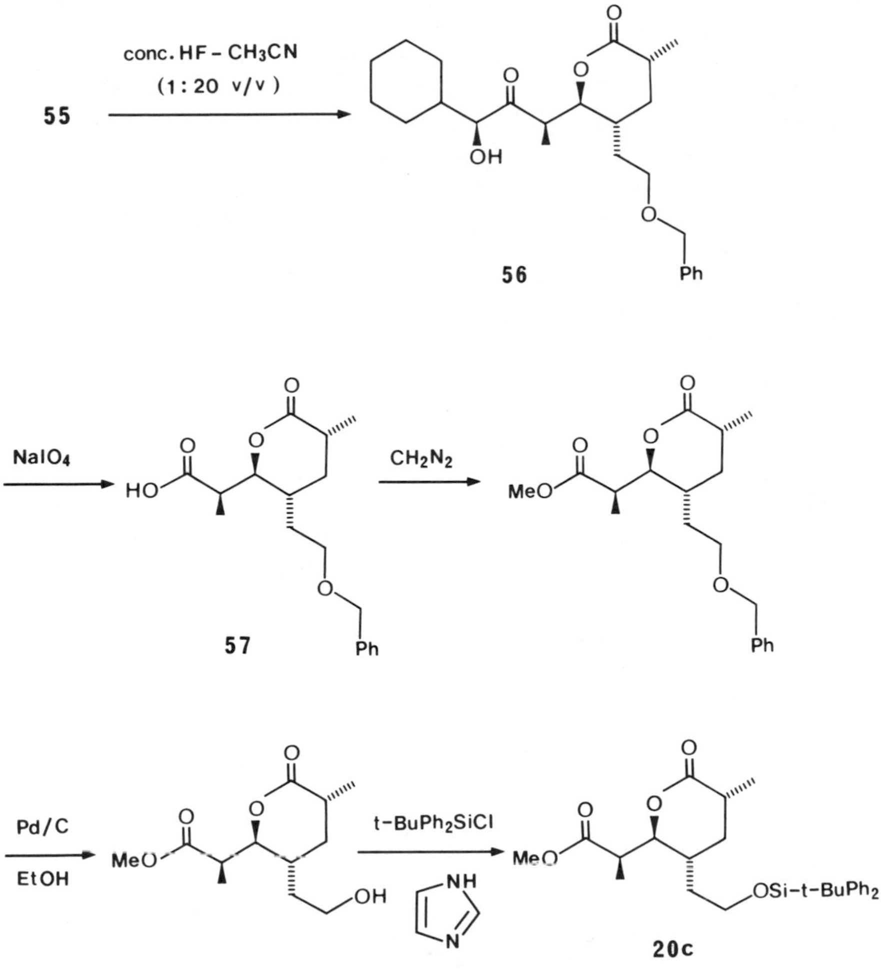

SCHEME 12

products created rather serious problems (such as low yield). The seleno reagents (X = SeMe, Se-*c*-C_6H_{11}, Se-*i*-C_3H_7, SePh) in combination with hindered bases (e.g., 2,6-di-*tert*-butyl-4-methylpyridine) generated the enolates. After the aldol reaction, the seleno group was readily and quantitatively eliminated by washing with a buffered solution (K_2HPO_4) containing benzenethiol. The overall selectivity of the aldol reaction with achiral primary

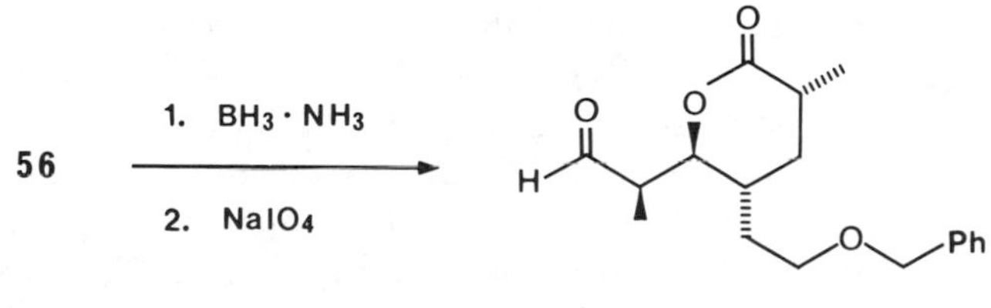

58

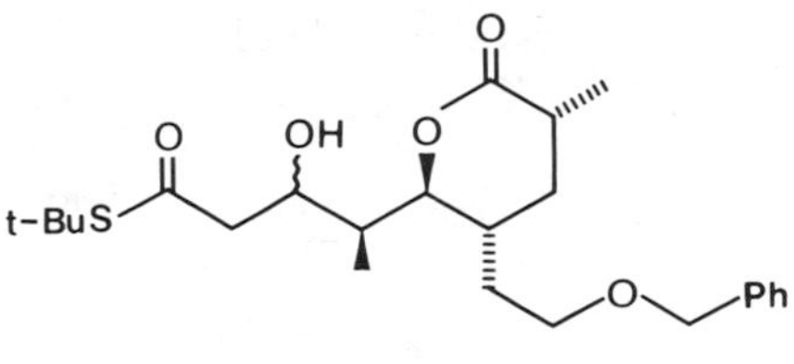

59

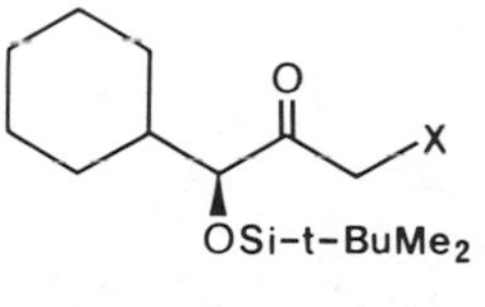

60

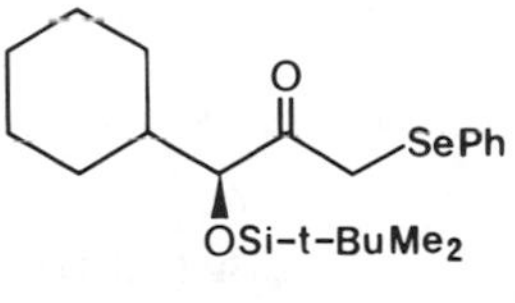

61

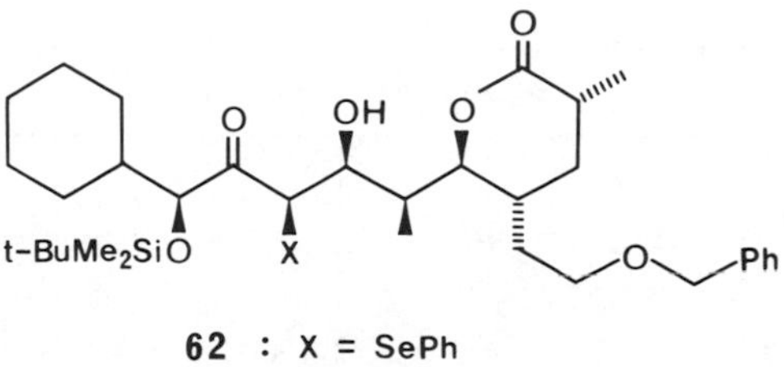

62 : X = SePh

63 : X = H

SCHEME 13

aldehydes, followed by removal of the seleno group, ranged between 20–50:1 with a 50–80% yield. However, the reaction with secondary aldehydes such as **58** became extremely sluggish and also caused other complications. Disappointingly, best results with **61**, for instance, were a modest (but separable) 2:1 ratio of desired **62** to its epimeric product, together with approximately 30% of the deselenated product, which was a separable ~4:1 mixture of the desired **63** to its undesired epimer. Deselenation of **62** was achieved in high yield, using benzenethiol in a phosphate buffer to afford **63**.

As there was a ~4:1 ratio of the desired to undesired deselenated product observed in the reaction cited above, we wondered if this was due to reaction with the deselenated boron enolate. To answer this question we prepared a series of protected methyl ketones (**64**) and examined the product ratios obtained after reaction of its boron enolate with aldehyde **58**, which is *chiral*. The results are shown in Table I. The best ratio was obtained with P = SiMe₃ (**64a**), which gave a 4:1 ratio of desired to undesired aldol products. We felt that this was a time for compromise. The ratio was clearly less than we had originally hoped to achieve; we accepted this reaction as a tentative solution mainly because of its simple execution.

Removal of the chiral auxiliary in the usual manner gave the corresponding acid, which was debenzylated and reprotected as the disilylated compound **65**. When the silylation was carried out with *tert*-butyldimethylsilyl triflate, substantial epimerization resulted at the carbon α to the lactone carbonyl group, presumably due to the lactone-ring opening. Consequently, silylation

TABLE I

PRODUCT RATIOS OBTAINED AFTER REACTION OF BORON ENOLATE OF **64** WITH **58**

P	Ratio (desired/undesired)
a) SiMe₃	4:1
b) SiPhMe₂	3.5:1
c) SitBuMe₂	3.3:1
d) SiEt₃	2.3:1
e) SitBuPh₂	1:1
f) Me	3.3:1
g) CH₂OCH₃	1.6:1

was achieved first by using *tert*-butyldimethylsilyl chloride with imidazole in tetrahydrofuran, followed by the same reagent in warm dimethylformamide. In this manner, **65** was obtained without epimerization. The acid **65** was converted to the thiol ester through the now standard procedure[8] (mixed anhydride formation, then TlS*t*Bu), followed by deprotection of the primary alcohol. This sequence afforded **15**, which was identical to the degradation product of **2** in every respect. Thus the absolute configuration of synthetic **15** was also secured.

65 : R = Si–t–BuMe$_2$

Conversion of **15** to **16** had already been effected (see above). The yield of this overall process was increased to 85% through the modified sequence of performing the oxidation with Collins' reagent and allowing the acid rearrangement to proceed for 20 h at room temperature. Except for a slight difference in isomeric ratios due to the C-6″ acetal, the synthetic intermediate was proved to have the correct stereostructure in comparison with the corresponding degradation product of **2**. This completed the synthesis of the right-hand fragment.

2.   *Preparation of the Left-Hand Fragment*

The next problem to be considered was the preparation of the left-hand fragment (**9**).[20] Compound **9** with P″ = Si*t*BuMe$_2$ and P‴ = THP was selected, since these protecting groups had been used during the degradation sequence. This portion of the molecule is rewritten as **9a**. Since the aldol approach to the right-hand fragment proved impressively effective, we planned to apply the same approach to prepare the left-hand fragment. Unfortunately, the 14,15-anti configuration of the two substituents at C-14 and C-15 (see **9a**) could not be obtained in optically active form via the boron-mediated aldol condensation. Looking at the problem retrosynthetically (Scheme 14), it can be seen that if we draw **9a** in an alternative form **9b**, the aldol reaction appears to be a viable prospect so long as "R" in **66** can be converted to an aldehyde. The most logical candidate for this transformation is an olefin and so "R" must be a vinyl equivalent, for which we choose R = CH$_2$CH$_2$SePh. Simple oxidation followed by syn elimination of benzeneselenol affords the olefin. Therefore, our next task was the preparation of **67**.

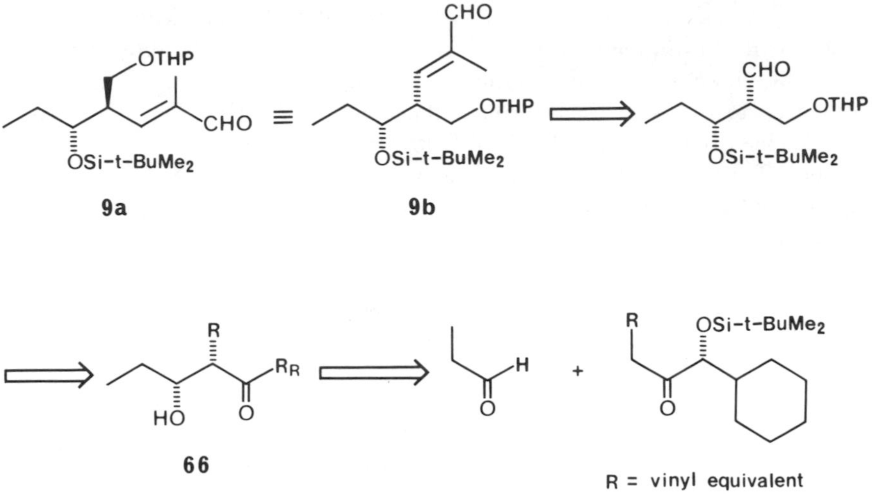

SCHEME 14

Initial attempts to alkylate the lithium enolate of **64c** with 1-bromo-2-phenylselenoethane failed disastrously. In fact, alkylation, even with ethyl iodide, proceeded only to the extent of 25% after one day at room temperature with 20% hexamethylphosphoramide as cosolvent. The lithium enolate of **64c** was quite unreactive.

The cyclopropyl ketone **68** was prepared from the corresponding acid [(R)-hexahydromandelic acid] and cyclopropyllithium followed by silylation of the hydroxyl group. Cleavage of the cyclopropyl ketone with lithium benzeneselenide in the presence of 12-crown-4 proceeded rapidly to give **67** in high yield.[21] As expected, reaction between the dicyclopentylborinic enolate of **67** and propanal at −78°C cleanly gave a 2,3-syn product as a single diastereomer (**69**). Desilylation and ozonolysis proceeded smoothly. However, when the selenoxide elimination was carried out without base, a large amount of selenium-induced cyclization took place to give what appeared to be **70**. When certain bases, e.g., triethylamine, were used, the elimination proceeded smoothly, but the terminal double bond that initially formed migrated to afford the corresponding conjugated enone **71**. Thus optimum conditions were brought about with the use of pyridine, a weaker base than triethylamine. No isomerization occurred and selenium-induced cyclization was kept to a minimum. Successive treatment with sodium meta-periodate, diazomethane, and *tert*-butyldimethylsilyl triflate afforded the protected ester **72**. Diisobutylaluminum hydride (DIBAL) reduction, followed by protection of the resulting hydroxyl group, ozonolysis (dimethylsulfide workup), and a Wittig reaction afforded the unsaturated ester **73**.

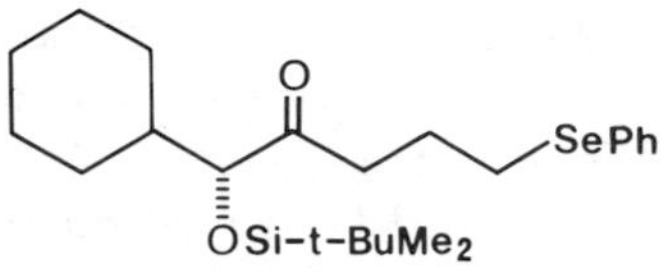

O

SePh

ÖSi–t–BuMe₂

67

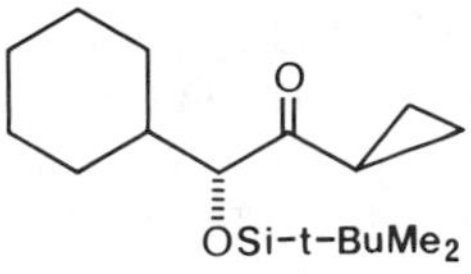

O

ÖSi–t–BuMe₂

68

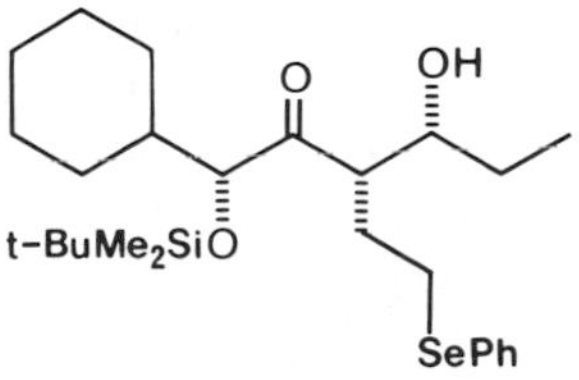

O OH

t–BuMe₂SiÖ

SePh

69

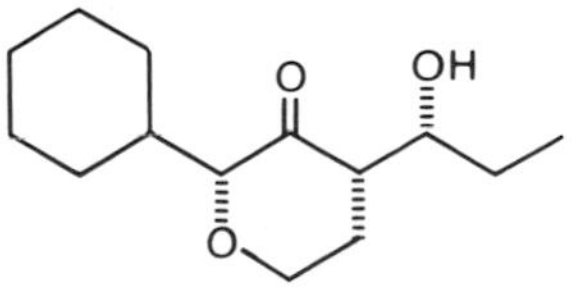

O OH

O

70

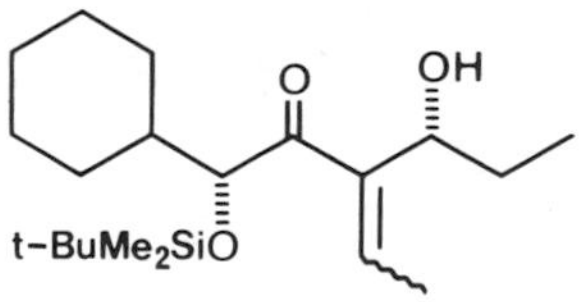

O OH

t–BuMe₂SiÖ

71

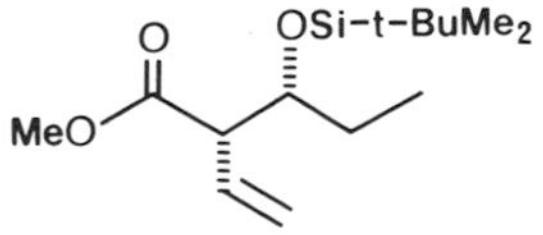

O OSi–t–BuMe₂

MeO

72

OTHP OSi–t–BuMe₂

O=C OEt

≡

O OEt

THPO

OSi–t–BuMe₂

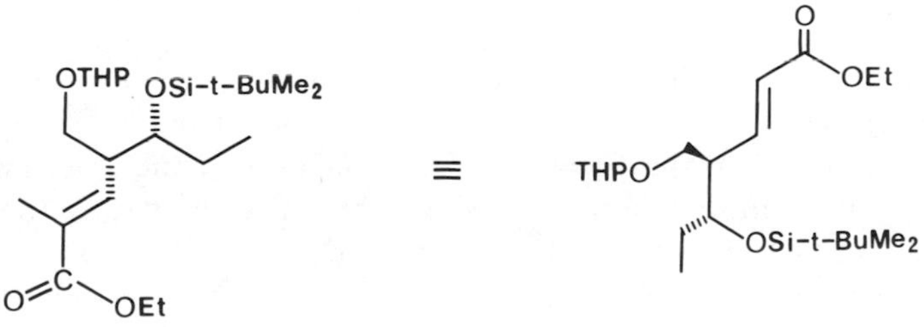

73

Further DIBAL reduction, followed by oxidation, afforded the left-hand fragment **9a**.

3. Coupling of the Right- and Left-Hand Fragments

During the coupling of the right- and left-hand fragments, the use of a bridging one-carbon unit (C-10) is necessary, and the most obvious way to join the two halves is by a Wittig-type reaction, which can introduce the double bond simultaneously. We had previously found that in analogous systems a Wittig reaction induced epimerization at the carbon (C-8) α to the carbonyl group during the condensation. We therefore sought an alternative method. We chose the Peterson reaction, which had been used successfully in the synthesis of narbonolide, which, in a way, served as a model for the present synthesis.[22] Thus the right-hand fragment **17** was first converted to the 2-pyridinethiol ester **74**, using the disulfide and triphenylphosphine. We were aware that the corresponding acid chloride could not be prepared because of the instability of the methyl hemiacetal. Conversion of **74** to the α-trimethylsilylmethyl ketone **75** with lithium bis[(trimethylsilyl)methyl]-cuprate proceeded rapidly. Regioselective generation of the enolate from

74 : R = -S―⟨pyridine⟩
75 : R = CH$_2$SiMe$_3$

75, as in the narbonolide synthesis, followed by addition of **9a**, afforded the coupled product **6e** in high yield. The base-sensitive C-3 silyloxy group remained virtually intact. Removal of both silyl protecting groups was necessary prior to cyclization because, as shown in our earlier degradation studies, all attempts at lactonization of seco-acid derivatives with the protected C-3 hydroxyl group had failed. Also, the work in the narbonolide synthesis[22] demonstrated that the phosphoric acid–mixed anhydride procedure for lactonization would be a method of choice. Thus the thiol ester **6e** was hydrolyzed with mercuric trifluoroacetate, followed by aqueous NaHCO$_3$ (100%), and then the C-3 and C-15 *tert*-butyldimethylsilyl groups were removed with pyridinium fluoride[23] to give **6b** (83%). Treatment of **6b** in tetrahydrofuran with triethylamine and diphenyl phosphorochloridate produced the mixed anhydride, which, after dilution with benzene, was added over a period of 8 h to warm (80°C) benzene containing 4-(dimethylamino)pyridine. The solution was refluxed for 10 h and then worked up in the usual manner. The yield of the lactone **76** was 32%.[24] Treatment of **76** with 70% aqueous acetic

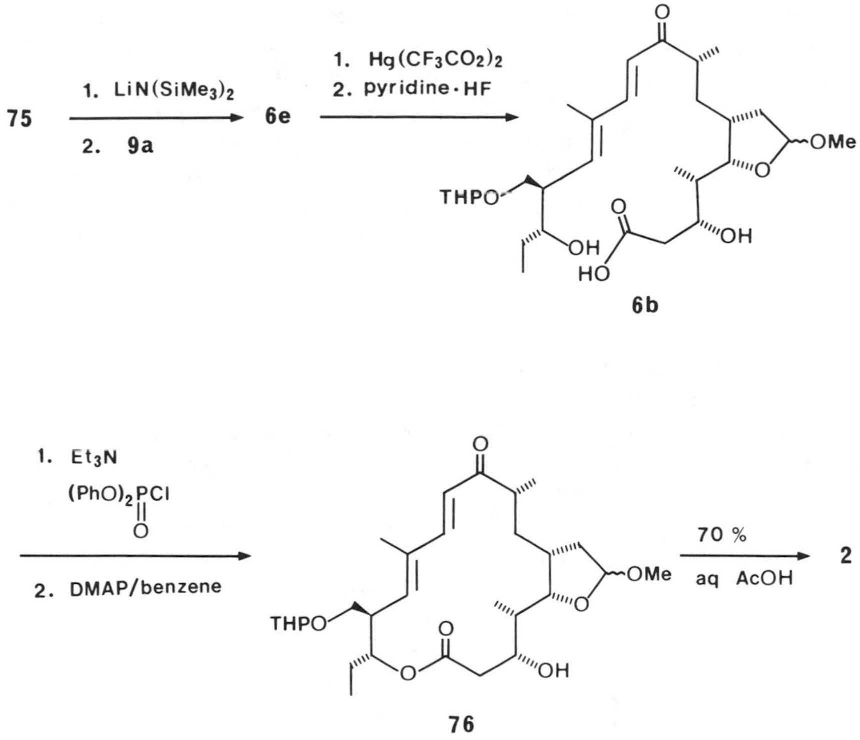

SCHEME 15

acid generated the C-6″ and C-14′ hydroxyl groups, and the synthesis of tylonolide (2), the intact and unmodified aglycone of tylosin, finally came to an end (Scheme 15).

V. Concluding Remarks

The present surge of macrolide syntheses[1] began in 1975 with the first synthesis of methymycin.[5] This synthesis, accomplished in our laboratories, was highly significant in that it demonstrated that medium- and large-sized lactones could be constructed from their corresponding seco-acids, contrary to the then prevailing view that had disfavored such intramolecular cyclization.[1] The seco-acid approach was thus established and paved the way to the subsequent explosive development of this field.

A second important problem, acyclic stereoselection, associated with macrolide synthesis was also clearly recognized in the methymycin synthesis.

During the approximately 3-year period, when the tylosin project was interrupted subsequent to the senior author's transfer to M.I.T., it had become increasingly clear that the "ring" approach used in the methymycin synthesis had its own limitations and proved to be rather inefficient for the construction of acyclic systems. As outlined in Section IV.A, this conventional approach makes use of the clearly defined cis and trans relationships of substituents on a cyclic system, and subsequent ring cleavage transfers this stereochemistry to the resulting acyclic system. In contrast, the aldol approach (see Section IV,B) greatly simplifies the synthetic design and creates directly two chiral centers in one operation. It so happens that our tylosin project has provided a unique opportunity to make a comparative evaluation of both approaches, since the project was resumed at about the time when 6-deoxyerythronolide B yielded to synthesis.[12]

The preceding sections not only disclose the actual course of the research involving, as usual, many technical flaws in the initial scheme and their remedies for the satisfactory execution of each step, but, more importantly, fundamental problems inherent in the two different approaches are clearly revealed. Readers are urged to refer to reference 4 for the concept of diastereofacial selectivity, since the clear recognition of this concept has led to the successful construction of acyclic systems.

Natural products continue to provide a test ground for concepts, strategies, and methods and further help identify new fundamental problems yet to be investigated. In this sense the tylonolide synthesis was indeed worth pursuing.

REFERENCES

1. For recent reviews of the chemistry and biochemistry see S. Masamune, G. S. Bates, and J. W. Corcoran, *Angew. Chem., Int. Ed. Engl.* **16,** 585 (1977). Also see S. Ōmura, "Macrolide Antibiotics." Academic Press; New York (in press).
2. The correct stereostructure was established only recently [S. Ōmura, H. Matsubara, and A. Nakagawa, *J. Antibiot.* **33,** 915 (1980)].
3. R. Morin and M. Gorman, *Kirk-Othmer Encycl. Chem. Technol. 2nd Ed.* **12,** 656 (1967).
4. The conceptual development of this and other macrolide projects as well as stereochemical descriptors pertaining to the acyclic system have now been fully documented [S. Masamune and W. Choy, *Aldrichim. Acta* **15,** 47 (1982)].
5. S. Masamune, C. U. Kim, K. E. Wilson, G. O. Spessard, P. E. Georghiou, and G. S. Bates, *J. Am. Chem. Soc.* **97,** 3512 (1975); S. Masamune, H. Yamamoto, S. Kamata, and A. Fukuzawa, *ibid.* 3513.
6. S. Masamune, Y. Hayase, W. K. Chan, and R. L. Sobczak, *J. Am. Chem. Soc.* **98,** 7874 (1976).
7. S. Masamune, H. Yamamoto, and J. Diakur, unpublished results.
8. S. Masamune and M. Sasaoka, unpublished results.
9. L. D.-L. Lu, *Tetrahedron Lett.* **23,** 1867 (1982).
10. P. A. Bartlett and W. S. Johnson, *Tetrahedron Lett.* p. 4459 (1970).
11. T.-L. Ho and G. A. Olah, *Angew. Chem., Int. Ed. Engl.* **15,** 774 (1976); M. E. Jung and M. A. Lyster, *J. Am. Chem. Soc.* **99,** 968 (1977).

12. S. Masamune, M. Hirama, S. Mori, Sk. A. Ali, and D. S. Garvey, *J. Am. Chem. Soc.* **103,** 1568 (1981).
13. S. Masamune, L. D.-L. Lu, W. P. Jackson, T. Kaiho, and T. Toyoda, *J. Am. Chem. Soc.* **104,** 5523 (1982).
14. P. E. Sonnett and R. R. Heath, *J. Org. Chem.* **45,** 3137 (1980).
14a. D. A. Evans and J. M. Takacs, *Tetrahedron Lett.* **21,** 4233 (1980).
14b. An improved version published after this work was completed has not been tested [D. A. Evans, M. D. Ennis, and D. J. Mathne, *J. Am. Chem. Soc.* **104,** 1737 (1982)].
15. H. C. Brown and N. M. Yoon, *Isr. J. Chem.* **15,** 12 (1976–1977).
16. H. C. Brown, N. R. Ayyangar, and G. Zweifel, *J. Am. Chem. Soc.* **86,** 397, 1071 (1964); G. Zweifel, N. R. Ayyangar, T. Munekata, and H. C. Brown, *ibid.* p. 1076; A Streitweiser, Jr., L. Verbit, and R. Bittman, *J. Org. Chem.* **32,** 1530 (1967); K. R. Varma and E. Caspi, *Tetrahedron* **24,** 6365 (1968); D. R. Brown, S. F. A. Kettle, J. McKenna, and J. M. McKenna, *Chem. Commun.* p. 667 (1967).
17. The highest % e.e. for the methallyl system is approximately 30% [H. C. Brown, P. K. Jadhav, and A. K. Mandal, *Tetrahedron* **37,** 3547 (1981); also see reference *14a.*
18. Cf. D. A. Evans and J. Bartroli, *Tetrahedron Lett.* **23,** 807 (1982).
19. M. Fetizon, M. Golfier, and J.-M. Louis, *Tetrahedron* **31,** 171 (1975).
20. S. Masamune, T. Kaiho, and D. S. Garvey, *J. Am. Chem. Soc.* **104,** 5521 (1982).
21. A. B. Smith, III and R. M. Scarborough, Jr., *Tetrahedron Lett.* p. 1649 (1978).
22. T. Kaiho, S. Masamune, and T. Toyoda, *J. Org. Chem.* **47,** 1612 (1982).
23. B. M. Trost and C. G. Caldwell, *Tetrahedron Lett.* **22,** 4999 (1981); K. C. Nicolaou, S. P. Seitz, M. R. Pavia, and N. A. Petasis, *J. Org. Chem.* **44,** 4011 (1979).
24. The α, β-unsaturated moiety [C(9)-C(11)] takes the s-cis-configuration (reference 2), whereas, the configuration of the corresponding chromophores in the seco-acids is very likely to be s-trans. This may be partly responsible for the low yield observed in the lactonization. Attempts to cyclize (**6b**) with the C-3 hydroxyl protected as the t-butyldimethylsilyl failed.

Chapter 6

PERICYCLIC REACTIONS IN ORGANIC SYNTHESIS AND BIOSYNTHESIS: SYNTHETIC ADVENTURES WITH ENDIANDRIC ACIDS A–G

K. C. Nicolaou
N. A. Petasis

Department of Chemistry
University of Pennsylvania
Philadelphia, Pennsylvania

I. Introduction: The Appealing Endiandric Acid Cascade

Nature is undoubtedly the most creative designer and builder of organic molecules. The amazing combination of rings and functional groups and the strictly defined stereochemistry present in natural products have always fascinated organic chemists. Furthermore, the unique pathways used by nature to construct these molecules are certainly unsurpassed by any synthetic chemist's ability. Nevertheless, several natural products have already been synthesized in the laboratory by successfully mimicking, at least to some extent, their formation in nature.

A very common structural feature of naturally occurring compounds is the presence of several chiral centers, which results in an overall optical activity. Synthesizing such complex molecules in enantiomerically pure form is one of the most challenging and currently very active areas of research in organic chemistry. Nature, of course, makes these dissymmetric molecules quite easily and efficiently by using enzymatic processes.

This enantiospecificity, however, usually observed by the presence of optical rotation, is not a rule for all natural products. A notable exception are the endiandric acids A, B, and C (Fig. 1) recently isolated from the Australian plant *Endiandra introrsa* (Lauraceae).[1-5] Although these compounds have eight chiral centers, they occur in nature in racemic form, as indicated by their lack of optical activity. To explain this unusual observation, the Australian workers have cleverly suggested that these natural products may be derived in nature from acyclic, achiral precursors by a series of *nonenzymatic* reactions.[2] Thus, as immediate precursors, they proposed the now-called endiandric acids E, F, and G (Fig. 2), which can form the tetracyclic acids A, B, and C, respectively, by intramolecular Diels–Alder reactions. These bicyclic acids, in turn, were assumed to be formed from the polyketide-derived acyclic polyenes I or II and III or IV via two consecutive electrocyclization reactions. According to this novel biogenetic hypothesis, there should be another possible bicyclic acid, namely, endiandric acid D (Fig. 2). which has a structure that does not allow any intramolecular Diels–Alder reaction. Indeed this acid was later isolated from the same plant.[6]

The frameworks of endiandric acids A, B, and C have not been encountered in nature before. Their condensed molecular structures, comprising two cyclohexenes, one cyclopentane, and one cyclobutane ring, and substituted with a phenyl ring and a short carboxylic acid-bearing chain, involve fascinating and rather complex molecular assemblies. Interestingly, the structure

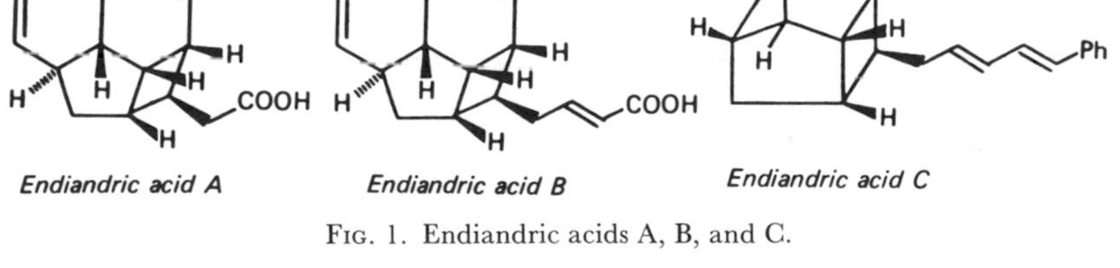

Endiandric acid A   Endiandric acid B   Endiandric acid C

FIG. 1. Endiandric acids A, B, and C.

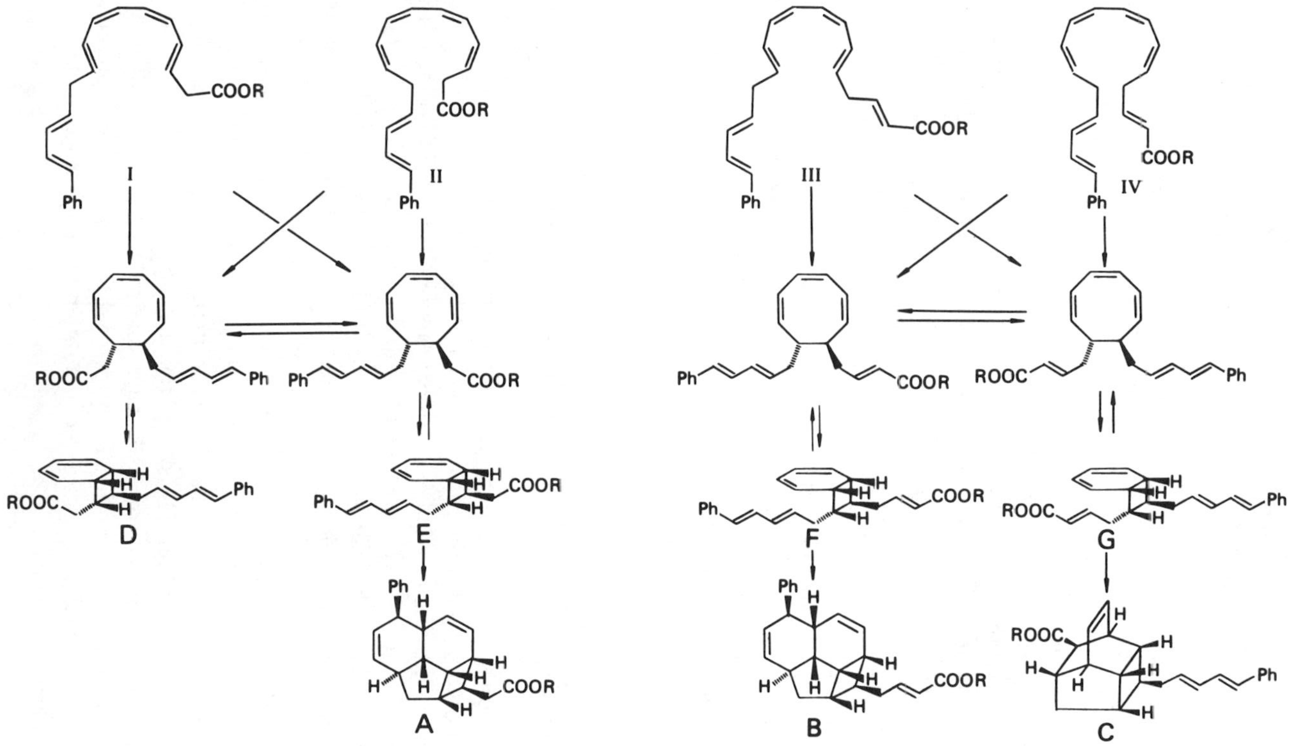

Fig. 2. The endiandric acid cascade.

Fig. 3. The similarity between basketene and endiandric acid C.

of endiandric acid C resembles that of basketene, (Fig. 3), a man-designed molecule, synthesized in the laboratory.[7]

Intrigued by these unique and novel structures and by the interesting hypothesis proposed for their biosynthesis, we decided to undertake investigations in this area directed primarily toward total synthesis and testing of the biogenetic hypothesis. Our work was also aimed at identifying the yet unclear biological and pharmacological significance of these substances and determining the stability of other members of the now-called endiandric acid cascade (Fig. 2).

An important component of both the biosynthetic scheme and our evolved synthesis is a group of three *pericyclic reactions* comprising two electrocyclizations and one intramolecular cycloaddition. As it turned out, the endiandric acids provided an excellent application of these important reactions in total synthesis, and these investigations revealed substantial support to their involvement in biosynthesis.[8-11]

II.   Retrosynthetic Analysis of Endiandric Acids A–G

The most logical first retrosynthetic disconnection of tetracyclic acids A, B, and C seemed to be the one suggested by the biosynthetic hypothesis, i.e., a retro-intramolecular Diels–Alder reaction (Fig. 4). This type of transformation is indeed well documented in the literature.[12-14] Another important aspect of this pathway is that it leads to bicyclic acids E, F, and G, which closely resemble the fourth acid anticipated to be a natural product at the time of our planning, namely, endiandric acid D. Thus as our synthetic targets became the four structurally related endiandric acids D–G, all of which could be easily seen to arise from a common intermediate, i.e., imaginary dialdehyde **8**, via two different Wittig-type reactions or a Wittig and an oxidation. Of course, the real precursor has to be a bicyclic molecule of the type shown in **9**, where one of the two substituents is the desired aldehyde and the other is a protected form of this functionality. Intermediate **9** can then be considered as a differentiated derivative of the symmetrical key intermediate **10**.

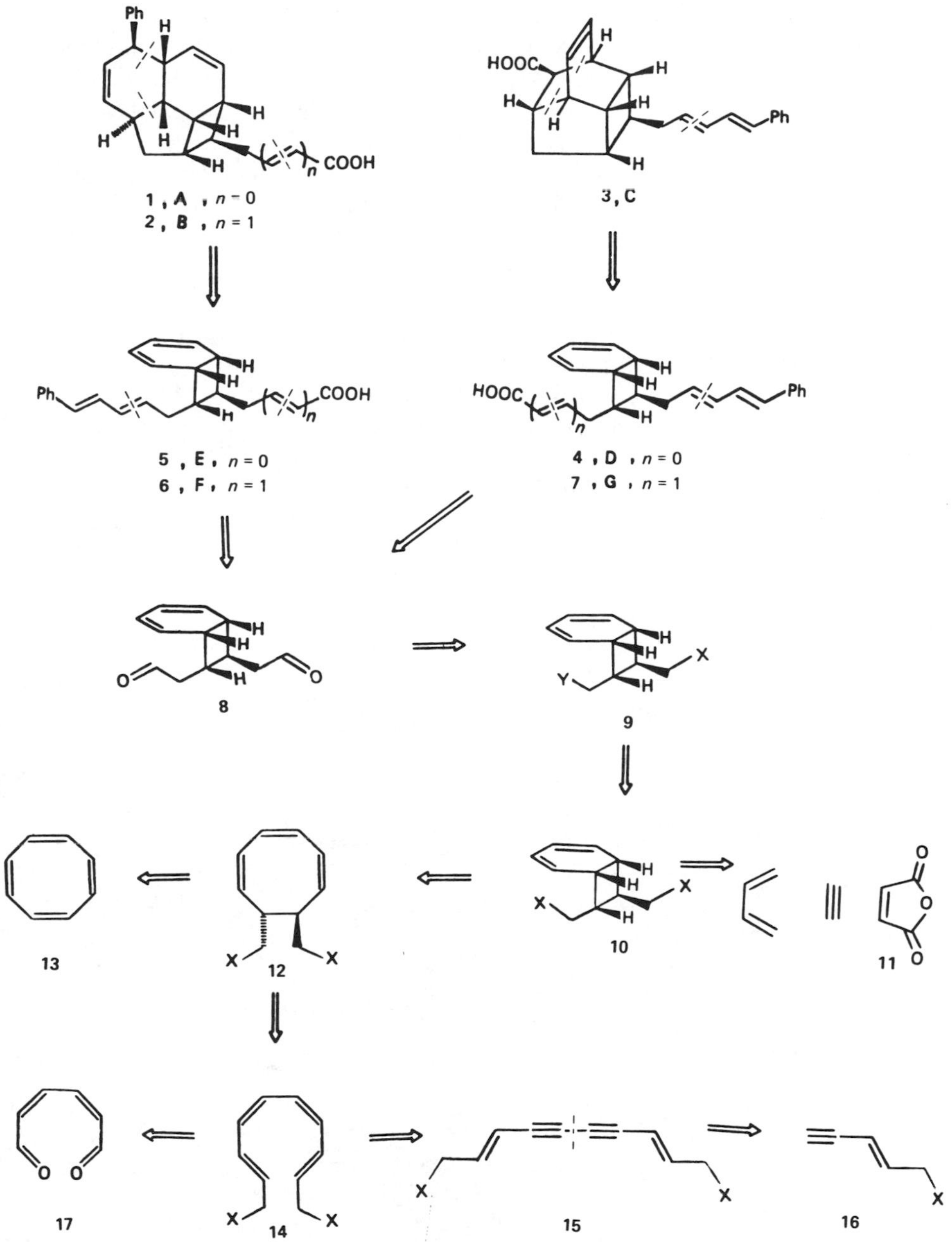

FIG. 4. Retrosynthetic analysis of endiandric acids A–G.

For the construction of the bicyclic intermediate **10**, several approaches were considered, the most promising being the following: (a) photolytic [2 + 2] cycloaddition of acetylene to maleic anhydride (**11**), followed by construction of the six-membered ring via a Diels–Alder reaction, (b) electrocyclization of the cyclooctatriene intermediate **12**, formed by derivatization of cyclooctatetraene (**13**), and (c) double electrocyclization of the conjugated tetraene **14** via the cyclooctatriene **12**. The scenario here was to obtain the required tetraene **14** by selective hydrogenation of the central diyne unit of precursor **15**, which is simply the product of symmetrical coupling of a 5-substituted pent-2-en-1-yne (**16**). The alternative construction of **14** via dialdehyde **17**, or its equivalents, seemed less promising because of the instability and inaccessibility of such precursors.

The required operations for the conversion of acyclic tetraene **14** to bicyclic diene **10** are quite interesting. They are: an 8πe conrotatory electrocyclization and a 6πe disrotatory electrocyclization. These pericyclic reactions are thermally allowed by the Woodward–Hoffmann rules;[15–21] and in order to obtain the trans-bicyclic product, the geometry of the tetraene should be either trans-cis-cis-trans or cis-cis-cis-cis. As shown in Fig. 5, when the tetraene has different substituents, both possible products can be obtained through the two cyclooctatriene intermediates, which are in equilibrium.

Although the foregoing pericyclic reactions are well known in the literature[22–24] to our knowledge, no applications to the construction of complex molecules, natural or unnatural, had been described up to the outset of our work.

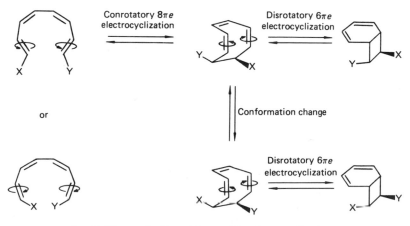

FIG. 5. Thermally allowed 8πe and 6πe electrocyclizations.

III. Synthesis of the Key Bicyclic Intermediate

In order to select the best synthetic route to the desired intermediate **10**, a literature search was undertaken in an effort to find precedents for the three alternative strategies described above. As a result, some interesting information was revealed, which was either in support of or against each of the three approaches.

A. THE PHOTOLYTIC APPROACH

This strategy begins with the photolytic [2 + 2] cycloaddition of acetylene to maleic anhydride (**11**) to give cyclobutene adduct **18** and is followed by a Diels–Alder reaction with butadiene to form **19** (Fig. 6). Both of these reactions are known in the literature, but they proceed with rather moderate yields.[25] Furthermore, the elaboration of **19** to **10** requires conversion of the olefin to a diene and epimerization of one of the two carbonyl substituents, making this sequence rather lengthy.

A more direct photolytic approach would be the addition of maleic anhydride to benzene but, unfortunately, the intermediate adduct obtained in this reaction undergoes a facile thermal cycloaddition of another molecule of maleic anhydride to give diadduct **20**[25] (Fig. 7).

Thus, from the foregoing evidence, the photolytic approach did not seem to be the method of choice, although it is a possible one.

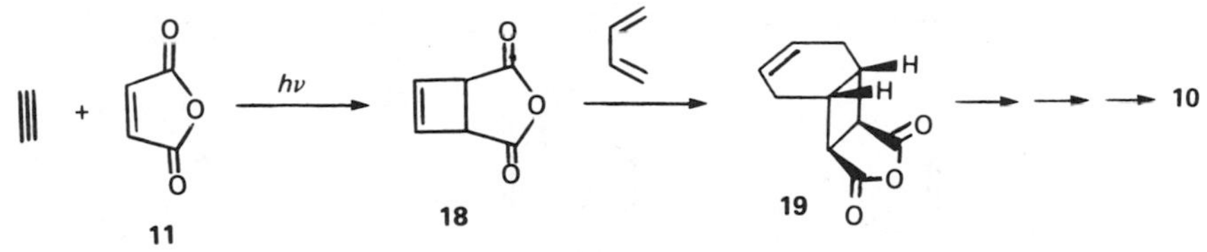

FIG. 6. Stepwise functionalization of maleic anhydride.

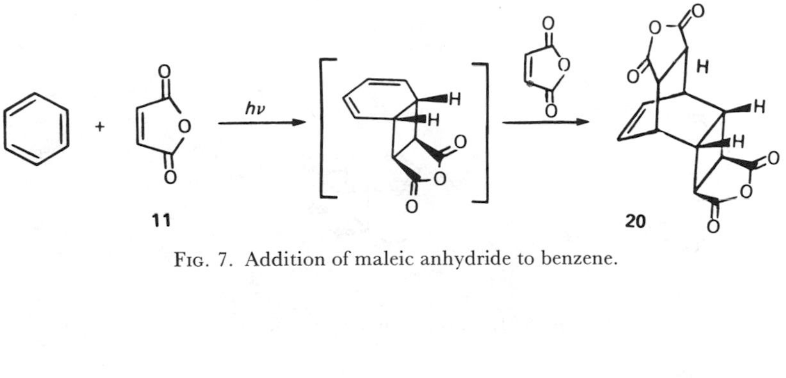

FIG. 7. Addition of maleic anhydride to benzene.

B. The Cyclooctatetraene Approach

Cyclooctatetraene (**13**) is known to undergo facile addition of electrophiles to give 1,2-disubstituted cyclooctatrienes (**21**) that are in equilibrium with the bicyclo[4.2.0]octadiene derivatives (**22**) (Fig. 8). For example, addition of bromine gives the trans dibromide **22a**.[26] Although this dibromide has the basic skeleton of the desired intermediate **10** (Fig. 4), it is necessary to replace the bromines with carbon substituents. An attempt, however, to convert **22a** to the dinitrile **22b**, using KCN, resulted in the formation of **23b**.[27] Apparently, the octatetraenes **23** are also in equilibrium with the octatrienes **21**, and their formation is favored when X is a group that extends the conjugation (Fig. 8). This is also confirmed by the conversion of **22a** to **23c** on treatment with the corresponding acetylenic Grignard reagents.[28,29]

We had hoped that treatment of **22a** with a cuprate reagent, bearing a saturated group, might provide us an access to a desired intermediate. Thus we prepared **22a** and treated it with dimethyl cuprate but, unfortunately, no **22d** could be obtained.

An alternative way of preparing 1,2-derivatives from cyclooctatetraene seemed more promising. Treatment of this annulene with an alkali metal (Li, Na, or K) gives the dianion, which is an aromatic species because it has 10 π-electrons. This dianion can then be quenched with an electrophile to give the corresponding 1,2-disubstituted cyclooctatriene **21**, which again is in equilibrium with **22** or **23**. Using methyl iodide as the electrophile, it was indeed possible to obtain **22d**, albeit in low yield.[30] With carbon dioxide, however, the major product obtained was the tetraenic diacid **23e**, probably

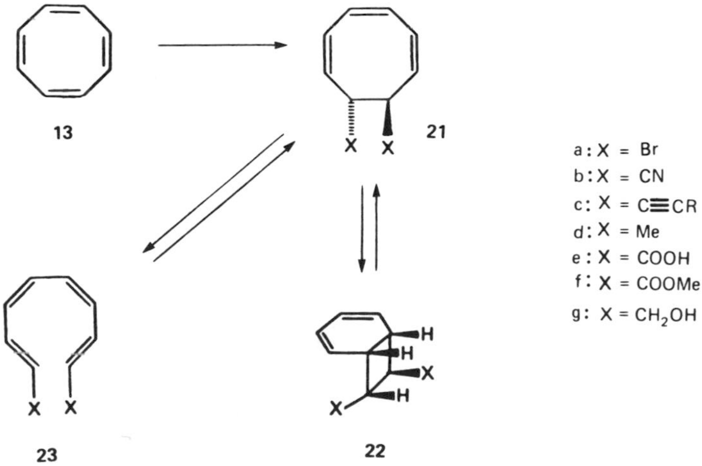
13
21
X X
23
X X
22
H
H
X
H
X
a : X = Br
b : X = CN
c : X = C≡CR
d : X = Me
e : X = COOH
f : X = COOMe
g : X = CH$_2$OH

FIG. 8. Derivatization of cyclooctatetraene.

because of stabilization from the extended conjugation.[31] Interestingly, when the diester **23f**, prepared from **23e**, was heated at 80°C it was converted to the bicyclic derivative **22f**.[31] Although **22f**, prepared in this fashion, seemed like a good starting point, the low overall yield from **13** was rather unattractive. Furthermore, another step was necessary for the conversion of **22f** to a useful intermediate, e.g., reduction to **22g**. In order to convert **13** to **22g** directly, we then attempted the quenching of the dianion of **13** with paraformaldehyde, but without any useful results.

Surprisingly, the cyclooctatetraene approach, despite being the most direct route to intermediates similar to **10**, for our purposes did not seem to be an efficient one, not to mention the high cost of the starting material.

C.   THE CONSECUTIVE ELECTROCYCLIZATIONS APPROACH

The disadvantages of the two above strategies having been realized, we then turned our attention to the third possibility, i.e., to prepare intermediate **22** (Fig. 9) by the two consecutive electrocyclization reactions described in Fig. 5. This approach not only seemed to be quite appealing, but it also seemed to be the most challenging one. Although some precedents exist in the literature, there are no synthetically viable examples. With **25d**, hydrogenation led to tetraene **23d**, which isomerized rapidly to **21d** and **22d** at room temperature.[23] In the case of the diester **25f**, however, the resulting tetraene **23f** had to be heated to higher temperatures for the electrocyclizations to occur.[24]

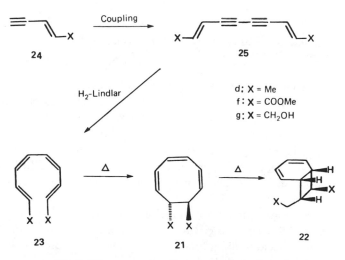

FIG. 9. The consecutive electrocyclizations approach.

We proceeded along this approach, starting from *trans*-pent-2-en-4-yn-1-ol (**24g**, Fig. 9), which is not only readily accessible[32] but also commercially available (Farchan Labs, Willoughby, Ohio). This acetylenic alcohol could be efficiently coupled to give **25g**, which was then subjected to numerous hydrogenation conditions. To our dismay, the commercial catalysts that were used could not give us appreciable amounts of useful products under a variety of conditions (solvents, temperatures, and poisons).

As an alternative to hydrogenation, we attempted using the hydroboration reaction with sterically hindered dialkyl boranes.[33] The presence of both the diyne unit and the two olefinic bonds, however, could not allow selective reduction of the two acetylenic bonds, neither with dicyclohexylborane nor with disiamylborane.

The solution to the problem of converting **25g** to **22g** came rather unexpectedly. During a discussion with Dr. John Partridge (Hoffmann–La Roche, Inc., Nutley, New Jersey), who was visiting us as a seminar speaker, he offered to provide us with a "super" Lindlar catalyst used in the Roche vitamin D synthesis. Indeed, a few days later, that catalyst arrived and, to our delight, it provided us with a respectable and reproducible 45–55% yield of **22g**! The tetraene **23g** or the cyclooctatriene **21g** could not be isolated under our hydrogenation conditions but, instead, the desired product **22g** was obtained directly, after chromatographic purification.

As it turned out, the two pericyclic reactions provided us with a powerful and convenient means of converting an acyclic polyene to a bicyclic intermediate having four chiral centers with precise relative stereochemistry and two of the desired rings.

IV. Functionalization of the Bicyclic Intermediate: Synthesis of Endiandric Acids D, E, F, and G

After our successful entry to the 7,8-disubstituted bicyclo[4.2.0]octadiene system (**22g**), we were faced with the problem of differentiating between the two hydroxyl groups of this diol. Initial attempts selectively to protect the exo OH, taking advantage of the seemingly higher steric hindrance of the endo OH, gave us inseparable 1:1 mixtures of the two possible monoprotected derivatives, even with bulky protective groups such as Si*t*BuPh$_2$, CPh$_3$, COPh, and CO*t*Bu. Selective deprotection of the diprotected derivatives was also unsuccessful.

Once it became clear that external reagents could not differentiate between the two hydroxyls in **22g**, we then turned our attention to possible internal protection. Indeed, Dreiding molecular models suggested a possible interaction of the endo OH with one of the olefinic bonds because of proximity reasons.

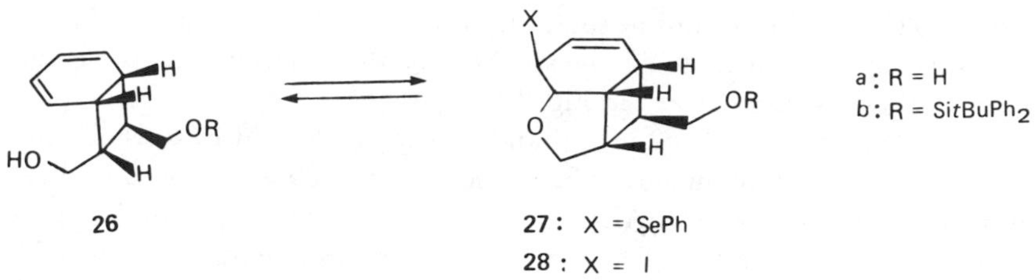

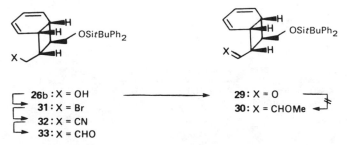

FIG. 10. Differentiation between the two OHs of diol **22g** (**26a**).

This observation led to the idea of using a reaction for temporarily engaging these two functionalities while protecting the exo OH. Being familiar with the cyclizations of unsaturated alcohols with phenylselenenyl chloride,[34] we first tried using the phenylselenoetherification reaction as a means of selectively blocking the endo OH. Thus treatment of diol **22g** (**26a**, Fig. 10) with PhSeCl gave in high yield the corresponding phenylselenoether **27a**. Reversal of this reaction, however, with Na–NH$_3$(l)[35] or with Me$_3$SiCl–NaI[36] could not be realized, probably because of interference from the olefinic bond.

Switching from selenium to iodine provided us with the solution. The iodoether **28a** could be prepared in almost quantitative yield from **26a** and, after protection of the exo OH as the *tert*-butyl diphenylsilyl ether (**28b**), the reversal of the iodoetherification could be achieved with Zn in acetic acid, thus providing us with the desired monoprotected derivative **26b**.

The one-carbon homologation of **26b**, although seemingly a simple operation, presented us with some more problems. Oxidation of **26b** with pyridinium chlorochromate proceeded with unforeseen attack on the diene system. Furthermore, although the aldehyde **29** (Fig. 11) could be obtained smoothly by Swern oxidation, the Wittig reaction to give **30**, the precursor of aldehyde **33**, did not work. Consequently, we focused on an alternative homologation approach via bromide **31** and nitrile **32**, which turned out to be a highly efficient sequence.

FIG. 11. Homologation of **26b** to **33**.

Nitrile **32** is a white crystalline solid, stable at room temperature, and it was used as the key intermediate for the synthesis of all endiandric acids. Its successful synthesis is summarized in Fig. 12.

The conversion of nitrile **32** to endiandric acids E (**5**) and F (**6**) is outlined in Fig. 13. Reduction of the nitrile **32** to the aldehyde **33** was done with diisobutylaluminum hydride (DIBAL), followed by acidic workup. Condensation of this aldehyde with the phosphorane derived from cinnamyl bromide (*trans*-PhCH=CHCH=PPh$_3$) resulted in the construction of the phenyldiene unit of **35**, but the newly formed olefinic bond was a 60:40 mixture $E:Z$. This ratio, however, was improved to $E:Z = 20:1$ when we replaced the phosphorane with the anion of the corresponding diethyl phosphonate [*trans*-PhCH=CHCH$_2$P(O)(OEt)$_2$, LDA]. Another phosphonate condensation was employed for the conversion of aldehyde **33** to α,β-unsaturated ester **34**, the precursor for the synthesis of endiandric acid C.

Removal of the protective group of **35** with fluoride anion formed alcohol

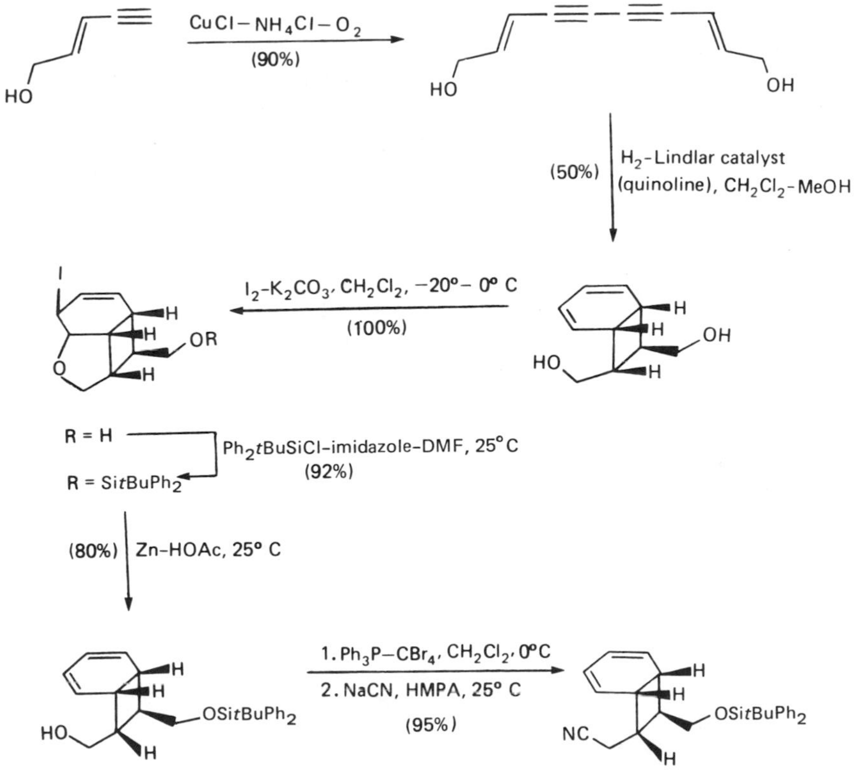

Fig. 12. Successful synthesis of the key intermediate nitrile **32**.

36, which was elaborated to aldehyde **39**, using our standard protocol. Jones oxidation of this aldehyde to the carboxylic acid did not work too well, apparently because of the presence of the acid-sensitive diene systems. Using the silver oxide oxidation, however, we arrived at endiandric acid E (**5**) in high yield. The "vinylog" of this acid, namely, endiandric acid F (**6**), was obtained by phosphonate condensation on aldehyde **39**, followed by hydrolysis of the resulting α,β-unsaturated ester.

The other two bicyclic acids, D (**4**) and G (**7**), were also prepared from nitrile **32**, as shown in Fig. 14. Deprotection of **32** and basic hydrolysis of the hydroxy nitrile **40**, followed by diazomethane treatment, gave hydroxy ester **42** in high yield. The same compound was initially obtained by reduction of **32** to the aldehyde **33**, Jones oxidation of **33**, esterification with diazomethane, and removal of the protective group. However, the yield of the oxidation step was again low (∼30%) because of the acid sensitivity of the substrate. Silver oxide oxidation of **33** also failed because of the insolubility of this compound in aqueous media.

Functional group manipulation of **42** to **46** proceeded uneventfully, as expected, and this aldehyde could be converted to endiandric acid D (**4**) by

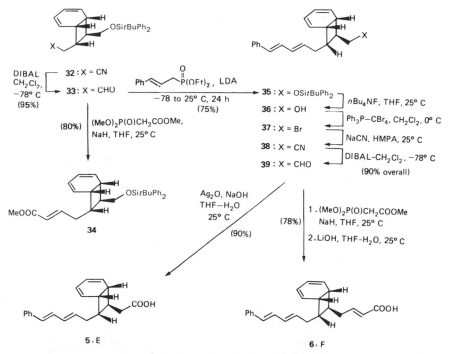

FIG. 13. Synthesis of endiandric acids E and F.

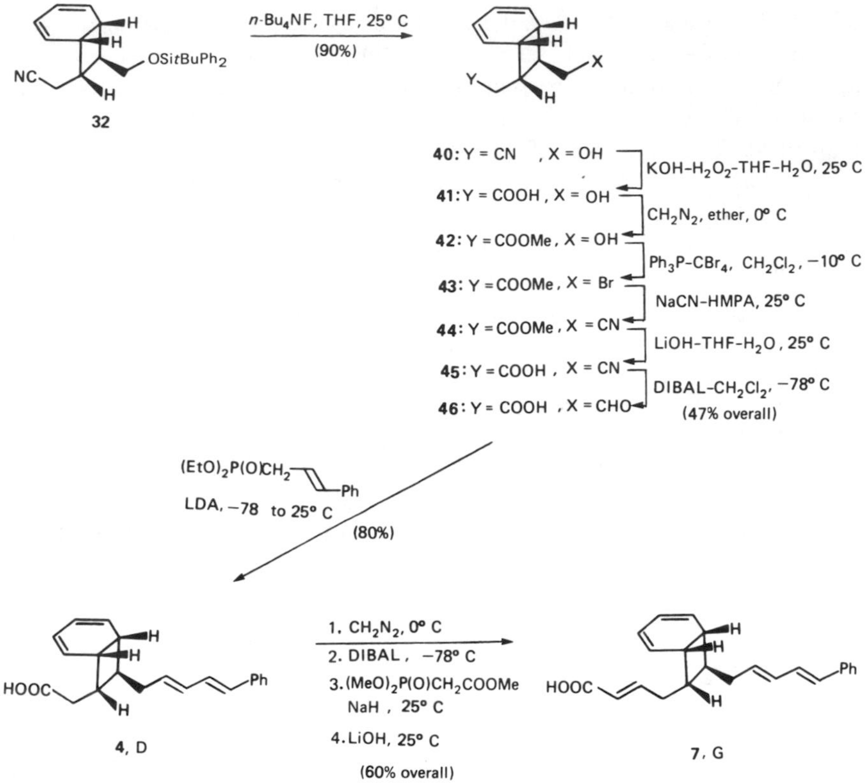

32

n-Bu₄NF, THF, 25° C
(90%)

40: Y = CN , X = OH ⎤ KOH-H₂O₂-THF-H₂O, 25° C
41: Y = COOH , X = OH ⎤ CH₂N₂, ether, 0° C
42: Y = COOMe , X = OH ⎤ Ph₃P-CBr₄, CH₂Cl₂, −10° C
43: Y = COOMe , X = Br ⎤ NaCN-HMPA, 25° C
44: Y = COOMe , X = CN ⎤ LiOH-THF-H₂O , 25° C
45: Y = COOH , X = CN ⎤ DIBAL-CH₂Cl₂, −78° C
46: Y = COOH , X = CHO ⎤ (47% overall)

(EtO)₂P(O)CH₂⟶Ph
LDA, −78 to 25° C
(80%)

HOOC ... **4, D**

1. CH₂N₂, 0° C
2. DIBAL , −78° C
3. (MeO)₂P(O)CH₂COOMe NaH , 25° C
4. LiOH, 25° C
(60% overall)

HOOC ... **7, G**

Fɪɢ. 14. Synthesis of endiandric acids D and G.

using the cinnamyl phosphonate procedure. At the time of our first synthesis of this compound, its natural occurrence had not yet been established, although a suspect compound had been isolated from the same plant. Thus we exchanged samples with the Australian workers (Professor D. St. C. Black, Monash University, Australia and, to our and their delight, the two compounds were identical in all respects (NMR, IR, MS, and mp).

Esterification of **4** and reduction to the aldehyde with DIBAL, followed by phosphonate condensation and ester hydrolysis, provided us with the "vinylog" of **4**, endiandric acid G (**7**, Fig. 14).

V.　The Intramolecular Diels–Alder Reactions: Synthesis of Endiandric Acids A, B, and C

The construction of the two tetracyclic skeletons of endiandric acids A, B, and C required another pericyclic reaction, namely, an intramolecular [4 + 2]

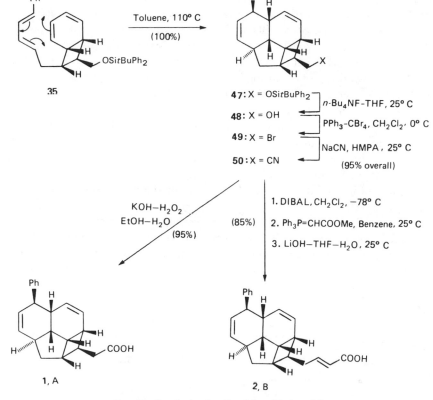

FIG. 15. Synthesis of endiandric acids A and B.

cycloaddition (Diels–Alder reaction). The cyclohexadiene system of the bicyclic precursors would provide the dienophile component in the case of acids A and B and the diene component in the case of acid C.

We first attempted the intramolecular cycloaddition for the synthesis of acids A and B with intermediate **35** (Fig. 13), which was heated to 110°C as a dilute solution in toluene in order to prevent intermolecular reaction or polymerization. The results of this experiment were really fascinating. Our expectations were indeed realized and the intramolecular Diels–Alder reaction proceeded nicely, forming intermediate **47** (Fig. 15) almost quantitatively and with the four new chiral centers having the desired stereochemistry.

From **47** we could prepare nitrile **50** in high overall yield, using again the standard chemistry described above and depicted in Fig. 15. Hydrolysis of this nitrile gave us endiandric acid A (**1**), whereas reduction to the aldehyde, followed by phosphorane condensation and ester hydrolysis, provided us with endiandric acid B (**2**). Comparison of these synthetic endiandric acids A and B with samples of the natural products furnished to us by the Australian workers (Professor D. St. C. Black, Monash University, Australia) showed identical physical and spectroscopic properties (NMR, IR, MS, and mp).

For the synthesis of endiandric acid C we used intermediate **34** (Fig. 13), which upon heating to 110° formed, as expected, the basketene-type skeleton of **51**, via an alternative Diels–Alder-pathway reaction (Fig. 16).

Transformation of **51** to **56** was performed again, using the standard methodology, and the final coupling with the anion of diethyl cinnamyl phosphonate worked nicely, as before, to give us endiandric acid C (**3**, Fig. 16). Comparison of the synthetic material with a sample of the natural product was once again a complete match.

The utilization of intermediates **34** and **35** for the preparation of **1–3**, was rather fortunate. As we found out later, heating of the esters of endiandric acids D, E, F, and G resulted in isomerizations, as shown in Fig. 2. The presence of only one possibility for intramolecular cycloaddition in **34** and **35**, assisted by the favorable positioning of the bulky silyl ether group in the exo position, apparently resulted in a more facile formation of the desired tetracyclic structures.

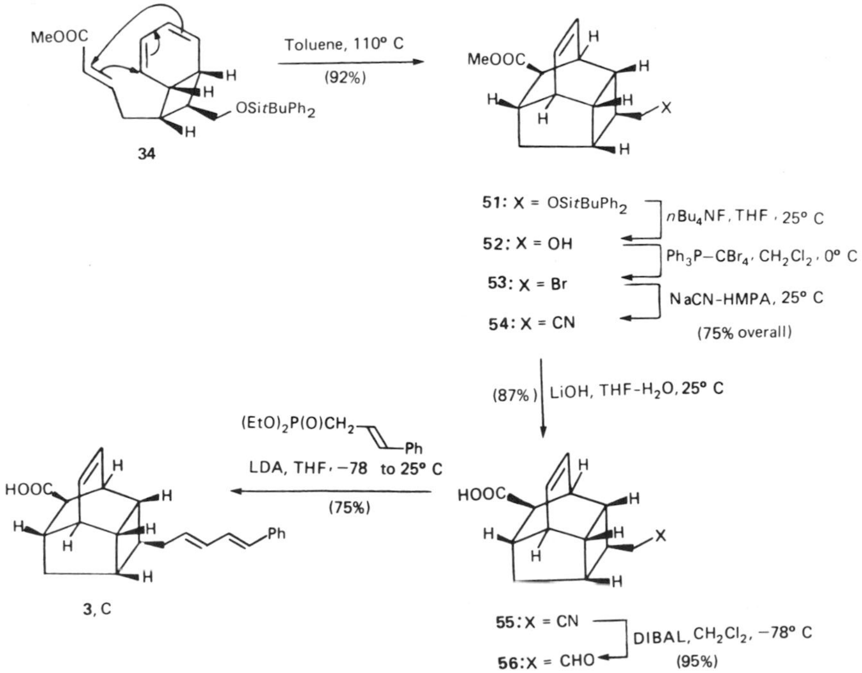

FIG. 16. Synthesis of endiandric acid C.

Interestingly, the thermal studies carried out on the esters D–G (Fig. 2, $R = Me$), uncovered the ease of isomerization between D↔E and F↔G, which resulted in the rather interesting conversions of D to A, F to C, and G to B (Fig. 2).

A more direct construction of the tetracyclic acids A–C from acyclic acetylenic precursors in essentially one step, combining the operations described above, was also carried out in our laboratories.[10,11] It should be noted that this intriguing approach parallels closely the hypothetical biogenetic scheme (Fig. 2).

VI. Conclusions

As the reader can see, the synthesis of the endiandric acids described in this article, despite the early obstacles, turned out to be rather short and efficient. The two-step conversion of a commercially available starting material to the bicyclic diol **22g** (Figs. 9 and 12), even with the moderate yield, is quite practical and represents a novel application of the two consecutive electrocyclizations (8πe and 6πe). The regiospecific monoprotection of the exo OH of this diol comprises a rather novel application of the iodoetherification reaction. The standard homologation sequence used, i.e., OH → Br → CN, provided us with the stable and versatile nitrile intermediates, which could be either reduced to the aldehydes or hydrolyzed to the carboxylic acids. Also, the use of the phosphonate condensation methodology allowed us to construct the (E,E)-phenyldiene unit in a highly stereoselective manner. Furthermore, the desired intramolecular cycloadditions worked out beautifully, giving rise to the tetracyclic skeletons with the proper stereochemistry and in high yield. Thus we were able to synthesize the four bicyclic endiandric acids D–G and the three tetracyclic ones A–C from a common precursor and by using similar reaction sequences. Noteworthy is the almost complete stereospecificity of these sequences with the stereochemical information encoded into the geometry of certain double bonds.

The availability of the bicyclic acids D–G prompted us to study their thermal behavior, resulting in the interesting observation that these molecules are isomerized readily, apparently via the equilibria indicated in Fig. 2, which lead to the tetracyclic acids.[8] This result can be used to explain the coexistence of the seemingly unrelated acids B and C in the same plant. Furthermore, the reasonable thermal stability of compounds E–G make them likely candidates as natural products.

Our findings from the investigations in the endiandric acid cascade demonstrate clearly the significance and powerful nature of the three pericyclic reactions involved, i.e., the 8πe electrocyclization, the 6πe electrocyclization, and the $(4 + 2)\pi e$ intramolecular cycloaddition (**Diels–Alder**), which proceed rather easily in these systems, even at ambient temperatures, suggesting their possible involvement in the biogenesis of these molecules as originally hypothesized.[2] Further applications of these reactions in organic synthesis are likely to appear in the future.

ACKNOWLEDGMENTS

We would like to extend our deep appreciation to our co-workers: R. E. Zipkin and Dr. J. Uenishi for their contributions to the endiandric acid project. Our appreciation is also expressed to Professor D. St. C. Black, Monash University, Australia, for samples of the natural products and helpful communications, and to Dr. J. Partridge of Hoffmann–La Roche, Nutley, N.J., for generous gifts of superior Lindlar catalyst. We would also like to thank Merck Sharp & Dohme, the A. P. Sloan Foundation, the Camille and Henry Dreyfus Foundation, and the National Institutes of Health for their generous financial support.

REFERENCES

1. W. M. Bandaranayake, J. E. Banfield, D. St. C. Black, G. D. Fallon, and B. M. Gatehouse, *J. chem. Soc., Chem. Commun.* pp. 162–163 (1980).
2. W. M. Bandaranayake, J. E. Banfield, and D. St. C. Black, *J. Chem. Soc., Chem. Commun.* pp. 902–903 (1980).
3. W. M. Bandaranayake, J. E. Banfield, D. St. C. Black, G. D. Fallon, and B. M. Gatehouse, *Aust. J. Chem.* **34,** 1655–1668 (1981).
4. W. M. Bandaranayake, J. E. Banfield, and D. St. C. Black *Aust. J. Chem.* **35,** 557–565 (1980).
5. W. M. Bandaranayake, J. E. Banfield, and D. St. C. Black, *Aust. J. Chem.* **35,** 567–579 (1982).
6. J. E. Banfield, D. St. C. Black, S. R. Johns, and R. I. Willing, *Aust. J. Chem.* **35,** 2247–2256 (1982).
7. S. Masamune, H. Cuts, and M. G. Hogben, *Tetrahedron Lett.* pp. 1017–1021 (1966).
8. K. C. Nicolaou, N. A. Petasis, R. E. Zipkin, and J. Uenishi, *J. Am. Chem. Soc.* **104,** 5555–5557 (1982).
9. K. C. Nicolaou, N. A. Petasis, J. Uenishi, and R. E. Zipkin, *J. Am. Chem. Soc.* **104,** 5557–5558 (1982).
10. K. C. Nicolaou, R. E. Zipkin, and N. A. Petasis, *J. Am. Chem. Soc.* **104,** 5558–5560 (1982).
11. K. C. Nicolaou, N. A. Petasis, and R. E. Zipkin, *J. Am. Chem. Soc.* **104,** 5560–5562 (1982).
12. G. Brieger and J. N. Bennett, *Chem. Rev.* **80,** 63–97 (1980).
13. A. Krantz and C. Y. Lin, *J. Am. Chem. Soc.* **95,** 5662–5672 (1973).
14. F. Naf and G. Ohloff, *Helv. Chim. Acta* **57,** 1868–1870 (1974).
15. R. B. Woodward and R. Hoffmann, "The Conservation of Orbital Symmetry." Academic Press, New York, 1970.
16. R. E. Lehr and A. P. Marchand, "Orbital Symmetry." Academic Press, New York, 1972.
17. M. J. S. Dewar and R. C. Dougherty, "PMO Theory of Organic Chemistry." Plenum, New York, 1975.
18. I. Fleming, "Orbital Symmetry." Wiley, New York, 1976.
19. T. L. Gilchrist and R. C. Starr, "Organic Reactions and Orbital Symmetry." Cambridge Univ. Press, London and New York, 1979.
20. A. P. Marchand and R. E. Lehr, eds., "Pericyclic Reactions." Vols. I and II. Wiley, New York, 1977.
21. E. N. Marvell, "Thermal Electrocyclic Reactions." Academic Press, New York, 1980.

22. H. Meister, *Chem. Ber.* **96,** 1688–1696 (1963).
23. R. Huisgen, A. Dahmen, and H. Huber, *J. Am. Chem. Soc.* **89,** 7130–7131 (1967).
24. E. N. Marvell, J. Sewbert, G. Vogt, G. Zimmer, G. Moy, and J. R. Siegmann, *Tetrahedron* **34,** 1323–1332, and references cited therein (1978).
25. W. Hartmann, H. G. Heine, and L. Schrader, *Tetrahedron Lett.* pp. 883–886, 3101–3104, and references cited therein (1974).
26. A. C. Cope and M. Burg, *J. Am. Chem. Soc.* **74,** 168–172 (1952).
27. H. Hoever, *Tetrahedron Lett.* pp. 255–256 (1962).
28. H. Straub, J. M. Rao, and E. Muller, *Liebigs Ann. Chem.* pp. 1339–1351 (1973).
29. S. J. Harris and D. R. M. Walton, *Tetrahedron* **34,** 1037–1042 (1978).
30. D. A. Bak and K. Conrow, *J. Org. Chem.* **31,** 3958–3965 (1966).
31. T. S. Cantrell, *J. Am. Chem. Soc.* **104,** 5480–5483 (1970).
32. L. J. Haynes, I. Heilbron, E. R. H. Jones, and F. Sondheimer, *J. Chem. Soc.* pp. 1583–1585 (1947).
33. G. Zweifel and N. L. Polston, *J. Am. Chem. Soc.* **92,** 4068–4071 (1970).
34. K. C. Nicolaou, R. L. Magolda, W. J. Sipio, W. E. Barnette, L. Lysenko, and M. M. Joullié, *J. Am. Chem. Soc.* **102,** 3784–3793 (1980).
35. K. C. Nicolaou, W. J. Sipio, R. L. Magolda, and D. A. Claremon, *J. Chem. Soc., Chem. Commun.* pp. 83–85 (1979).
36. D. L. J. Clive and V. N. Kale, *J. Org. Chem.* **46,** 231–234 (1981).

Chapter 7

PLATO'S SOLID IN A RETORT: THE DODECAHEDRANE STORY

Leo A. Paquette

Evans Chemical Laboratories
The Ohio State University
Columbus, Ohio

I. Introduction

For many centuries the aesthetically pleasing features of the dodecahedron have attracted the attention of learned individuals. Driven by the mathematical elegance of this Platonic solid, the ancient Greek mathematicians viewed the object as the building block of heavenly matter.[1] The seventeenth century was witness to its extraordinary use by the astronomer Johannes Kepler in a model of the universe.[2] Invariably, the vogue has been to regard this most complex of the symmetric convex polyhedra with the highest esteem. Typically, organic chemists have, in more recent times, been moved to regard its molecular transliteration, dodecahedrane $(C_{20}H_{20})$, as the "Mount Everest of Alicyclic Chemistry."[3] Thus, history did early set forth before us the formidable challenge of constructing a polyquinane[4] hydrocarbon so symmetric (I_h) that its ^{1}H- and ^{13}C-NMR spectra would individually be characterized by a single line.

Although many investigators have sought to develop synthetic protocols capable of arriving at dodecahedrane and have in the process made several beautiful and important experimental contributions,[5] the target has remained elusive until recently. To our good fortune, these laboratories were witness to the realization of this achievement last year.[6] Our intent in this overview of the successful synthesis is to outline the chronology with which ideas and experimentation evolved, including the many unforeseen and unexpected developments that ultimately led to resolution of the problem.

II. The Decision against Convergency and Building of the Cornerstone

The contributions of convergency to any synthetic endeavor of major proportions are always tantalizingly attractive. In the present instance, however, we soon came to regard an approach of this type as having serious steric and entropic disadvantages. Thus the dimerization of triquinacene (**A**), favored by Woodward,[7] Eaton's capping of peristylane (**B**),[8] and our own planned installation of additional interconnective C—C bonds in $(\pm)$-bivalvane (**C**)[9] all suffer intrinsically from alignment problems. Whereas

A

B

C

any two desired segments can certainly be properly conjoined by means of a C—C bond, it is an entirely different matter to coax the substrate into proper conformational alignment (as idealized in **A–C**). For the stated reasons, the two structural components strive to be as mutually distal as possible.[10]

Accordingly, we opted early to employ a serial synthesis that we hoped,

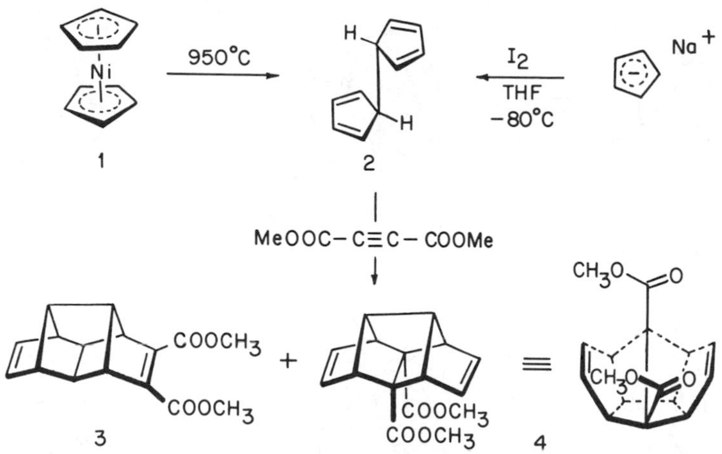

SCHEME 1

however, could be significantly abbreviated by taking advantage of the enormously high symmetry of the target molecule. This decision was made with the added optimistic expectation that arrival at dodecahedrane might well be achieved in laboratory steps fewer than, or equal to, the number of constituent carbon atoms.

Rapid access to a cornerstone intermediate was encouraged by Hedaya's finding that flash vacuum pyrolysis of nickelocene (1) at 950°C produced cyclopentadienyl radicals[11] whose dimerization to 9,10-dihydrofulvalene (2) was established by chemical conversion to a 1:1 mixture of 3 and 4 (Scheme 1).[12] Particularly relevant to us was the fact that the C_{2v}-symmetric 4 already contains four five-membered rings interlocked in a manner that fixes the six methine hydrogens in an all-cis relationship. The expense and practical limitations of the original method were overcome by making recourse to the oxidative coupling of cyclopentadienide anion with iodine[13] and effecting the domino Diels–Alder reaction at low temperatures.[14,15] This operation, when conducted with total exclusion of oxygen, routinely provides 4 or the related diacid (depending on workup) in 15–20% yield.

III. A Retrosynthetic Glimpse and Symmetry Considerations

Although many informal dodecahedrane degradation schemes can be written, the key features of our original strategy were to incorporate the highest element of symmetry within each precursor and to avoid the need for "one-carbon insertion" into an almost fully elaborated C_{19} structure. As seen in

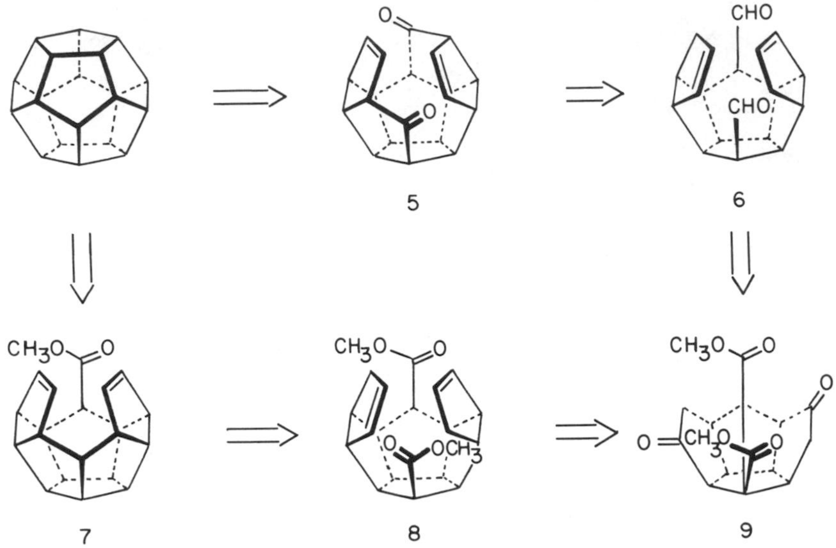

SCHEME 2

Scheme 2, both C_2 (e.g., **5, 6,** and **8**) and C_s (e.g., **7**) C_{20} molecules comply with this concept. A protocol of this type should not only simplify the derived spectroscopic data and structural assignments, but should also allow chemical transformations to be carried out concurrently on like functional groups. Consequently, the conversion of **4** to **9** was next investigated.

Recourse to hydroboration–oxidation did indeed lead to **9**. However, the isomeric C_s diketo diester also is formed and in relatively larger amounts.[16] Although **9** could be readily accessed by this procedure, we chose on larger scale to avoid the need for chromatographic purification by effecting highly efficient iodolactonization of **10** and preserving during three additional steps the cross-corner oxygenation pattern that materializes therein (Scheme 3).[17]

Although **4** and **9** feature an unwanted transannular C—C bond, its disengagement is to be deferred for two reasons. First, its presence locks the left- and right-handed portions of these molecules into conformationally

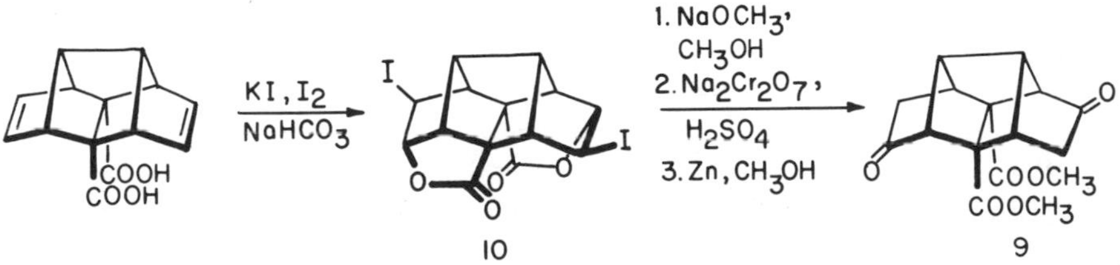

SCHEME 3

rigid norbornyl units and essentially guarantees full stereoselectivity during the planned appendage of additional rings. Second, cleavage at a later stage, when the level of convexity has been heightened, should disallow the entry of solvent molecules into the cavity and result only in protonation from the exterior surface as ultimately required. It should be emphasized here that the tactical elaboration of dodecahedrane requires at every stage that the methine carbons be constructed contrathermodynamically, with the large group facing the molecular interior and the small hydrogen positioned exo. This very important detail sooner or later can be expected to introduce very high levels of nonbonded steric strain, which must be carefully dealt with.

IV. Arrival at the "Closed" and "Open" Dilactones

The domino Diels–Alder reaction described above has the noteworthy feature of setting 14 carbon atoms into proper position. To our mind, introduction of the remaining six as early as possible would best serve our purposes. Consequently, **9** was allowed to react with excess diphenylcyclopropyl-sulfonium ylide. Following efficient arrival at the C_2-symmetric bis(spiro-cyclobutanone) **11**, our laboratory operations were necessarily directed to enhancing the quinane level of the substrate. Ideally, not only was axial symmetry to be preserved, but reagents were to be properly selected such that

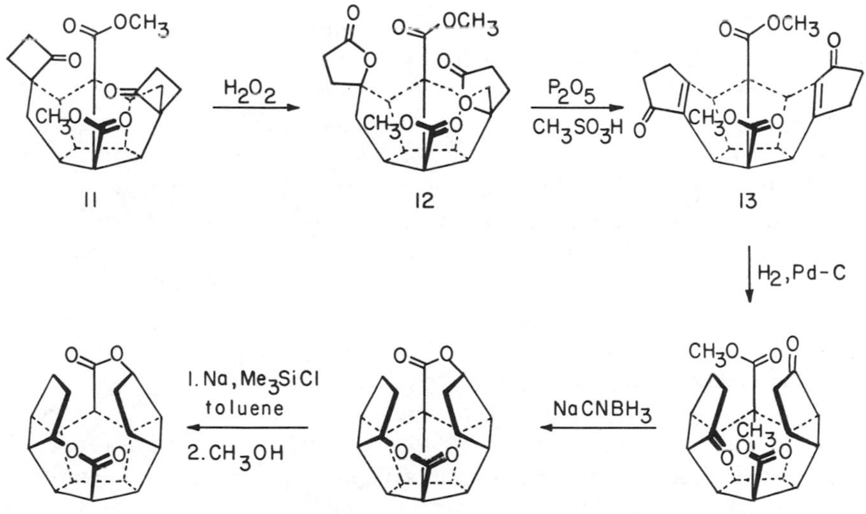

SCHEME 4

additional unwanted substituents would not become appended. Baeyer–Villiger oxidation of **11**, accomplished by the action of hydrogen peroxide, followed by isomerization of the bis(spirolactone) so produced (**12**) under strongly acidic conditions, delivered **13** (Scheme 4).[17] Catalytic hydrogenation of **13** was controlled completely by exo approach and served to position four additional methine hydrogens on the exterior of the developing sphere and to complete the construction of two additional cis,syn-fused cyclopentane units. Controlled reduction of **14** with sodium cyanoborohydride afforded the so-called closed dilactone **15**.

The reasonably rigid conformational features of **15** so fix the stereoalignment of the two lactone carbonyl groups relative to the interconnective bond that 1,4 reduction proceeds readily. Also, the sphericality of the framework is sufficiently developed so that protonation from the exo direction occurs exclusively to furnish "open" lactone **16**, the structural assignment to which was confirmed by X-ray analysis.[18]

To this point, our strategem had proceeded very directly forward. As constituted, the triseco molecule **16** has 14 cis-locked methine hydrogens, all 20 carbon atoms predisposed in the proper fashion, and a single type of functional group to facilitate subsequent chemical changes. To arrive at dodecahedrane, it remained to develop a workable dehydrative retro Baeyer–Villiger sequence and ultimate threefold C—C bond formation. Our efforts to resolve these practical issues uncovered many unexpected complications and made much chemical improvisation necessary.

V. Chemical Reactivity of the "Open" Dilactone

Although lactones have frequently been dehydrated to cycloalkenones in strongly acidic solution, **16** does not follow this precedent and is transformed instead to its isomer **17**.[17] More characteristically, **16** is extremely susceptible to unwanted transannular cyclization under both alkaline and acidic conditions. Thus heating with sodium hydride in toluene triggers Dieckmann condensation and formation of **18**. Conversion to internal ketal **19** occurs upon exposure to methanolic hydrogen chloride or trimethyloxonium tetrafluoroborate. Similarly, reaction with triphenylphosphine dibromide in refluxing acetonitrile and subsequent methanolysis afforded **20** and not the anticipated ring-opened dibromo diester.[17] Although **16** could be reduced under carefully controlled conditions to dilactol **21**, this substance proved unsuitable for further study because of a comparable marked tendency for transannular bonding with loss of C_2 symmetry.

As a consequence, we were forced to consider the implementation of a blocking scheme. Of the several conceivable possibilities, recourse was first

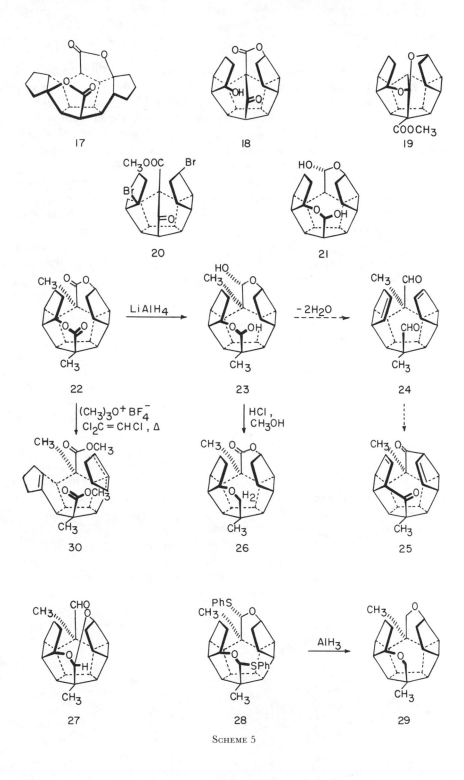

17   18   19

20   21

22   23   24

30   26   25

27   28   29

SCHEME 5

made to **22** because its structural features were most closely allied to those of our target. This dimethyl lactone, conveniently available by reductive methylation of **16**, was readily converted to dilactol **23** (Scheme 5)[19] whose projected role was to serve as precursor to diene dialdehyde **24** (compare **6**). Our purpose in arriving at **24** was to examine the feasibility of effecting a twofold Prins cyclization as a route to the highly valued intermediate **25**. However, all attempts to produce **24** led instead to intramolecular oxidation–reduction and isolation of **26**. Dilactol **23** does experience dehydration upon melting, but the product is the unsymmetrical acetal **27**. While our expectations in arriving at **28** and **29** were successfully realized, suitable cleavage of their heterocyclic rings was never achieved. When the observation was made that Meerwein's reagent acts on **22** only at more elevated temperatures and with combined ring opening–prototropic migration (see **30**), a conscious decision was made to examine alternative tactics.

VI. Functionalization Reactions of the "Closed" Dilactone

In view of the virtually complete absence of solvation within the cavities of **16** and **22** and the ultraproximity of transannular centers, it is not surprising that a reactive functional group, once generated, acquires hyperreactive tendencies under these circumstances. While the closed dilactone also suffers from equally severe nonbonded steric interactions, we believed that the presence of an interconnective bond spanning the two α-carbonyl sites might well serve as a temporary protective device. Consequently, our efforts were directed back to **15**, which, in its own right, made evident facets of unexpected chemical behavior. For example, treatment of **15** with 2 equivalents of trimethyloxonium tetrafluoroborate, followed by a methanolic solution of sodium methoxide, gave monolactone monoester **31** but not a doubly cleaved product, even under the most forcing conditions tolerable. Evidently, the substantial conformational change that accompanies the conversion of **15** to **31** leads to improper stereoalignment about the remaining lactone functionality and renders repetition of the first step difficult. Identical conformational factors appear to gain importance in the reaction of **15** with methanolic hydrogen bromide, which delivers **32**. Heating **32** with additional reagent at

31

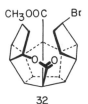

32

the reflux temperature for more than 1 week led to no additional chemical change.[17]

At about the same time, the observation was made that **15** could be converted in relatively low yield to dibromo anhydride **33** upon heating with triphenylphosphine dibromide in acetonitrile solution. Although this compound proved not to be useful,[17] we were led to consider the possibility that closer timing of the lactone cleavage steps could result in twofold ring opening as desired. The well-known lower nucleophilicity of chloride relative to bromide ion prompted reaction of **15** with methanolic hydrogen chloride at room temperature. Consistent with the operating hypothesis, axially symmetric dichloro diester **34** was isolated in 62% yield. Disappointingly,

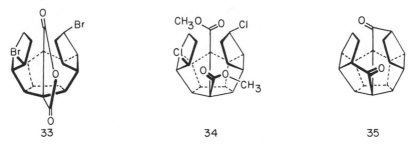

33   34   35

however, conditions for dehydrochlorinating **34** or for effecting its conversion to diketone **35** by reductive cyclization in the presence of metals or organometallic reagents were not uncovered.[17] The failure of these wishful formalisms caused us to shelve **34** for a period of approximately 30 months while other aspects of this investigation were being pursued. However, this substance, which can be prepared in a minimum of nine laboratory steps, will be shown below to be a truly pivotal intermediate.

VII. (C_2)-Dioxa-C_{20}-octaquinane, a Heterocyclic Trisecododecahedrane

Greater control of the reactivity of **15** was achieved by its hydride reduction to lactol **36**, which can also be prepared directly from **14** under comparable conditions. As in the case of **21** and **23**, the indicated structure is not the result of kinetic control; rather, equilibration of the hydroxyl groups to the more stable exo environment occurs during workup. When dissolved in thionyl chloride, **36** is converted quantitatively to chloroether **37** with preservation of the aldehyde oxidation level (Scheme 6).[20,21] Our original intention was to capitalize on the inherent stability of oxonium ions as a means of causing ring opening of **37** to **38** with the aid of silver salts. At this point, our naiveté about the proclivity of such molecules for carbocation-induced rearrangement

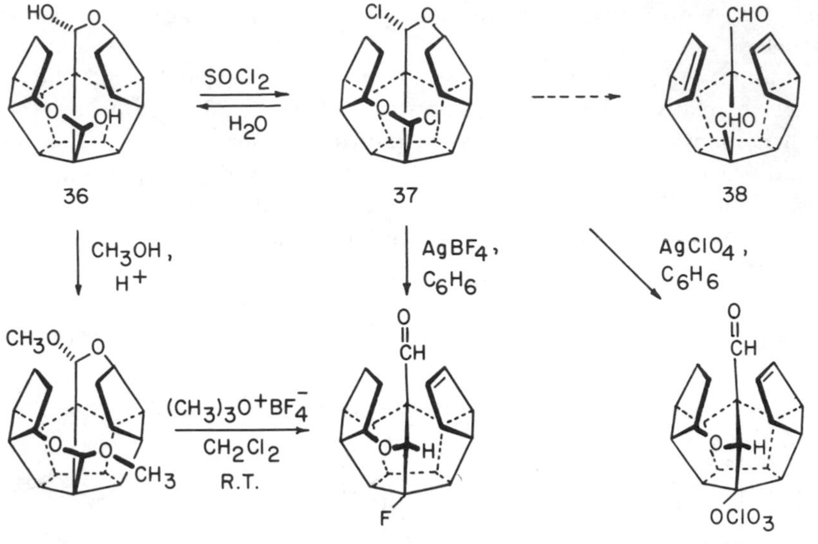

SCHEME 6

surfaced again. When **37** was treated with anhydrous silver tetrafluoroborate or perchlorate in dry benzene, the products isolated in good yield proved to be **39** and **40**. The identical reaction course is followed upon exposure of **41** to Meerwein's reagent. To convince ourselves that **40** was indeed a covalently bound tertiary perchlorate, an X-ray crystal structure analysis was performed.[18] Although the precise timing of these rearrangements have not been unequivocally established, we are of the opinion that departure of the first chloride ion is accompanied by a 1,2-shift of the central bond (with substantial relief of steric strain) to generate a tertiary cationic center that captures the only anion available (generally very nonnucleophilic). Only subsequently is the desired fragmentation believed to occur in the other half of the molecule.

The turn of events culminating in the isolation of **39** and **40** obviously did not conform to our single-minded goal of maintaining axial symmetry. However, the complication was neatly circumvented by reductive cleavage of the 1,4-dichloride part structure of **37** with sodium in liquid ammonia. An appreciation of the ease with which the resulting bis(dihydropyran) (**42**) can enter into transannular bonding is gained by the formation of **43** and **44** (Scheme 7).[21] Consequently, we were relegated to deploying **42** only in reactions where positive charge buildup was low or negligible. When its diepoxide (**45**) became available, it was found to be shelf stable despite its α,β-epoxy ether nature and the close transannular proximity of the two

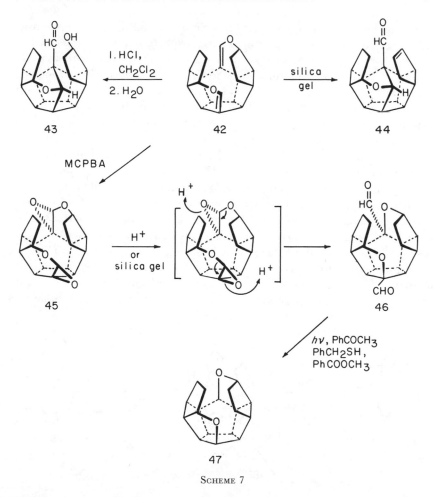

SCHEME 7

strained three-membered rings. In the presence of a wide variety of acidic reagents, including silica gel, however, **45** underwent rapid isomerization to afford ring-contracted dialdehyde **46**. The formation of **46** is believed to arise by oxirane ring opening toward the tertiary carbon center with concurrent 1,2 oxygen migration, as in the illustration. When the di-*tert*-butylperester derived from **46** failed to undergo decarboxylation, attempts to decarbonylate **46** directly were given attention. When a satisfactory response was not realized under traditional conditions, **46** was irradiated with an intimate mixture of acetophenone, benzyl mercaptan, and ethyl benzoate under an argon atmosphere. With the isolation of **47**, acquisition of the most advanced polyquinane structure known to that time had been realized [20,21] and we believed the stage to be set for construction of the carbon analog.

For this purpose, **42** was subjected to Simmons–Smith cyclopropanation and converted efficiently to **48**. However, our expectations of this intermediate were not vindicated, as it shares with its precursor a marked proclivity for kinetically controlled transannular bonding. As seen by the conversions to **49** and **50** (Scheme 8), the choice of conditions is not important, with one significant exception. Heating of **48** with *N*-bromosuccinimide results in smooth cyclopropane ring cleavage to give **51**, whose bromine atoms can be removed conventionally by tri-*n*-butyltin hydride reduction. During our further examination of the chemistry of **52**, it was quickly determined to follow unprofitable channels earlier seen with **23** and **28**.

The preceding evidence left serious doubts about the viability of carbocation chemistry in the continued elaboration of dodecahedrane. Certainly, nonbonded steric compression was to become substantially more serious before all of the framework construction was completed. Since carbocations are most apt to isomerize along those available avenues that would minimize this stress and because little control can be exerted otherwise to redirect this course of events, we chose to concentrate on photochemical methods for two relevant reasons. First, the energy supplied by this technique to substrate molecules

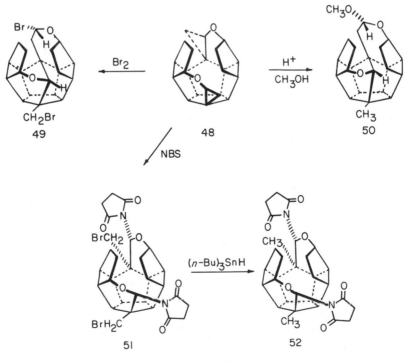

SCHEME 8

should be ample to overcome those incremental enhancements in strain, which would surely be encountered. Second, unlike many polar reactions, biradical processes are not usually accompanied by rearrangement. Also, they exhibit minimal geometric constraints to C—C bond formation and excellent thermodynamic driving force to proceed in this direction.

VIII.   The Pilot Photochemical Experiment

Ester carbonyl groups are rarely photoactive and **15, 16, 22,** and **34** proved not to be exceptions to this general rule. In considering the preparation of a suitably photoreactive ketonic species, we opted for **53.** The intent was to control the regioselectivity of C—C bond formation by taking advantage of the well-known ability of α-diketones for photoinduced hydrogen transfer and cyclization. In unconstrained systems, γ-hydrogen abstraction is overwhelmingly preferred and 2-hydroxycyclobutanones are formed exclusively. In **53,** only δ-hydrogens are available in close spatial proximity to the $n \rightarrow \pi^*$ excited carbonyl groups. Our expectation was that **52** would follow the

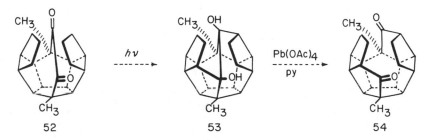

| | 52 | 53 | 54 |

example of several known monocarbonyl compounds that yield cyclopentanols upon irradiation and be transformed to **53** by twofold cyclization. It will be noted that **53** contains an interconnective bond improperly predisposed for arrival at dodecahedrane. Cleavage with lead tetraacetate should rectify this problem and provide an expedient route to triseco diketone **54.**

As outlined in Scheme 9, **52** could be readily prepared by sequential tri-n-butyltin hydride reduction of **34,** reductive methylation of **55** to give **56,** acyloin condensation, and ferric chloride oxidation.[22,23] Why introduce the two methyl groups? The topology of the dianion of **55** is more than adequate to guarantee entry of the methyl groups from the exterior face. The carbomethoxy substituents consequently become locked into the interior of the molecule, as is necessary, and epimerization is no longer possible. Furthermore, it had become quite apparent during past studies, such as those found in Schemes 5 and 8, that the ^{1}H-NMR absorption to these tertiary methyl protons serves as a beacon that signals the maintenance or loss of axial symmetry. In the event, X-ray analysis confirmed the structure of the beautifully crystalline yellow α-diketone and indicated the carbonyl groups to be canted apart at a dihedral angle of 19.7°, a state of affairs that should render them highly responsive to photoexcitation.

Proper photochemical activation of **52** did give rise to a diol whose ^{13}C- and ^{1}H-NMR spectra showed that C_2 symmetry had been maintained. Relevantly, this product was cleaved to a diketone, which, however, was not **54** because of its inability to undergo α-deuterium–hydrogen exchange and its unprecedented facile pinacolization in the presence of a slight excess of lithium diisopropylamide.[22,23] We quickly determined that **57** had been produced instead. Any mechanism that can be written to proceed from **52** to **57** necessarily violates established precedence at one or more stages. The pathway that materializes operates as the direct result of steric congestion, which is sufficiently extreme to force radical rearrangement to gain prominence.

Despite these insurmountable complications, we were encouraged to proceed in a photochemical mode. Since the "staging" bond interconnecting the two carbonyl groups in **52** may well be the principal steric impediment to appropriate homo-Norrish closure, attention was given to ketone candidates that lacked this complication.

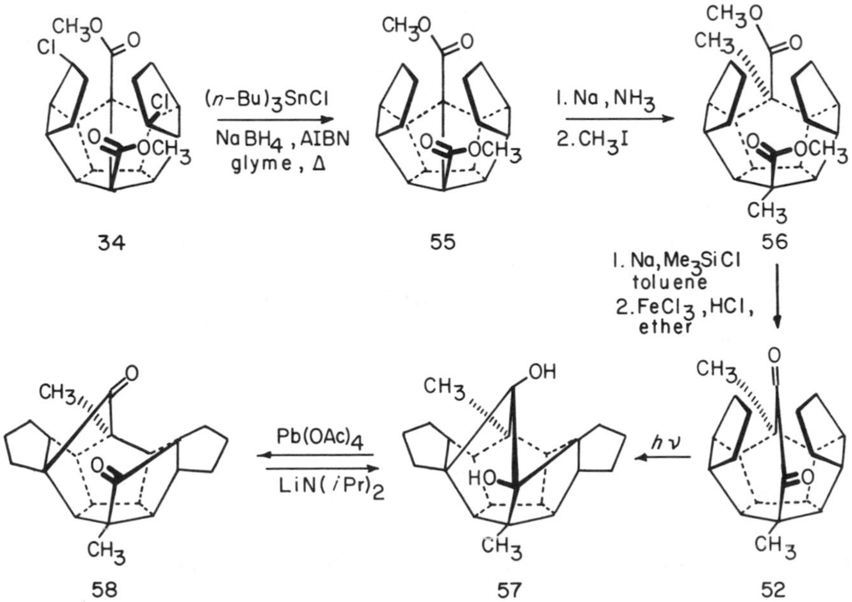

SCHEME 9

IX. Trisecododecahedranes via Dichloro Diester Reduction

Earlier mention was made of the fact that formation of carbanionic centers at the chlorinated carbon atoms of **34** led to simple reduction (formation of **55**). In contrast, electron transfer to **34** should result in formation of radical anion **59**, where the most electropositive center has been reduced (Scheme 10). If such does materialize, electronic overlap with the transannular carbonyl group should be sterically disfavored and allow for intramolecular S_N2 displacement of chloride ion and formation of **60**. Continued reduction of the latter, it was reasoned, should, again on the basis of standard half-wave potentials, involve attack at the ketone carbonyl and lead to central bond cleavage. Since the resulting ester enolate is not properly oriented to displace the second chlorine, independent reductive cleavage should eventuate there. In the final analysis, only a single cyclization can result from this protocol. Nonetheless, the transformation has the undeniable value of catapulting the synthetic scheme (only 10 steps from cyclopentadienide anion!) to the tetraseco stage of construction.

Under the influence of lithium or sodium in liquid ammonia, **34** splendidly experiences reduction to generate dienolate **61**, which is directly transformed to **62** upon treatment with excess methyl iodide.[23,24] In this way, transannular

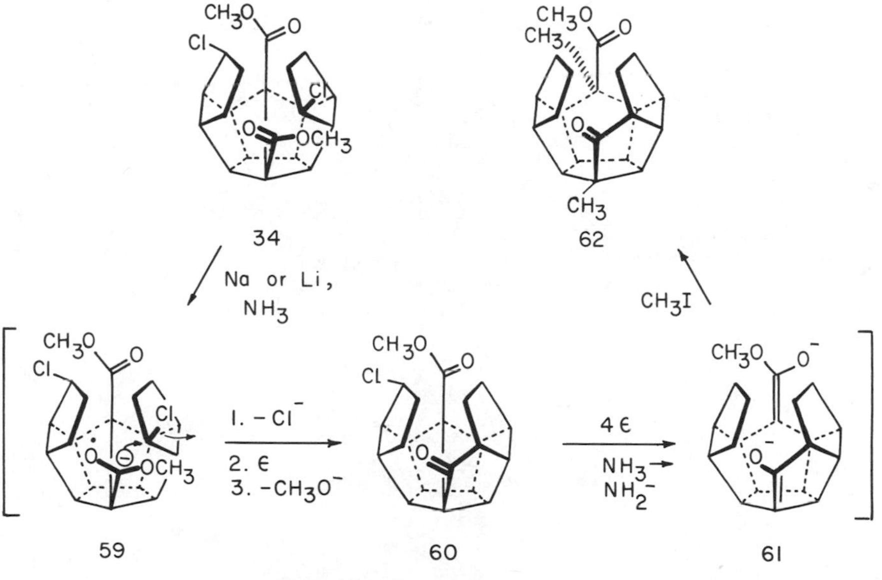

SCHEME 10

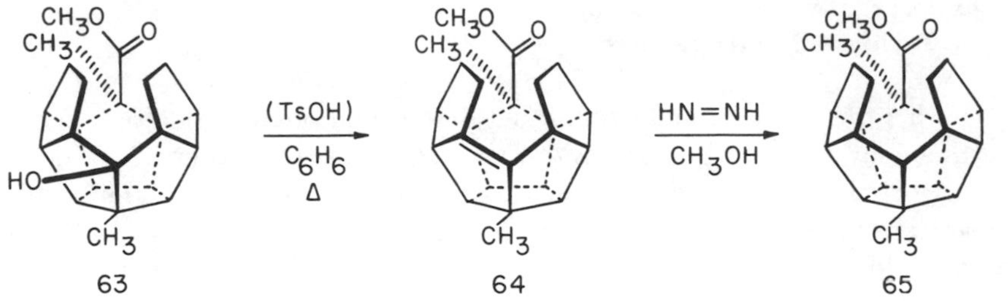

SCHEME 11

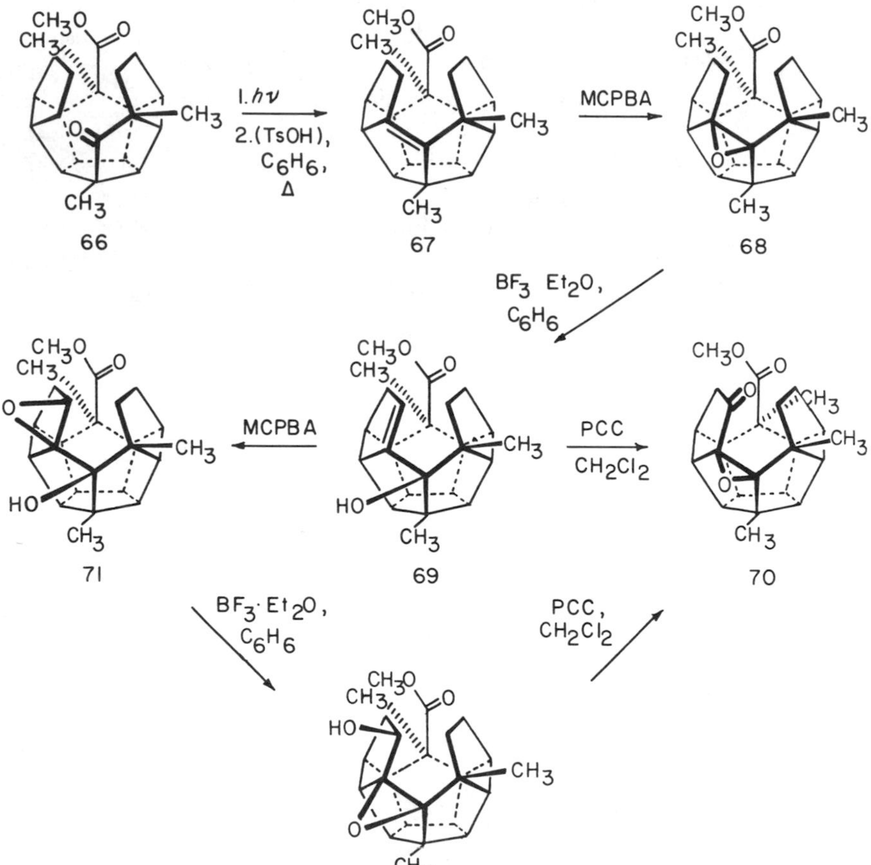

SCHEME 12

cyclization is precluded and the endo orientation of the important carbonyl groups is conveniently set. Light-induced cyclization of **62** took place only at the ketone site without framework rearrangement to give **63** (Scheme 11). In turn, the dehydration of this tertiary alcohol could be routinely achieved under acidic conditions without evidence of Wagner–Meerwein methyl migration. Although **64** resisted catalytic hydrogenation in a Parr apparatus, its double bond succumbed readily to diimide reduction. With the acquisition of **65**, planar symmetry (C_s) was accessed for the first time.

Because the opposed methylene groups in **62** are unfunctionalized, a brief attempt was made to correct this situation. For this purpose, recourse was made to the trimethyl keto ester **66**, which had accumulated because it is a by-product formed in the production of **62**. Following conversion to **67** as before, epoxide **68** was prepared (Scheme 12). On exposure to boron tri-fluoride etherate, remarkably clean isomerization to **69** was observed. This tertiary alcohol is not responsive in the normal manner to chromium-based oxidants. Instead of undergoing conversion to the transposed enone, **69** gives rise to epoxy ketone **70**. Confirmation of this chemical change was realized by perepoxidation of **69**, Lewis acid-catalyzed epoxy alcohol–epoxy alcohol rearrangement leading from **71** to **72**, and a final oxidation step.[23] Molecular models clearly show that the **71** → **72** isomerization is sterically driven. Whereas the three-membered ring in **71** does little to alleviate those non-bonded steric interactions present in the gap, the pinching effect exerted by the epoxide ring in **72** serves to enlarge the seam of the molecule. However, this structural distortion effectively deters **70** from being responsive to photoactivation.

Circumstances such as this caused us to defer consideration of this problem to a later stage of the synthesis.

X. The Monoseco Level of Elaboration

The tantalizing issue that had to be addressed was the involvement of the oxygenated carbon atom in **65** in dual cyclopentane ring formation. Because it was already clear that the ester group required modification, we made the decision to proceed with aldehyde **73** (Scheme 13). This substance was arrived at in excellent yield by sequential diisobutylaluminum hydride reduction and pyridinium chlorochromate oxidation.[23,24] The fact that the CHO unit was bonded to a fully substituted carbon now had to be dealt with, since precedent left no doubt that these structural features are highly conducive to photode-carbonylation. Although **73** is certainly prone to carbon monoxide extrusion, we succeeded, by operating at low temperatures, in achieving 29% conversion

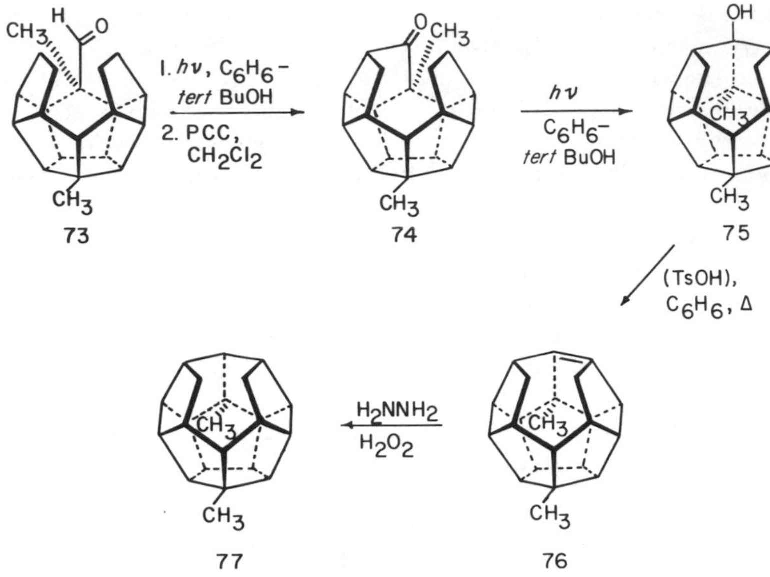

SCHEME 13

to the cyclopentanol. With subsequent oxidation, the desired diseco diketone **74** was obtained. Needless to say, we were delighted to discover that **74** experiences photocyclization with high efficiency and that removal of the tertiary hydroxyl group in **75** and saturation of the double bond in **76** are simple, good-yielding steps.

The eight-line ^{13}C-NMR spectrum of **77** clearly signaled a return to C_{2v} symmetry status. Its infrared spectrum features an unusually high absorption at 3150 cm^{-1} assumed to arise by virtue of the enormously compressed opposed methylene hydrogens. In fact, X-ray structure analysis of **77** revealed the nonbonded C----C distance to be 3.03 Å and the pronounced internal stresses resulting from this intramolecular contact to result in substantial bond-angle distortions and flexure of the cyclopentane rings at the gap. In reality, the molecule is shaped like a flattened pouch.[25]

XI. Installation of the Final Bond:
Formation of 1,16-Dimethyldodecahedrane

The very gratifying results just described were initially marred by our inability to effect the extrusion of molecular hydrogen from **77**, a feat we originally believed might occur readily. However, whereas **77** is formally a dihydrododecahedrane, **76** is isomeric with the target molecule. Accordingly, attempts to effect double-bond isomerization and ultimate ring closure were next undertaken. Disappointingly, a variety of protic acids failed to alter **76** chemically. It was found, however, that trifluoromethanesulfonic acid in dichloromethane solution at room temperature causes the rapid disappearance of this olefin. A mixture of products results, from which the most highly symmetric isomer (**78**) can be isolated in 27% yield. Because these extreme acid conditions also convert **75** to **78**, a total of 19 steps are required to arrive at this spherical hydrocarbon.[23,26]

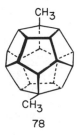

CH₃

CH₃

78

The unexpected occurrence of a methyl migration during this installation of the final framework bond was substantiated by [1]H- (three signals) and ^{13}C-NMR (four resonance lines) as well as by X-ray analysis.[25,26] Some of the more unique physical properties of **78** include its reluctance to melt (discoloration begins above 350°C), its density (1.412 g/cm³), its relatively short transcavity distance (0.9 Å), and its marked resistance to fragmentation in the mass spectrometer.

The precise timing of the alkyl migration remains to be established and is currently the subject of experimental scrutiny.

XII. Further Developments

At this point, pursuit of the parent $(CH)_{20}$ hydrocarbon became still more irresistible. Utilization of **78** as a serviceable precursor was not viewed favorably because of the inoperability of Schleyer's catalyst systems[27,28] under the high-vacuum conditions necessary for its volatilization[28] and the unlikely workability of chemospecific methyl oxidation and ensuing decarboxylation. Also, the entire synthetic sequence would be protracted by the inclusion of these additional proposed transformations. As a result, we sought to adapt the synthesis just described in as efficient a manner as possible.

Highly pertinent to our aims was the discovery of methodology for transforming dichloro diester **34** to tetraseco keto alcohol **79**. Treatment with approximately 12 equivalents of lithium metal in liquid ammonia, followed by an ethanol quench, gave the best results. The ester group having been reduced

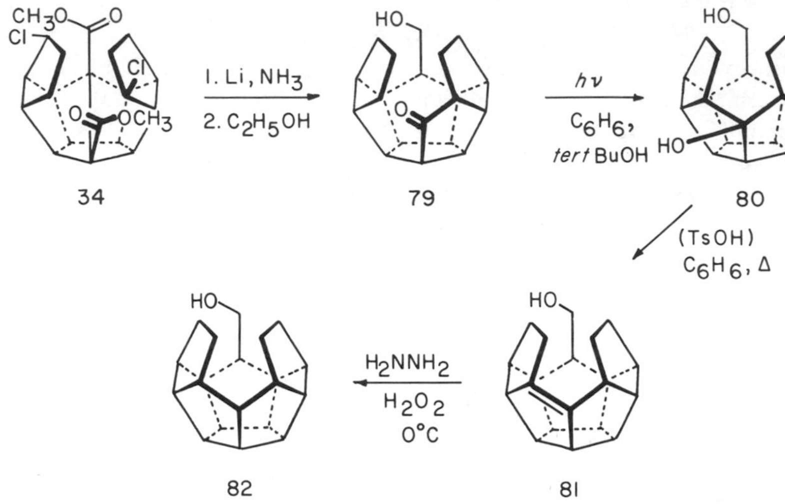

SCHEME 14

as in **79** (Scheme 14), we did not run the risk of transannular cyclization.[27,28] Fortunately, **62** and **66** proved to be realistic prototypes for the homo-Norrish photocyclization of **79**, whose conversion to **80** was characteristically efficient. Subsequent chemospecific dehydration of the tertiary hydroxyl group in **80** to give **81**, followed by diimide reduction, led to **82** in good yield. A return to C_s symmetry was signaled by the simplified 12-line ^{13}C-NMR spectrum. Were **82** to be capable of straightforward oxidation to C_s-aldehyde **83**, our goal of extreme brevity would easily have been met.

As matters turned out, a great deal of experimentation showed that we had grossly misjudged the actual state of affairs in this most sterically congested molecule. Many oxidizing agents acted on **82** to give predominantly the α,β-unsaturated aldehyde **85**; others, principally those that act at the mechanistic level by activating the hydroxyl substituent by conversion to a good leaving group, transformed **82** uniquely to olefin **86**. Under the conditions of Jones oxidation, smooth conversion to norketone **87** occurred.[27,28] This evidence suggested that the first-formed product in those reactions that led to **85** was indeed **83**. However, because of the appreciable steric crowding around the periphery of the central cavity in this molecule, substantial levels of enolization, as in **84**, takes place. This intermediate can be expected to be more readily oxidized than **82** and to serve as precursor to **85**. Whatever the case, an alternative route to **83** by reduction of **85** under dissolving metal conditions was investigated. Treatment of **85** with lithium and *tert*-butyl alcohol in liquid ammonia did indeed yield the saturated aldehyde in its

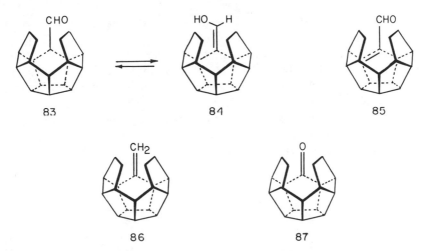

enol form (**84**). Unfortunately, however, **84** proved to be extremely susceptible to air oxidation, being completely converted to **85** in a short time.

For this reason, the decision was made to engage **34** in reduction–monoalkylation in order to install a blocking group that would effectively prevent enolization of the aldehyde functionality, yet prove readily removable at a later stage. It became imperative, therefore, that the anionic center α to the carbomethoxy group in **61** be more reactive toward electrophiles than that α to the ketone carbonyl. Preferential electrophile capture at the ketone enolate site would seriously frustrate our aims. Because little precedent exists concerning this question, we sought to test our synthetic concepts by examining the feasibility of preparing methyldodecahedrane.

XIII.   Controlled Monosubstitution and Isolation of an Isododecahedrane

Addition of **34** to a solution of lithium (6 equivalents) in liquid ammonia at −78°C followed by 1 equivalent of methyl iodide afforded a mixture of **88** (46%) and **89** (24%). Equally encouraging was the expeditious manner in which methyl secododecahedrene (**90**) was fashioned from this intermediate (Scheme 15).[29,30] In earlier work, a dimethyl seco olefin related to **90**, i.e., **76**, was found to be subject to rapid acid-catalyzed isomerization and cyclization. Consequently, we immediately proceeded to treat **90** with trifluoromethanesulfonic acid in dichloromethane as before. A complex mixture of hydrocarbons was formed, among which methyldodecahedrane was clearly not dominant. Repeated recrystallization afforded the major component (49%)

in a pure state. Identification as the polyquinane **91** was achieved by X-ray analysis. Evidently, the substantial steric compression present in the protonated form of **90** is more than adequate to foster a most remarkable transannular electrophilic substitution. The twinned norbornyl character of the methano bridges in **91** was seen to project the associated internal hydrogens well beyond intramolecular contact range.

This unexpected development demonstrated once again the unreliability of carbocation chemistry in these systems and forced upon us the development of a more generally applicable and universal approach to dodecahedrane ring construction.

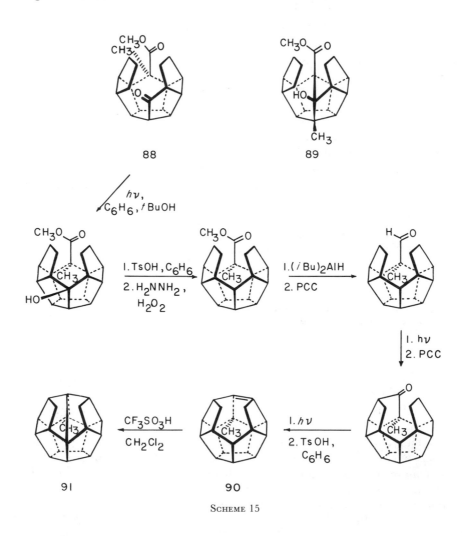

SCHEME 15

XIV. Bypassing the Unwanted Reaction Pathway

Saturated hydrocarbon **92** was prepared by diimide reduction of **90** to enable full examination of catalytic dehydrogenation as a means of introducing the final carbon–carbon bond. When an intimate mixture of **92** and 10% palladium-on-carbon was heated in a sealed stainless steel reactor at 250°C for 30 min, reaction occurred as judged by capillary gas chromatographic analysis. Among the products formed were several compounds that appeared to contain sites of unsaturation. Presumably, dehydrogenation had materialized by removal of hydrogen from the methine units. Repetition of the process under an atmosphere of argon did not alter the results.

It occurred to us at this point that it might be possible to suppress the olefin-forming process by performing the dehydrogenation under a hydrogen atmosphere. Indeed, success was realized upon heating (250°C) an intimate mixture of **92** with 50 times its weight of 10% palladium-on-carbon, previously exposed to 50 psi of hydrogen, for 7 h in a sealed stainless steel chamber as before. Although capillary gas chromatography indicated that a variety of products had formed, the olefinic components earlier in evidence were now insignificant. From ^{1}H-NMR analysis of the mixture, monomethyldodecahedrane (**93**) was determined to be present to the extent of 35–40%. By recrystallization of the reaction mixture from benzene, a 28% yield of **93** was realized.[29,30]

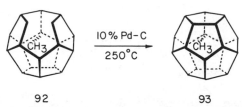

| 92 | | 93 |

Access to dodecahedrane by this procedure first required that a proper blocking group be introduced. The selection of this side chain had to be made judiciously because it had to survive several bond-stitching reactions. Also, we were limited to its introduction via S_N2 methodology. Most importantly, the substituent must not foster photodecarbonylation of the aldehyde nor enter into bonding with this photoexcited moiety. As will be seen in the sequel, chloromethyl phenyl ether satisfied all of these requirements.

XV. (CH)$_{20}$: The Parent Dodecahedrane

As shown in Scheme 16, reductive alkylation of **34** with $ClCH_2OC_6H_5$ gave **94** in quite acceptable yield. The phenoxymethyl side chain causes no interference during the three steps required to construct the other side of the sphere as in **95**. Nor does it become entangled with proximate reactive functional groups as **95** is modified to the aldehyde level and subjected to photochemical cyclization. At this point, the risk of complications due to enolization α to an aldehydic carbonyl site was behind us and it became appropriate to direct attention to the deblocking maneuver. To this end, sequential Birch reduction of **96** proceeded to make keto aldehyde **97** cleanly available. Retroaldol cleavage of **97** in alkaline solution provided diseco ketone **98** in 37% overall yield.[27,31] Recourse to photochemistry again permitted installation of the penultimate bond. Following arrival at secododecahedrane, the saturated hydrocarbon was heated as before with hydrogen-presaturated 10% palladium-on-carbon at 250°C for ≥4.5 h. Under these conditions, dehydrogenative conversion to **99** occurred with 40–50% efficiency. Dodecahedrane could be readily separated in 34% yield by simple recrystallization from benzene.

In line with expectation, the ^{1}H- and ^{13}C-NMR spectra of **99** in CDCl$_3$ solution are characterized by singlets, the former at δ 3.38 and the latter at 66.93 ppm. The relatively low chemical shift of the dodecahedryl protons, which suggests that at least one tensor of the proton spin is strongly accented, is not unique to the fully spherical molecule but is seen in all topologically

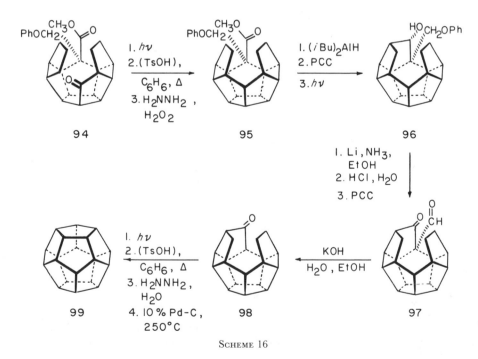

SCHEME 16

related molecules that have acquired reasonable levels of sphericality. The ^{13}C-H coupling constant of 134.9 compares quite favorably with those previously determined for **78** (131.2 and 135.0 Hz).

The vibrational frequencies exhibited by **99** (120 identity operations) agree fully with a highly rigid network of interlinked methine units. Only three infrared-active bands are observed at 2945, 1298, and 728 cm^{-1}. As presently obtained, dodecahedrane happens to be optically isotropic, having crystallized in the cubic system as quite well-formed octahedra and cubeoctahedra. This would imply that **99**, in this phase, almost certainly possesses a face-centered cubic unit cell. By means of a standard microcrystallographic technique, the refractive index of dodecahedrane was determined to be $n_c = 1.670$, $n_D = 1.674$, and $n_F = 1.685$.[31]

Because dodecahedrane showed no visible evidence of melting upon being slowly heated in sealed capillaries up to 450°C (at which point a brown coloration was obvious), a special microscope hotstage was built and calibrated. With this apparatus, the melting point of **99** was determined to be 430 ± 10°C, the relatively low precision being forced upon us by the quite rapid rate of heating that was required to achieve the measurement. At these temperatures and atmospheric pressure, dodecahedrane sublimes very rapidly. Nonetheless, a very fleeting liquid phase was observed.

In summary, the total synthesis of dodecahedrane has been achieved in 23 steps from cyclopentadienide anion. Although the sequence is brief, quantities of **99** for ongoing studies will prove scarce pending future improvements, which most certainly will surface.

ACKNOWLEDGMENT

This research, which was most capably executed by those highly skilled co-workers named in the references, was made possible by the generous financial support of the National Institutes of Health.

REFERENCES

1. T. Heath, "A History of Greek Mathematics," Vols. I and II. Oxford Univ. Press, London and New York, 1976.
2. J. Kepler, "Continens Misterium Cosmografica." 1596.
3. P. Grubmüller, Ph.D. Dissertation, Friedrich-Alexander-Universität, Erlangen-Nürnberg, West Germany (1979).
4. L. A. Paquette, *Top. Curr. Chem.* **79**, 43 (1979).
5. P. E. Eaton, *Tetrahedron* **35**, 2189 (1979).
6. R. J. Ternansky, D. W. Balogh, and L. A. Paquette, *J. Am. Chem. Soc.* **104**, 4503 (1982).
7. R. B. Woodward, T. Fukunaga, and R. C. Kelly, *J. Am. Chem. Soc.* **86**, 3162 (1964); see also I. T. Jacobson, *Acta Chem. Scand.* **21**, 2235 (1967).
8. P. E. Eaton and R. H. Mueller, *J. Am. Chem. Soc.* **94**, 1014 (1972); P. E. Eaton, R. H. Mueller, G. R. Carlson, D. A. Cullison, G. F. Cooper, T.-C. Chou, and E.-P. Krebs, *ibid.* **99**, 2751

(1977); P. E. Eaton, G. D. Andrews, E.-P. Krebs, and A. Kunai, *J. Org. Chem.* **44,** 2824 (1979).

9. L. A. Paquette, W. B. Farnham, and S. V. Ley, *J. Am. Chem. Soc.* **97,** 7273 (1975); L. A. Paquette, I. Itoh, and W. B. Farnham, p. 7280; L. A. Paquette, I. Itoh, and K. Lipkowitz, *J. Org. Chem.* **41,** 3524 (1976).

10. J. Clardy, B. A. Solheim, J. P. Springer, I. Itoh, and L. A. Paquette, *J. Chem. Soc., Perkin Trans. 2* p. 296 (1979).

11. E. Hedaya, D. W. McNeil, P. O. Schissel, and D. J. McAdoo, *J. Am. Chem. Soc.* **90,** 5284 (1968).

12. D. McNeil, B. R. Vogt, J. J. Sudol, S. Theodoropulos, and E. Hedaya, *J. Am. Chem. Soc.* **96,** 4673 (1974).

13. W. E. Doering, *in* "Theoretical Organic Chemistry—The Kekulé Symposium," p. 45. Butterworth, London, 1959; E. A. Matzner, Ph.D. Thesis, Yale University, New Haven, Connecticut (1958).

14. L. A. Paquette and M. J. Wyvratt, *J. Am. Chem. Soc.* **96,** 4671, (1974).

15. L. A. Paquette, and M. J. Wyvratt, H. C. Berk, and R. E. Moerck, *J. Am. Chem. Soc.* **100,** 5845 (1978).

16. L. A. Paquette, M. J. Wyvratt, O. Schallner, D. F. Schneider, W. J. Begley, and R. M. Blankenship, *J. Am. Chem. Soc.* **98,** 6744 (1976).

17. L. A. Paquette, M. J. Wyvratt, O. Schallner, J. L. Muthard, W. J. Begley, R. M. Blankenship, and D. Balogh, *J. Org. Chem.* **44,** 3616 (1979).

18. P. Engel and W. Nowacki, *Z. Kristallogr., Kristallgeom., Kristallphys., Kristallchem.* 4 (1977).

19. D. W. Balogh and L. A. Paquette, *J. Org. Chem.* **45,** 3038 (1980).

20. D. Balogh, W. J. Begley, D. Bremner, M. J. Wyvratt, and L. A. Paquette, *J. Am. Chem. Soc.* **101,** 749 (1979).

21. L. A. Paquette, W. J. Begley, D. Balogh, M. J. Wyvratt, and D. Bremner, *J. Org. Chem.* **44,** 3630 (1979).

22. D. W. Balogh, L. A. Paquette, P. Engel, and J. F. Blount, *J. Am. Chem. Soc.* **103,** 226 (1981).

23. L. A. Paquette and D. W. Balogh, *J. Am. Chem. Soc.* **104,** 774 (1982).

24. L. A. Paquette, D. W. Balogh, and J. F. Blount, *J. Am. Chem. Soc.* **103,** 228 (1981).

25. G. G. Christoph, P. Engel, R. Usha, D. W. Balogh, and L. A. Paquette, *J. Am. Chem. Soc.* **104,** 784 (1982).

26. L. A. Paquette, D. W. Balogh, R. Usha, D. Kountz, and G. G. Christoph, *Science* **211,** 575 (1981).

27. R. J. Ternansky, D. W. Balogh, and L. A. Paquette, *J. Am. Chem. Soc.* **104,** 4503 (1982).

28. L. A. Paquette, D. W. Balogh, R. J. Ternansky, W. J. Begley, and M. G. Banwell, *J. Org. Chem.* **48,** 3282 (1983).

29. L. A. Paquette, R. J. Ternansky, and D. W. Balogh, *J. Am. Chem. Soc.* **104,** 4502 (1982).

30. L. A. Paquette, R. J. Ternansky, D. W. Balogh, and W. J. Taylor, *J. Am. Chem. Soc.* **105,** 5441 (1983).

31. L. A. Paquette, R. J. Ternansky, D. W. Balogh, and G. Kentgen, *J. Am. Chem. Soc.* **105,** 5446 (1983).

Chapter 8

THE SYNTHESIS OF FOMANNOSIN AND ILLUDOL

M. F. Semmelhack
Department of Chemistry
Princeton University
Princeton, New Jersey

I. Introduction

The synthesis of fomannosin began for the author with the publication of the structure (**1**) in 1967[1,2] and was completed in 1980.[3] The opportunity to describe the project in this chapter in addition to the terse presentation in the *Journal of the American Chemical Society* is welcome, because the origin of the

ISBN 0-12-450280-6

problem and evolution of the successful pathway reflect personal interactions as much as detached scientific logic. These factors are important in most scientific endeavors, more so those that require a team effort, such as a complex synthesis, and are seldom described in print.

The structure was attractive from the moment of reading the initial report: it was (and still is) the only natural methylenecyclobutene known, a rare functional group in general; it is unstable on standing at room temperature, and the structure was inferred from detailed X-ray diffraction analysis of a dihydro derivative (2).[1,4] It was simple enough to allow optimism that a many-year, many-person effort was not required and that a few general problems needed to be solved. A satisfying bonus was the reported antibiotic activity published at the same time as the structure.[2] The molecule was the subject of idle paper chemistry for several years, while the main labors of the research group involved organo-transition metal synthesis methodology. The initiation of the project required a special provocation.

A new graduate student indicated she was willing to join the research group, but she wanted nothing to do with the esoteric transition metal reagents! A plan for fomannosin was devised, without obvious recourse to elements in the middle of the periodic table, and the project was begun in 1974.

II.   The Target

Fomannosin (1) is a sesquiterpene metabolite of the wood-rotting fungus *Fomes annosus*, which affects pine tree stands in the Southeastern United States.[2] More recently, isolation from *Fomitopsis insularis* has also been reported.[5] It was isolated by chloroform extraction of a still culture and purified by silica gel chromatography. The pure material is reported as an unstable semisolid and was shown to be toxic toward two-year-old *Pines tadae* (loblolly pine) seedlings, *Chlorella pyrenoidosa,* and some bacteria. Controlled hydrogenation produced dihydrofomannosin (2), with disappearance of two olefinic proton signals in the ^{1}H-NMR spectrum compared to 1. The X-ray diffraction analysis on a derivative of 2 together with IR and NMR data on 1 and 2 led to the establishment of the structure of 1. The absolute configuration was later defined by a second X-ray study on the ester of dihydrofomannosin (2) with (−)-camphanic acid.[4] There were no published accounts of synthesis nor biosynthesis studies at the inception of this project.

The central problems in the synthesis in order of anticipated difficulty were: (a) introduction of the methylenecyclobutene double bonds, (b) introduction of the two chiral centers with correct relative configurations, and (c) manipulation of the three proximate oxygen functionalities, which differ in oxidation state. Potential pitfalls such as acid- or base-catalyzed epimerization at C-9 and decomposition of the methylenecyclobutene unit also play a large role in choice of strategy.

An obvious approach to the four-membered ring is the photochemical [2 + 2] cycloaddition of two alkenes or of an alkyne and an alkene (giving a cyclobutene directly). This process [Eq. (1)] generates one chiral center (at C-7), which must be controlled relative to the configuration at C-9. In the most direct precursor [**A**, Eq. (1)], little control is expected. An application of

the direct strategy was reported in 1977[6] as an extension of a study of photochemical addition of alkenes to unsaturated lactones. Photocycloaddition of ethylene to the lactone in Eq. (2) produced two diastereomers in approximately equal amounts. Both isomers were carried through several synthetic steps before an opportunity to separate the desired series was taken. In the end, dihydrofomannosin acetate was obtained. Greater control was attempted, again with an ethylene photocycloaddition, utilizing a rigid bicyclic system.[7]

Kosugi:

The price to be paid for greater stereochemical control is the additional steps necessary to turn the tricyclic adduct [**C**, Eq. (3)] into the fomannosin skeleton. More specific drawbacks are the low selectivity in the photocycloaddition (3:1 ratio of desired/undesired direction of approach) and the use of ethylene, which leaves no handle for introducing the cyclobutene double bond. The desired isomer **C** was carried on to a close precursor of dihydrofomannosin [Eq. (3)]. Neither of these approaches appears flexible enough to

Miyano:

$$(3)$$

utilize or recycle the wrong stereoisomers that were obtained, nor to allow introduction of the cyclobutene double bond. A related general method of cyclobutane synthesis with a different stereochemical requirement is the thermal or Lewis acid-catalyzed cycloaddition of a ketene with an alkene or alkyne. This was the basis of our first plan.

III. The Ketene Cycloaddition Approach to Fomannosin

A. STRATEGY

The key retrosynthetic disconnection is shown in Eq. (4). The penultimate intermediate (**3**) would undergo an intramolecular condensation (Knoevenagel, Wittig, etc.), as employed before.[7] The key intermediate is **4**, in which the stereochemical relationships have been established. The cyclobutenone unit would be generated by cycloaddition of a ketene (**5**) with acetylene or an equivalent [Eq. (5)]. There is no reason to expect stereospecific formation of **4** in preference to the diastereomer **6**. The plan to produce a stereospecific process included two main features. First, the ketene addend can be made more rigid, where the two faces of the ketene C=C unit are more obviously differentiated. A simple possibility is the rigid, bicyclic ketal **7**; approach of a substituted alkene or alkyne from the less hindered (exo) face would lead to **8**;

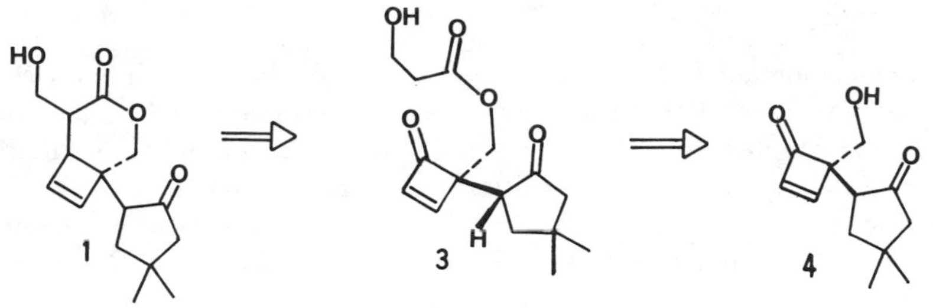

(4)

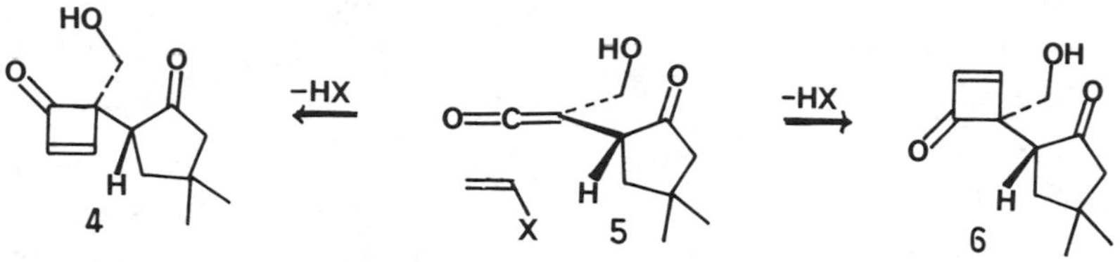

manipulation of the functionality in the four-membered ring would generate the desired cyclobutenone **9**. The isomeric cycloadduct **10** might also be converted to **7** by a different set of manipulations. Then, if the cycloaddition were not stereospecific, the isomers (e.g., **8** and **10**) could be separated and converted to **9**. The first stage of the synthesis is preparation of **7** or its obvious precursor, the ester **11**, which, in turn would arise from keto ester **12**.

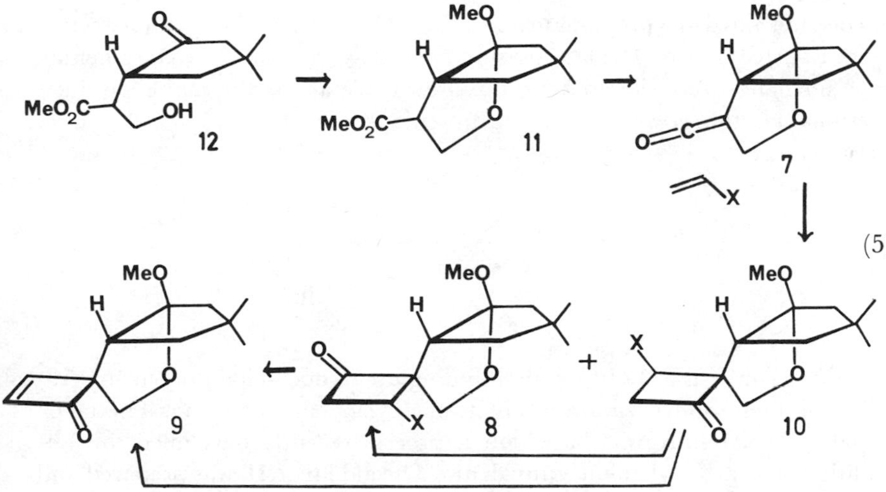

(5)

B. Synthesis

A simple approach to **12** is the alkylation of a differentiated malonate ester (**13**) with α-ketotosylate **14a** to give **15a**. Unfortunately, the alkylation [Eq. (6)] is complicated by formation in significant amounts of the undesired isomer (**15b**), apparently from initial elimination of HOTs in a process related to the Favorskii reaction. Reaction of the more difficulty accessible α-bromo and α-iodo ketones (**14b** and **14c**) with thallium diethyl malonate gave selective formation of **15a**, but in modest yield.

$$\tag{6}$$

14   13

a: X = OTs   15a   15b
b: X = Br
c: X = I

Nucleophilic ring opening of epoxide **16** with excess diethyl sodiomalonate in methyl alcohol at reflux produced the promising intermediate **17**, but very slowly and ultimately in low yield [Eq. (7)]. The use of an enolate–aluminum species[8] to open cyclohexene epoxides gave hope that similar modifications would be effective here. However, treatment of **16** with the product from dimethyl lithiomalonate and diethylaluminum chloride (presumably the diethylaluminum derivative of dimethyl malonate) led to slow reaction and a low yield of **17** [R = $CH(CO_2Me)_2$]. Similarly, the reaction of the dilithium salt of ethyl 3-hydroxypropionate (with Et_2AlCl) and of *tert*-butyl lithioacetate (with Et_2AlCl) failed. The latter system, exactly as described by Danishefsky, gave unchanged epoxide and *tert*-butyl acetoacetate as the major product. Presumably, the *gem*-dimethyl substituents are putting up a substantial steric barrier.

$$\tag{7}$$

16   17

A third synthesis of **12** began by building up an open-chain precursor (**18**) with the intent of cyclization to create the cyclopentanone ring. The tactic tested most carefully was based on a process recently developed for 1,4-addition of a carbonyl anion equivalent.[9] The aldehyde **18** was prepared and treated with benzylthiazolium chloride (**19**) and triethylamine in DMF at

100°C for more than 2 days [Eq. (8)]. From a quite complicated mixture, the desired cyclopentanone (20) was isolated in 30% yield. Although this first example of the intramolecular version of Stetter's reaction may have general significance, the yield could not be improved, and an alternate path was developed.

$$\text{18} \xrightarrow{\text{19}} \text{20} \tag{8}$$

An obvious possibility is the functionalization of 3,3-dimethylcyclo-pentanone, for example, through alkylation of enolate anion 21. However, we and others[6] had demonstrated that the undesired enolate 22 could not be

$$\text{22} + \text{21} \longrightarrow$$

avoided. A solution was presented to us by Professor David Cane during a visit at Brown University. He proposed that 3-methyl-2-cyclopentenone can be converted to the kinetically favored enolate anion by reaction with lithium diisopropylamide and alkylated regiospecifically. We prepared the enolate anion at $-78°C$ in THF and added this solution to a solution of excess methyl bromoacetate in THF at $-25°C$. After a simple chromatography, keto ester 23 was obtained in 48% yield. Then reaction with dimethylcopper lithium at $-60°C$ produced 24 in 90% yield. After protection of the ketone unit as the ethylene ketal, the ester side chain was hydroxymethylated (base, then CH_2O) and equilibrated to the internal ketal 11 (mixture of epimers, 76% yield) by treatment with acidic methyl alcohol [Eq. (9)].

$$\text{23} \longrightarrow \text{24} \xrightarrow{} \text{11} \tag{9}$$

To prepare for the critical ketene cycloaddition studies, the free acid 25a was prepared by KOH hydrolysis as colorless crystals and converted, via

reaction of its salt with oxalyl chloride at 25°C, to the acid chloride **25b**.
Reaction with triethylamine in carbon tetrachloride at 0°C led to the appear-
ance of a precipitate ($Et_3N \cdot HCl$) and, presumably, the ketene **7** in solution.
The ketene was insufficiently stable to allow isolation. With a series of electron-
rich alkenes that are known to react with ketenes to give cyclobutanones[10,11]
[e.g., 1,1-dimethyoxyethylene by Scarpati, phenyl vinyl thioether and *tert*-
butyldimethyl(vinyloxy)silane by Hasek], added after or during generation
of the ketene, no trace of the desired cyclobutanone product was detected. In
general, discrete 1:1 adducts could not be isolated.

An encouraging change in this pattern was observed with ethoxyacetylene
as the ketenophile (in carbon tetrachloride with triethylamine at 0°C for 40 h;
for a related example, see ref. 12). Not only did the ^{1}H-NMR spectrum of the
crude product indicate the presence of a 3-ethoxycyclobutenone (vinyl protons
at δ 4.85, 4.91), but each of the two major components gave the proper high
resolution mass spectral molecular weight, infrared C=O (1754 cm^{-1})
and C=C (1567 cm^{-1}) stretching frequencies, and UV spectral data (λ_{max}
236 μm), in close agreement with data on related structures.[12] The main
by-products were high molecular weight materials, presumably from self-
reaction of the ketene. After careful purification, one isomer was obtained as a
solid (mp 43.5–45.5°C) in 23% yield, whereas the other remained an oil at
25°C in 19% yield. The β-ethoxycyclobutenone unit offers important simpli-
fications in planning the utilization of both isomers. Starting from **26a**, hydride
reduction of the ketone, followed by acid-catalyzed rearrangement, would
produce the desired cyclobutenone **9**. The functional equivalence of the C-1
and C-3 positions in **26a** and **26b** means that they can be interconverted
chemically, which amounts to inversion of configuration at C-6. It is expected
that each could be converted to **9**, using a different sequence of operations.
However, the first question is which isomer is which?

Evidence for the relative configurations of the isomers was available from
spectroscopic analysis and chemical behavior. The configurations of spiro-
cyclobutanones have been correlated with ^{13}C- and ^{1}H-NMR chemical

TABLE I
^{13}C-NMR DATA FOR ISOMERS **26a** AND **26b**

| Carbon[c] | δ^a ^{13}C | | Carbon | δ^{13}C | |
	23a	**23b**		**23a**	**23b**
1	13.88 q[b]	13.98	9	40.26 t	40.81
2	69.01 t	69.28	10	40.26 s	40.63
3	184.72 s	184.30	11	46.96 t	47.25
4	106.63 d	107.16	12	120.96 s	120.90
5	187.50 s	188.04	13	50.20 q	50.27
6	70.37 s	71.94	14	29.74 q	26.69
7	69.50 t	69.52	15	28.62 q	28.36
8	53.14 d	52.13			

[a] In ppm downfield from tetramethylsilane.
[b] Multiplicity in SFORD spectrum: s = singlet; d = doublet; t = triplet; q = quartet.
[c] The numbering scheme used in this table is from the drawing of structure **23b**.

shifts.[13] The relevant data are given in Table I, showing the deshielding of protons in the C-9 methylene group of the solid isomer, suggesting proximity to the carbonyl unit, and the ^{13}C signals for C-8 and C-3 appear relatively downfield; consistent with simple analogs[13] structure **26a** can be suggested for the solid. The high R_f on silica gel chromatography and a much slower rate of reaction with sodium borohydride in methyl alcohol for the liquid isomer may reflect a sterically encumbered carbonyl group, consistent with structure **26b**. While these data allowed tentative structure assignments, a more definite assignment was crucial at this stage of the synthesis. The solid isomer was subjected to X-ray diffraction analysis, which verified the structural assignment (**26a**) as shown in the ORTEP representation (Fig. 1).

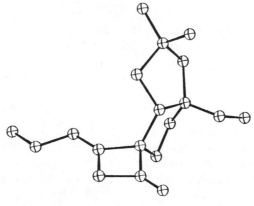

FIG. 1.

The lack of stereoselectivity in the ketene cycloaddition was disappointing and not easily rationalized, since the diastereotopic faces of the ketene C=C unit appear from models to have very different steric environments. Stepwise addition via a dipolar intermediate (**27** in particular) seems to require preferential formation of **26a** from steric considerations. However, ketene addition via a $_2\pi_s + {}_2\pi_a$ concerted pathway requires the orthogonal approach [14,15] of the reacting species (as in **28a** and **28b**). The steric interaction originates primarily from the acetylenic proton and the difference in energy for transition states resulting from **28a** and **28b** is likely to be small.

Given the formation of both stereoisomers in similar amounts, two strategies were considered. At some stage, different reaction sequences might carry both **26a** and **26b** to a simple cyclobutenone such as **9**. Alternately, one pure isomer (e.g., **26**) might be equilibrated to a mixture of **26a**–**26b**, the desired isomer obtained by chromatography, and the undesired isomer recycled through the equilibration step. The latter idea was developed through reaction of **26b** with aqueous sodium hydroxide to give **29** (mixture of tautomers), followed by methylation with diazomethane to produce **30a**–**30b** as a 3:2 mixture in 95% yield [Eq. (10)]. In this way, pure **26a** (**30a**) was obtained from the cycloaddition products in 20–25% yield overall (based on acid **25a**).

$$(10)$$

In a series of preliminary reactions, further conversions of **26b** were explored. Reaction with diisobutylaluminum hydride, followed by delicate acid hydrolysis (oxalic acid in a water–ether two-phase system), produced **31** in 64% yield. This product proved difficult to work with in attempts to add the final three-carbon unit. Careful hydrolysis of the internal ketal **30b** produced **32**, which was acetylated and thioketalized to provide a precursor (**33**) for cyclization, but efforts to carry on with intermediates **31–33** have not been successful. There are several reasonable schemes for proceeding from **26**, but the moderate yield in ketene cycloaddition and the lack of stereoselectivity made the general approach unattractive.

It is easy to become enmeshed in a synthetic strategy when the obviously beautiful original plans cannot be executed in a direct and efficient way. Partial success is sometimes harder to overcome than clear-cut failure. A broader view beyond the narrow immediate problems and recognition of better general solutions often come out of discussions with colleagues and visitors. We had the lucky chance of a visit by Professor Diulio Arigoni at Cornell in 1975, during which he made us aware of special biosynthesis questions presented by the structure of fomannosin (**1**).

IV. Synthesis of Illudol and Fomannosin from a Common Intermediate

A. THE BIOSYNTHESIS CONNECTION

A biosynthetic connection can be proposed between the fomannosane and illudane skeletons.[16,17] Fomannosin (**1**) is not a product from simple cationic cyclizations of farnesyl derivatives. The detailed pathway has now been elucidated,[18,19] verifying the earlier speculation. Through double ^{13}C-labeling experiments, a pathway was established from farnesyl pyrophosphate (**35**) through humulene (**36**) and the protoilludane skeleton (**37**) to fomannosin (Scheme 1). Although the detailed pathway from **37** to **1** has not been defined, the connectivity implied in Scheme 1 points to a synthesis strategy for fomannosin and leads naturally to a consideration of the synthesis of illudol (**34**) from a common intermediate related to **37** in the biosynthesis scheme. An important virtue is the rigid skeleton of **37**, where the critical stereocenters can be generated in a predictable framework.

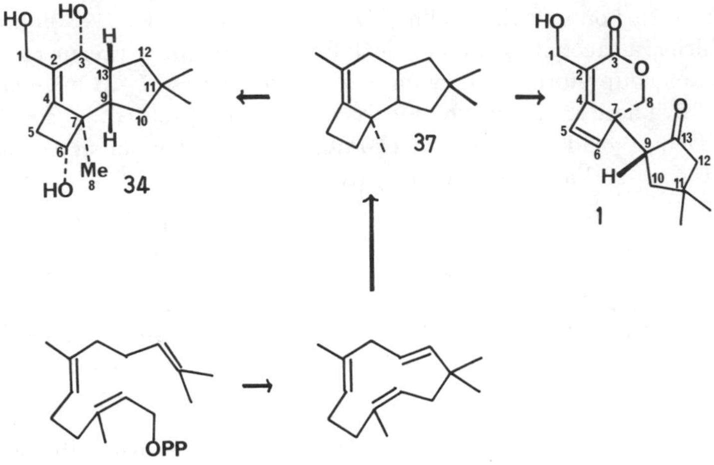

SCHEME 1. Biosynthesis connection.

B. ILLUDOL

The structure of illudol (**34**), a metabolite of *Clitocybe illudeus*,[20] was proposed in 1967 on the basis of spectral and chemical evidence[21] but the absolute configuration has not been determined. A synthesis was reported in 1971 based on a nonstereospecific photochemical [2 + 2] cycloaddition of 1,1-diethoxyethylene with a 2-cyclopentenone to form the crucial cyclobutane unit.[22]

Illudol (**34**) has five contiguous chiral centers: two of them (C-3 and C-6) bear hydroxyl groups where selective carbonyl reduction will generate the proper configuration; one (C-13) is adjacent to a potential carbonyl (C-3) and therefore easily epimerized to the more stable (assumed) cis 4–6 ring fusion; the relative chirality at the C-7–C-9 pair is the same as required for fomannosin and must be controlled during ring construction of the common intermediate.

C. OVERALL STRATEGY

The strategy inspired by the biosynthetic relationships is outlined in Scheme 2. The skeleton is to be assembled through cycloaddition of a cyclobutene carboxylate (**38**) with a diene (**39**). The ester unit in **38** is important in accelerating the cycloaddition and also in providing the oxygen functionality at C-8 in proposed intermediate **40** and later becomes the oxygen in the lactone

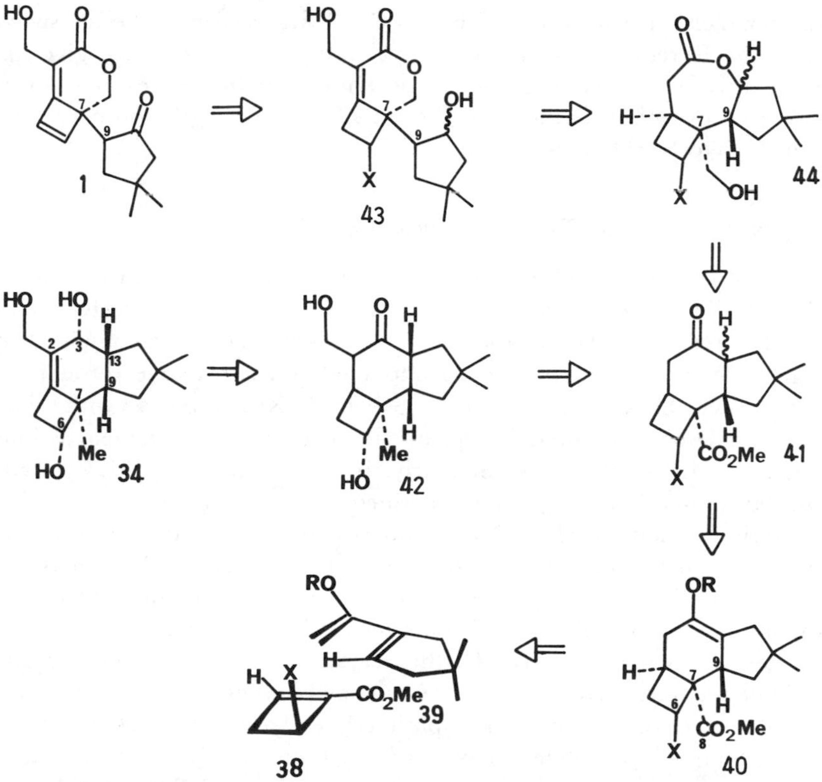

SCHEME 2. Fomannosin and illudol from a common intermediate.

ring of fomannosin. However, the ester unit must be reduced to a methyl group for illudol, a process invariably involving brutal conditions that should be planned for an early stage. Baeyer–Villiger oxidation of **41** requires selective migration of the secondary carbon (C-13), consistent with the usual pattern for this reaction. One carbon must be added to **41**, adjacent to a ketone carbonyl (to give **42**) and adjacent to a lactone carbonyl (to give **43**). The group X in **38** must be stable enough to persist to the last stages of the fomannosin sequence, but be eliminated under mild enough conditions to minimize decomposition of the very delicate, electron-deficient methyl-enecyclobutene unit in **1**. At the same time, X must appear as a hydroxyl group in illudol.

The critical stereochemical relationships for fomannosin are fixed during the cycloaddition. As shown in Scheme 2, endo addition should be favored and would give the proper $7R,9S$–$7S,9R$ configuration. Illudol has three additional chiral centers (C-3, C-6, and C-13); the earlier synthesis[22] showed that hydride reduction of the carbonyl groups at C-3 and C-6 gives predominantly the correct configurations. The remaining center at C-13 will arise from hydrolysis of the enol ether in **40**; the cis ring fusion may not be strongly favored thermodynamically.

D. Synthesis of the Common Intermediate 41

This phase begins with the problem of preparing cyclobutenecarboxylate **38** and diene **39**. Several versions of **39** were prepared [R = CH$_3$, Si(CH$_3$)$_2$(tBu), Si(CH$_3$)$_3$] from ketone **45**, which, in turn, was obtained on a large scale from isobutyraldehyde and methyl vinyl ketone through 4,4-dimethylcyclohexanone, according to Scheme 3. Reaction of **45** with lithium diisopropylamide followed by quenching with methyl p-toluenesulfonate produced **39a** in 40% yield. Similarly, by use of the chlorotrialkylsilanes, **39b** (99%) and **39c** (96%) were obtained.

A series of model Diels–Alder reactions with **39a** were disappointing and led nearly to abandonment of the basic strategy. No cycloadducts were obtained with simple acrylate derivatives, neither under thermal reaction conditions nor in the presence of Lewis acid and transition metal catalysts. At elevated temperatures (>120°C) the acrylates polymerized and the diene could be recovered. Nevertheless, it was hoped that cyclobutene carboxylate **38** would be more reactive than simple acrylates, based on relief of bond angle strain. Simple cyclobutenes have appeared in the Diels–Alder reaction infrequently,[23,24] and we are unaware of examples closely related to **38**. Similarly, it was hoped that the siloxydienes **39b** and **39c** would be more

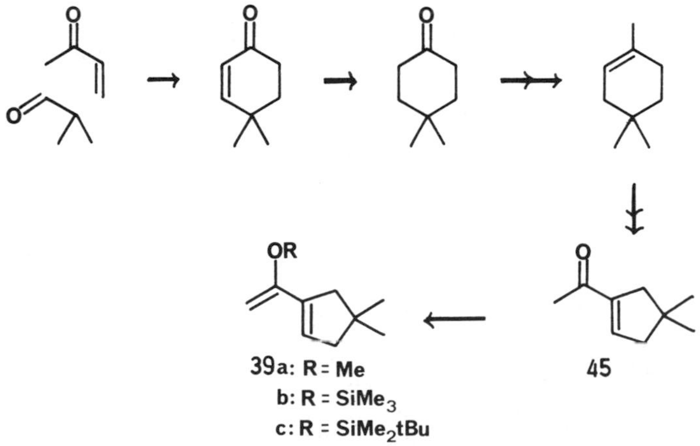

39a: R = Me
b: R = SiMe$_3$
c: R = SiMe$_2$tBu

45

Scheme 3. Preparation of dienes **39a–c**.

reactive then **39a** for electronic and steric reasons. The required cisoid arrangement would be favored at equilibrium by large R groups.

The highly functionalized cyclobutene **38** is of a type that has recently been prepared by Lewis acid-catalyzed [2 + 2] cycloaddition of propiolic esters with electron-rich alkenes.[25,26] After initial studies, using various Lewis acids, we found that a simple thermal reaction of ethyl propiolate with 1,1-diethoxyethylene produces an appropriate cyclobutene (**46**) at 50°C in dichloromethane (65% yield). As expected, ring opening of **46** occurs rapidly at 90°C ($t_{1/2}$ = 0.5 h) to give diene **47** in high yield. This reactivity bodes ill

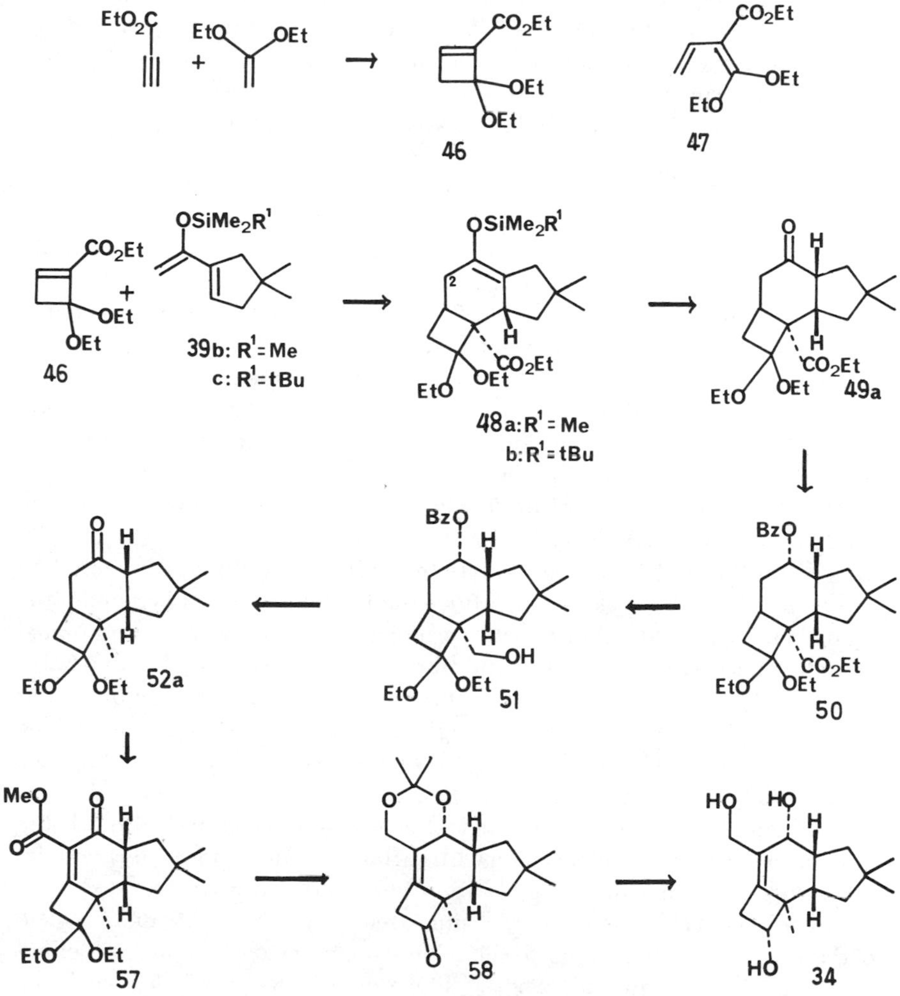

SCHEME 4. Synthesis of illudol.

for the desired Diels–Alder reaction with **39**, but under carefully controlled conditions (minimum reaction temperature, extended time), 1:1 adducts were obtained in 70–75% yield and >95% selectivity (regio, stereo), after purification by column chromatography. Structures **48a** and **48b** were assigned to the adducts, based on spectral data. However, the stereochemical assignment was secure only after conversion to illudol (Scheme 4), which requires the configuration in **40**.

Hydrolysis of **48a** and **48b** with either fluoride anion or gentle base produced a mixture of the desired cis ring fusion isomer (**49a**) and the corresponding trans isomer (**49b**) in similar amounts. The pure cis isomer was converted to the equilibrium mixture of **49a–49b** (60:40) upon treatment with sodium methoxide in methyl alcohol. However, desilylation of **48a** could be achieved with 3-Å molecular sieves as catalyst in methyl alcohol to give exclusively the cis product **49a**, apparently by kinetically controlled protonation.

E. Synthesis of Illudol

The formation of the C-7 methyl substituent in illudol (**34**) from **49a** involves reduction of the ketone carbonyl to an alcohol, protection of the hydroxyl group as the benzyl ether (**50**), and then reduction of the ester to the primary alcohol (in **51**) (Scheme 4). Following a general procedure,[27] the alcohol was activated as the phosphorodiamidate and the carbon–oxygen bond was cleaved with lithium metal in ethylamine; the benzyl protecting group was also cleaved. Then oxidation with Collins' reagent gave the ketone **52a** in 56% yield overall from **48a**.

An alternative preparation of **52a** which avoided the benzyl protecting group, was developed through application of Ireland's reduction method directly on the *tert*-butyldimethylsilyl ether **48b**. The reduced compound **53** was obtained in 74% yield. Desilylation could not be accomplished under the mild conditions that allow kinetic protonation; fluoride anion-promoted desilylation led to a mixture of cis and trans ring-fusion isomers (**52a** and **52b**), which could be separated easily by chromatography. The trans isomer was equilibrated in base to recycle it through to the desired cis arrangement. The combined yield of **52a** after two equilibrations was 89%, corresponding to a yield of 65% overall from **48b**.

Addition of a one-carbon unit at C-2 of **48b** was accomplished by carboxylation of the kinetic enolate anion with carbon dioxide (followed by methylation with diazomethane to give **55**). The anion from **55** reacted with phenylselenenyl chloride to afford a single selenenyl ketone (**56**, in 37% yield overall from **48b**). Oxidation to the selenoxide provoked spontaneous elimination at 20°C to give **57** (96%). The stereochemistry of **56** is assigned consistent with the required syn-elimination pathway.[28]

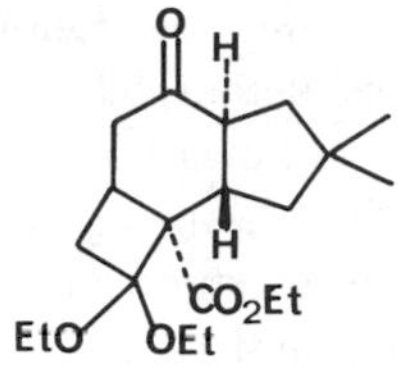

The keto ester **57** was employed in the previous synthesis of illudol,[22] and we made no effort to improve on the published procedure. Reduction of **57** with excess sodium bis (2-methoxyethoxy)aluminum hydride produced a diol that reacted with acetone (catalytic *p*-toluenesulfonic acid) to give keto acetonide **58**. Hydride reduction and acid-promoted hydrolysis of the acetonide gave a single product, identified as racemic illudol (**34**) by comparison of NMR spectral data and chromatographic properties with a sample from nature, which was generously provided by Dr. M. Anchel of the New York Botanic Garden.

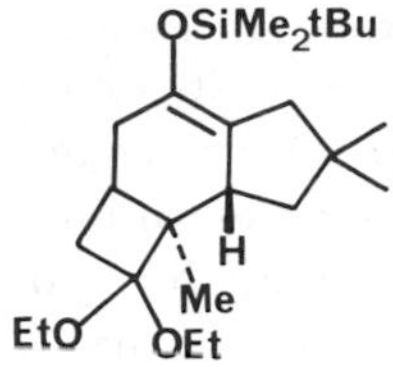

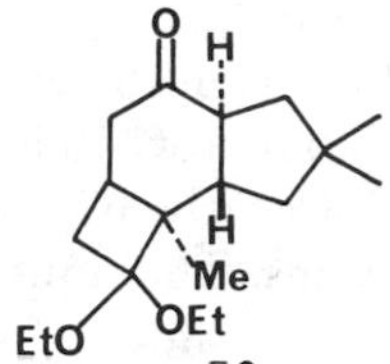

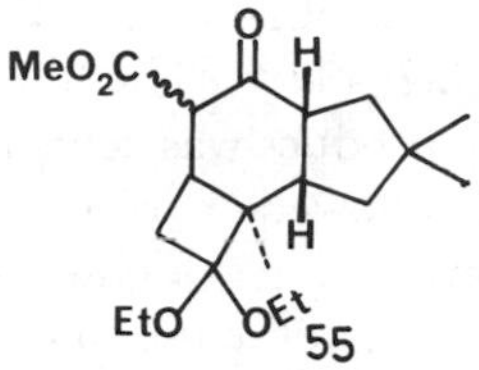

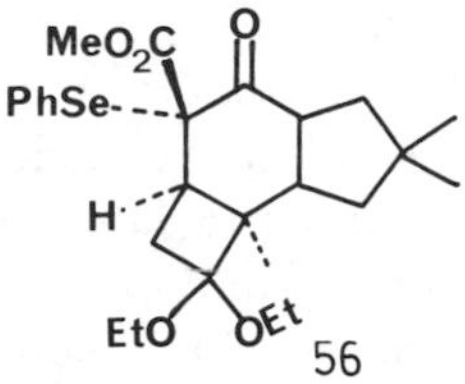

F. SYNTHESIS OF FOMANNOSIN

The strategy for conversion of **48** to fomannosin (**1**) is outlined in Scheme 5. There are four crucial steps: (a) oxidative cleavage at C-3–C-13, (b) introduction of the hydroxymethyl unit, (c) trans lactonization to give the fomannosin skeleton, and (d) introduction of the reactive cyclobutene double bond.

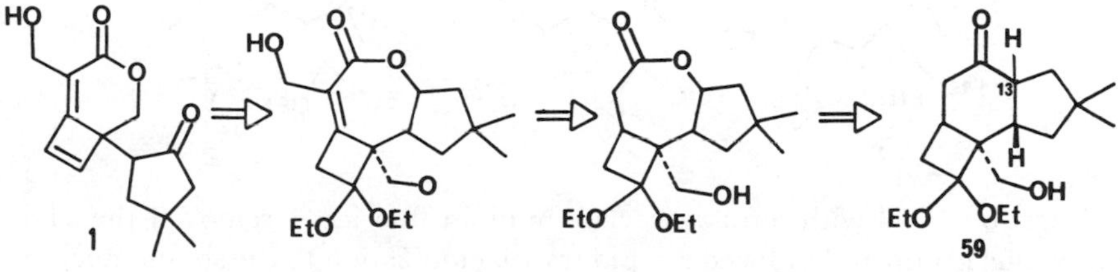

SCHEME 5. Strategy for fomannosin from **59**.

The initial sequence tested began with reduction of ester **49a**, followed by hydrolysis of the enolsilyl ether unit, to give hydroxy ketone **59**. The configuration at C-13 is not important since that carbon will eventually become part of a carbonyl group, but it was amusing to find a single isomer, exclusively the trans. The trans isomer is now the favored kinetic product, presumably because of internal proton delivery from the hydroxymethyl substituent at C-8. Baeyer–Villiger oxidation of **59**, using buffered *m*-chloroperbenzoic acid, produced a single isomer (**60**), and the primary hydroxyl group was protected by reaction with *tert*-butylchlorodimethylsilane (**61**; 90% yield from **59**). Formaldehyde was the source of the hydroxymethyl side chain at C-2, using a sequence parallel with that for illudol (Scheme 4). The lithium enolate of **61** was added to excess formaldehyde; the product was converted to the dianion with lithium diisopropylamide and allowed to react with phenylselenenyl chloride. Under optimum conditions, a single product was obtained (**62**), but the yield was only 48%, and 40–45% of the intermediate (**63**) was recovered. Attempts to force complete conversion (excess reagents, longer time, and higher temperature) did not improve the absolute yield. The stereochemistry of **62** is again based primarily on easy selenoxide elimination (86% yield, 20°C, 0.3 h) to give **64**. The pattern of reactivity of enolate anions derived from **61** and from **49** indicates preferential electrophile addition syn to the proton at C-4, presumably due to steric effects.

Reaction of **64** with tetra-*n*-butylammonium fluoride[29] removed the silyl protecting group and induced a rapid translactonization to the six-membered lactone (**65**; 47% yield) appropriate for fomannosin (**1**). However, at this stage, serious problems were encountered in the hydrolysis of the ketal unit leading to the cyclobutanone (in **66**). All attempts at acid-catalyzed hydrolysis on **65** and **64** and related derivatives failed to give a cyclobutanone; preliminary characterization of two by-products (**67** and **68**) suggests ring cleavage as the primary process. Other conditions were tested unsuccessfully, including trimethylsilyl iodide.[30]

An alternative ordering of the steps avoided the ring cleavage (Scheme 6). Baeyer–Villiger oxidation directly on **58** gave **69** (96% yield) and ketal hydrolysis at −20°C produced the cyclobutanone **70** (91%). At higher temperature, lactone cleavage products were observed. After protection of the primary hydroxyl group (**71**), the ketone carbonyl group was reduced with sodium borohydride and the resulting secondary hydroxyl (in **72**) was converted to a tetrahydropyranyl ether (**73**). The configuration of the hydroxyl group at C-6 was tentatively assigned as shown, based on least-hindered approach of hydride and potential internal delivery after initial coordination of the borohydride reagent with the oxygen at C-8. A single isomer was obtained after column chromatography, in 85% yield from **70**. Introduction of the hydroxymethyl group and the C-2–C-4 double bond followed the procedure used earlier, except that intermediate **74** (obtained in 48% yield) was treated with pyridinium tosylate in methyl alcohol to free the secondary hydroxyl (**75**). Then selenoxide elimination gave the unsaturated lactone **76** in 71% yield. Selective acetylation of **76** at the primary hydroxyl could not be achieved, but reaction with ethyl vinyl ether and 1 mole equivalent of pyridinium tosylate at −22°C afforded **77** in 87% yield. Reaction of **77** with methanesulfonyl chloride served to protect the secondary hydroxyl (in **76**) and to provide the activation needed later for introduction of the cyclobutene double bond. Desilylation of **78** with hydrofluoric acid allowed trans lactonization to occur, producing diol **79**, in 72% overall yield from **77**. The primary hydroxyl was silylated carefully at −22°C, and the secondary hydroxyl was oxidized with chromium trioxide to give **80** in 88% yield from **79**. Treatment of **80** with 1 mole equivalent of fluoride anion in tetrahydrofuran at 25°C gave racemic fomannosin (**1**). After rapid chromatography on silica gel, the pure

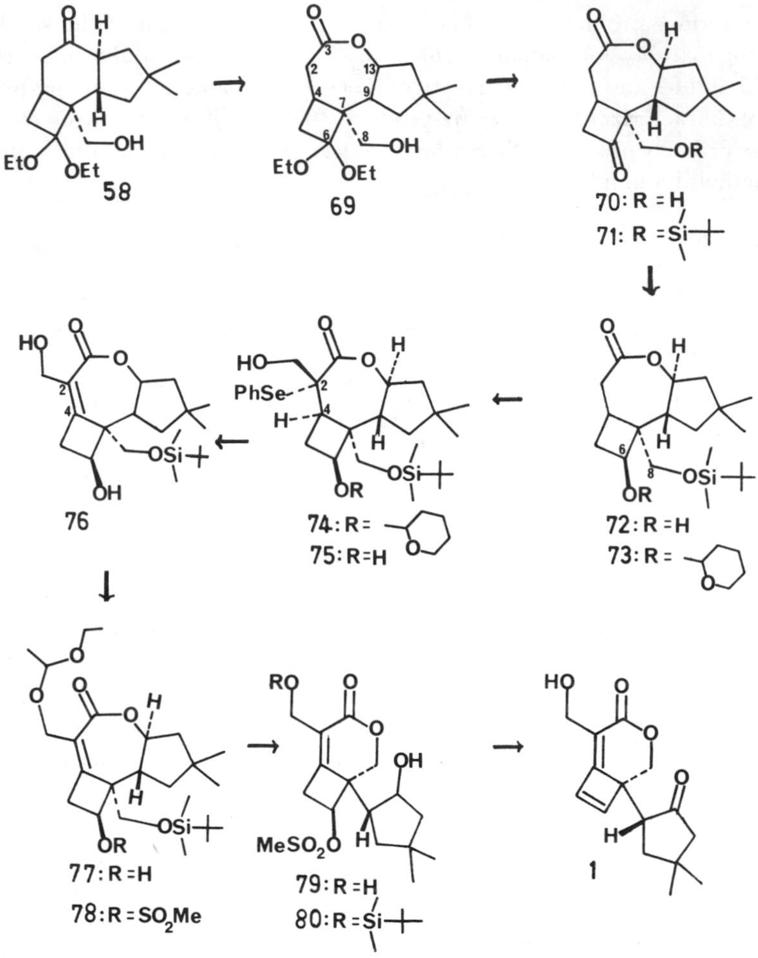

SCHEME 6. Synthesis of fomannosin.

sample was obtained as a colorless, unstable semisolid, in 81% yield. Comparison with a sample of natural (−)-fomannosin shows identical TLC properties in two solvent systems and identical ^{1}H-NMR and IR spectral data. The ^{13}C-NMR spectral data matched well with reported values. The comparison samples of natural material were generously provided by Professor David Cane. The compound shows a remarkable sensitivity toward acid and base. Attempted epimerization to obtain the isomer with opposite configuration at C-9 could not be achieved because of rapid polymerization. Similarly, a solution of 1 in deuterochloroform deteriorates over a few hours at 25°C to produce an insoluble film on the NMR sample tube.

V. Summary

The first total synthesis of fomannosin has been completed. A rigid tricyclic intermediate was constructed by a special stereospecific Diels–Alder reaction of a cyclobutene carboxylate. From this intermediate, a Baeyer–Villiger oxidative cleavage provided the basic skeleton, and an efficient trans lactonization gave the precise tricyclic arrangement. The delicate cyclobutene double bond was introduced at the very last stage through elimination of a secondary methanesulfonate ester. Although the mesylate was readily eliminated in base, it served as a protecting group during the preceding oxidation step. Cleavage reactions of the four-membered ring caused serious roadblocks when the exo-cyclobutene double bond was introduced before the final manipulations at the C-6 oxygen substituent. However, the problem was solved by introduction of the exo double bond at an intermediate stage and then delicate protecting group manipulations to establish the hydroxyl at C-6.

The strategy was particularly attractive because the same tricyclic intermediate could be converted to another sesquiterpene cyclobutane derivative, illudol, in a relatively short and stereospecific sequence. The original idea for the common intermediate arose from the recognition that the biosyntheses of illudol and fomannosin are closely related, apparently proceeding through a common intermediate or a pair of very similar structures.

ACKNOWLEDGMENTS

Organic total synthesis is accomplished in the laboratory. Analysis and "paper" chemistry are important to outline the possibilities and estimate probability of success, but progress requires careful experimentation, and many of the essential ideas arise out of experimental observations. Impressive advances in organic theory notwithstanding, questions of selectivity, such as changing a 1:1 mixture of products to a 9:1 ratio, can be done only by systematic experimentation. Therefore, it is very important to recognize the co-workers who contributed ideas and all of the executions of the ideas: Dr. Shuji Tomoda, Dr. Ken Hurst, Ms. Susan Boettger (now Dr.), and Dr. Hiroto Nagoaka.

REFERENCES

1. J. A. Kepler, M. E. Wall, J. E. Mason, C. Basset, A. T. McPhail, and G. A. Sim, *J. Am. Chem. Soc.* **89,** 1260 (1967).
2. C. Basset, R. T. Sherwood, J. A. Kepler, and R. B. Hamilton, *Phytopathology* **57,** 1046 (1967).
3. M. F. Semmelhack, S. Tonoda, H. Magaoka, S. D. Bottger, and K. M. Hurst, *J. Am. Chem. Soc.* **104,** 747–759 (1982).
4. D. E. Cane, R. B. Nachbar, J. Clardy, and J. Finer, *Tetrahedron Lett.* 4277 (1977).
5. S. Nozoe, H. Matsumoto, and S. Urano, *Tetrahedron Lett.* pp. 3125–3128 (1971).
6. H. Kosugi and H. Uda, *Chem. Lett.* p. 1491 (1977).
7. K. Miyano, Y. Ohfune, S. Azuma, and T. Matsumoto, *Tetrahedron Lett.* 1545 (1974).
8. S. Danishefsky, T. Kitahara, M. Tsai, and J. Dynak, *J. Org. Chem.* **41,** 1669 (1976).
9. H. Stetter, *Angew. Chem. Int. Ed. Engl.* **15,** 639 (1976).

10. R. Scarpati and D. Sica, *Gazz. Chim. Ital.* **92,** 1073 (1962).
11. R. H. Hasek, P. G. Gott, and J. C. Martin, *J. Org. Chem.* **29,** 1239 (1964).
12. H. H. Wasserman, J. U. Piper, and E. V. Dehmlow, *J. Org. Chem.* **38,** 1451 (1973).
13. B. M. Trost and P. H. Scudder, *J. Am. Chem. Soc.* **99,** 7601 (1977).
14. L. Ghosez, R. Montaigne, A. Roussel, H. Vanlierde, and P. Mollet, *Tetrahedron* **27,** 615 (1971).
15. R. B. Woodward and R. Hoffmann, "The Conservation of Orbital Symmetry," pp. 163–168 Academic Press, New York, 1971.
16. W. Parker, J. S. Roberts, and R. Ramage, *Q. Rev., Chem. Soc.* **21,** 331 (1967).
17. T. K. Devon and A. I. Scott, "Handbook of Naturally Occurring Compounds," Vol. 2. Academic Press, New York, 1972.
18. D. E. Cane and R. B. Nachbar, *Tetrahedron Lett.* p. 2097 (1976).
19. D. E. Cane and R. B. Nachbar, *J. Am. Chem. Soc.* **100,** 3208 (1978).
20. M. Anchel, A. Hervey, and Robbins, *Proc. Natl. Acad. Sci. U.S.A.* **36,** 300 (1950).
21. T. C. McMorris, M. S. R. Nair, and M. Anchel, *J. Am. Chem. Soc.* **89,** 4562 (1967).
22. T. Matsumoto, K. Miyano, S. Kagawa, S. Yu, J. Ogawa, and A. Ichibara, *Tetrahedron Lett.* p. 3521 (1971).
23. V. V. Plemenkov and V. P. Kostin, *J. Org. Chem. USSR (Engl. Transl.)* **15,** 1086 (1979).
24. T. R. Kelly and R. W. McNutt, *Tetrahedron Lett.* p. 285 (1975).
25. B. B. Snider, D. J. Rodini, R. S. E. Conn, and S. Sealfon, *J. Am. Chem. Soc.* **101,** 5283 (1979).
26. R. D. Clark and K. G. Untch, *J. Org. Chem.* **44,** 248 (1979).
27. R. E. Ireland, D. C. Muchmore, and U. Hengartner, *J. Am. Chem. Soc.* **94,** 5098 (1972).
28. H. J. Reich, F. Chow, and S. K. Shah, *J. Am. Chem. Soc.* **101,** 6638 (1979).
29. E. J. Corey and B. B. Snider, *J. Am. Chem. Soc.* **94,** 2549 (1972).
30. M. E. Jung and M. A. Lyster, *J. Am. Chem. Soc.* **99,** 968 (1977).

Chapter 9

EVOLUTION OF A SYNTHETIC STRATEGY: TOTAL SYNTHESIS OF JATROPHONE

Amos B. Smith III*

Department of Chemistry
Laboratory for Research on the Structure of Matter
and The Monell Chemical Senses Center
University of Pennsylvania
Philadelphia, Pennsylvania

* Camille and Henry Dreyfus Teacher Scholar, 1978–1983; and National Institutes of Health (National Cancer Institute) Career Development Awardee.

223

I. Introduction

In the fall of 1976, I initiated at Penn a research program that set as principle goal the stereocontrolled total synthesis of jatrophone (**1**), an architecturally interesting macrocyclic diterpene first isolated in 1970 by the late Professor S. Morris Kupchan from extracts of *Jatropha gossypiifolia* L (Euphorbiaceae).[1] Jatrophone merited consideration as a synthetic target for two reasons. First, the unique structure, which combined the features of a spiro-3(2*H*)-furanone ring system with an eleven-membered macrocyclic ring provided, at that time, a significant synthetic challenge. Second, jatrophone displayed significant inhibitory activity against a variety of cell lines including sarcoma 180, Lewis lung carcinoma, P-388 lymphocytic leukemia, and Walker 256 intramuscular carcinosarcoma.[1] During the next five years this venture proved to

| Jatrophone | Normethyljatrophone | Epijatrophone |
| 1 | 2 | 3 |

be the central focus of my laboratory. In retrospect, it is now obvious that the unique architecture and functionality of the jatrophone molecule directed the evolution of a major portion of the research that we undertook during those years. The outcome proved not only to be the first (and to this point the only) total synthesis of jatrophone, but also the total synthesis of a number of related natural products, as well as the development of several new methods and synthetic strategies. The list below summarizes to date the major accomplishments that have emanated from the jatrophone venture.

1. The stereocontrolled total synthesis of (±)-jatrophone (**1**), (±)-normethyljatrophone (**2**), and (±)-epijatrophone (**3**).[2]
2. The stereocontrolled total synthesis of the pentenomycins, their epimers and dehydropentenomycin.[3]
3. The total synthesis of geiparvarin and isogeiparvarin.[4]
4. The synthesis of two analogs of jatrophone: *cis*- and *trans*-normethyl-jatropholactone.[5]
5. The development of a general protocol for construction of the 3(2*H*)-furanone ring system involving a convenient synthesis of 1,3-diketones.[6]
6. A study of the chemistry of the 3(2*H*)-furanone ring system including alkylation, conjugate addition, and reaction with alkyl thiols.[6]
7. Development of a versatile latent α-ketovinyl anion equivalent.[7]
8. Isolation and structure elucidation of three new jatrophone derivatives: the hydroxyjatrophones A, B, and C.[8]
9. A study of the interplay of the structure and conformation of the jatrophone nucleus vis-a-vis antitumor activity.[9]

In keeping with the informal spirit of this collection of essays on the development of synthetic strategy, I will attempt to present here in chronological order a fairly complete account of the evolution of the ideas, spin-offs and setbacks of what I now refer to as the jatrophone era.

Before beginning our discussion, it is important to note that the central structural feature of the jatrophone molecule, namely the 3(2*H*)-furanone ring, is common to an increasing number of antitumor agents, including such diverse systems as the eremantholides A, B, and C (**4a–c**)[10] and geiparvarin (**5**).[11] Thus the development of a viable synthetic strategy for jatrophone was anticipated to have considerable impact on the synthesis of related 3(2*H*)-furanone antitumor agents.

Eremantholide

4a: R = *i*Pr
4b: R = *sec*-Bu
4c: R = CMe=CH$_2$

Geiparvarin

5

II. General Background

The structure and absolute stereochemistry of jatrophone, initially assigned on the basis of spectroscopic data, was confirmed through aegis of a single-crystal X-ray analysis.[1] The central features of the derived structure proved to be the spiro-3(2H)-furanone ring system, the macrocyclic ring, and the cross-conjugated trienone functionalities. In addition, two chiral centers were present, at C-2 and C-15.

Concomitant with the original isolation and structural studies, Kupchan, in an investigation of the chemistry of jatrophone [12] obtained several interesting results vis-a-vis the possible chemical mechanism for the observed cytotoxic effects. For example, treatment of jatrophone (1) with thiols, including the

amino acid cysteine, led to conjugate addition at C-9, followed by transannular bond formation between C-8 and C-12 to yield novel, highly unstable adducts such as **6**. Sulfhydryl groups present on proteins were found to react in an

cis-Normethyljatropholactone
7

trans-Normethyljatropholactone
8

9: R¹ = OH, R² = Me
10: R¹ = Me, R² = OH

11

analogous fashion.[12] Similar transannular bond formation was also observed on treatment of jatrophone (1) with either hydrochloric or hydrobromic acid or on acid-catalyzed ketalization.[1] These observations led Kupchan to suggest that jatrophone and related compounds exert their cytotoxic effect by alkylation and thereby inactivate biological macromolecules involved in growth regulation.[12] As will be discussed in the latter part of this chapter, we have expanded the original Kupchan study of the chemistry of jatrophone vis-a-vis antitumor activity to include several closely related synthetic analogs of jatrophone [i.e., normethyljatrophone (2) and *cis*- and *trans*-normethyl-jatropholactones (7) and (8)] as well as three new naturally occurring jatrophone derivatives, [i.e., hydroxyjatrophones A–C (9–11)].[8]

III. A Strategy for the Total Synthesis of Jatrophone

From the retrosynthetic perspective, three major structural problems are immediately discernible in the jatrophone molecule. The first two are stereochemical in nature, the third concerns the general difficulty encountered in construction of a macrocyclic ring. Let us consider first the stereochemical problems inherent in the jatrophone molecule.

Jatrophone
1

Any viable synthetic approach to jatrophone must consider the problem of the relative orientation of the secondary methyl substituent at C-2, in addition to orchestrating configurational control at the C-5—C-6 and C-8—C-9 olefinic linkages. Significantly, the C-2 methyl substituent is vinylogously α to the ketone at C-7 and therefore, formally at least, is capable of epimerization via conjugate enolization. Such a process could also lead to isomerization of the C-5—C-6 olefinic bond. Assuming that structure 1 represents the most stable isomer of jatrophone, stereochemical control at C-2 and at the C-5—C-6 olefinic bond could be effected near the end of a synthetic strategy. At the outset, however, no experimental information concerning the thermodynamic stability of jatrophone was available to support such a possibility. Furthermore, molecular-model studies indicated that upon equilibration neither isomer at C-2 would predominate. Finally, for enolization and thereby equilibration to occur, the dienone system (C-2 to C-7) must be capable of assuming a coplanar conformation. Here, molecular-model studies were more definitive, indicating that such a coplanar conformation would be energetically unlikely. Thus, it appeared that any approach to jatrophone must control *from the outset* the relative stereochemistry at C-2 and C-15 as well as the configurations at the C-5—C-6 and C-8—C-9 olefinic bonds.

With these structural problems in mind, we initiated our retrosynthetic analysis by considering advanced intermediates **12–14**. First, we envisioned that introduction of the trans olefinic linkage at C-8—C-9 in **14** could be accomplished via a hydroxyl-directed reduction[13] of **13** (e.g., lithium aluminum hydride). Subsequent mild oxidation (MnO$_2$ or Collins)[14,15] of the derived allylic alcohol and deketalization would then yield jatrophone.

12   DIBAL   **13**

                                                              LiAlH$_4$

Jatrophone   Oxidation   **14**
1

Second, it appeared that **13** would be readily available from **12** via reduction of the C-7 carbonyl group. Here a prudent choice of reducing reagents would be necessary in order to avoid the potential problem of 1,4-reduction of the α,β-unsaturated system. For this transformation, DIBAL[16] or NaBH$_4$ in the presence of CeCl$_3$[17] appeared to be the reagents of choice. The strategic selection of the acetylene functionality at C-8—C-9 was seen as the cornerstone of our approach, in that the linear nature of this functionality would provide the configurational control required at C-5—C-6. That is, molecular model studies of **12** revealed that the Z configuration at C-5—C-6 should be considerably more stable than that of the corresponding E isomer. Alternatively, models of jatrophone (**1**) suggested that both configurational isomers at C-5—C-6 are feasible molecular systems, the natural Z isomer being only slightly less strained.

This detailed stereochemical analysis of intermediates **12–14** led quite naturally to the formulation of **15** as the required advanced precursor prior to cyclization. Aldol condensation followed by stereospecific conversion of the acetylenic linkage in **16** to a trans double bond would then afford jatrophone.

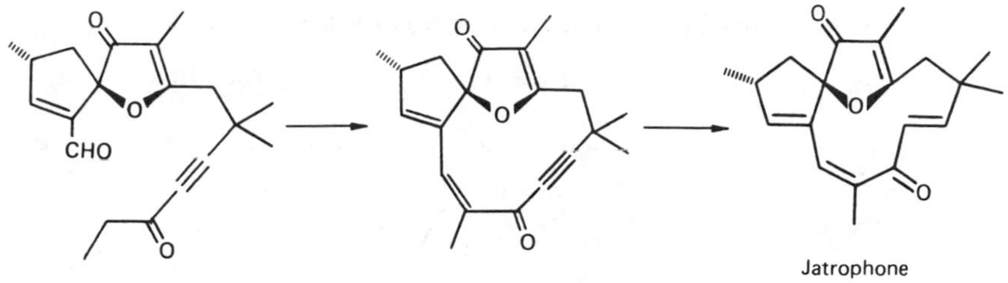

15	**16**	Jatrophone **1**

Toward this end, we demonstrated early on in this research program that model acetylenic alcohol **17**, prepared as shown below, undergoes, in nearly quantitative yield, stereoselective trans reduction to **18** upon treatment with $LiAlH_4$ in ether.[13] Likewise, reduction of ketone **19** with $CrSO_4$[18] afforded ketone **20**, in 53% yield, as the only product. These transformations provided the required precedent for the final conversion of the acetylenic linkage to a trans olefin in our jatrophone strategy.

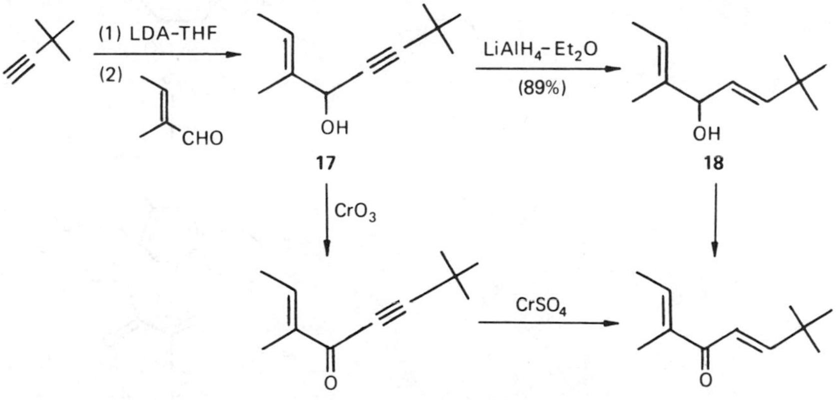

Assuming the validity of the above scenario, we devised a three-phase plan for the jatrophone program. Each phase of the plan had as its goal a major synthetic target. An overview of the plan is illustrated below. The first phase called for the development of a viable synthetic approach to the $3(2H)$-furanone ring system, compatible with the spiro nature of the furanone ring in jatrophone, and then exploitation of this methodology for the construction of the model furanone **21**. The second phase set as the principle goal the synthesis of normethyljatrophone (**2**). Required here would be the development of a method for closure of the eleven-membered macrocyclic ring found in jatrophone. Macrocycle **23** was thus selected as the initial advanced target for phase II. The latter, in conjunction with the previously developed protocol

THREE-PHASE PLAN FOR CONSTRUCTION OF JATROPHONE

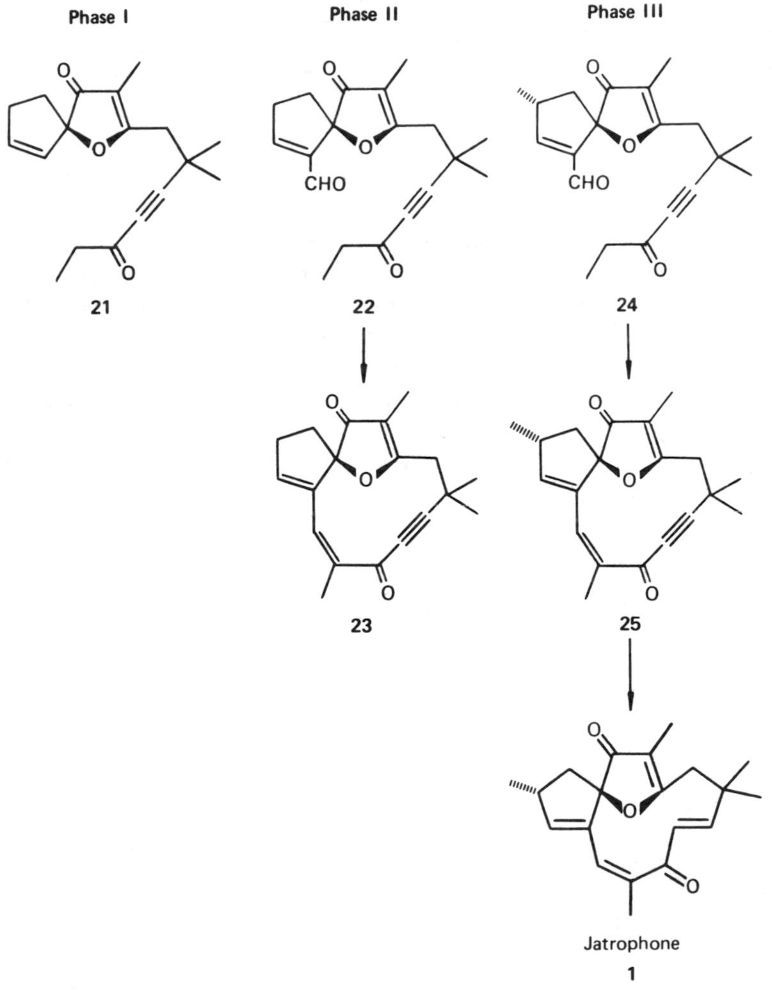

Jatrophone

1

for the regiospecific conversion of the C-8—C-9 acetylenic bond to a trans olefin would then allow elaboration of normethyljatrophone (**2**). Finally, in

Normethyljatrophone
2

phase III, we would turn our efforts to the synthesis of jatrophone; the key intermediate here was envisioned to be **24**.

As alluded to above, selection of jatrophone as a synthetic target demanded the availability of an efficient (and hopefully general) strategy for construction of the 3(2*H*)-furanone ring system. No general approach to this system, however, was available at the outset of this synthetic venture. In fact, the chemistry of this increasingly important heterocycle had been little explored. Thus, the development of a viable synthetic approach to the 3(2*H*)-furanone ring system, as well as an exploratory study of the reactions of this interesting heterocycle, was considered consistent with the overall goal of this research program.

IV. Synthesis of Simple 3(2*H*)-Furanones: The First Major Spin-Off of the Jatrophone Program

In 1971 Margaretha[19] reported that the sodium hydride-catalyzed acylation of hydroxy ketone **26** with ethyl formate, followed by an acid-promoted dehydration, led to furanone **27** in 50% yield. Unfortunately, all attempts to exploit this transformation for synthesis of 5-substituted 3(2*H*)-furanones, through utilization of an alkyl ester in place of ethyl formate, proved fruitless.

Further consideration suggested that an intramolecular version of the Margaretha transformation might offer a general solution. Indeed, as early

as 1965 Lehmann[20] reported that α-acyloxy ketones such as **28**, when treated
with sodium dimsyl anion in Me$_2$SO, afford a mixture of α-hydroxyfuranone
29 and butenolide **30**, the latter being the major product. The predominance
of **30**, however, was troublesome in view of the well-known difference in
acidity of ketones versus esters ($\sim$4–5 pK$_a$ units).[21]

as 1965 Lehmann

To explore the Lehmann transformation in detail, we subjected α-acyloxy
ketone **31** to the NaH–Me$_2$SO reaction conditions; the result, albeit in low
yield, was butenolide **32**. Even kinetic deprotonation (LDA–THF at $-78°$C),
followed by an acidic workup, did not afford the desired furanone (i.e., **33**) in
useful amounts. Instead, a complex mixture containing both **32** and **33**, along
with a number of unidentified components, resulted.

Collectively, these results suggest (see Scheme 1) that an equilibrium is
established between the ketone and ester enolates (**31a** and **31b**) at a rate that
is fast relative to addition of the ketone enolate to the ester carbonyl. Furthermore, the greater facility with which ketones are known to undergo nucleophilic addition relative to ester carbonyls[22] leads to the prediction that
butenolide **32** should in fact be the major product. However, if the system is
constructed (see Scheme 2) so as to prevent elimination of water from the
favored intermediate (**34**), the course of the reaction is expected to shift to
formation of the 3(2H)-furanone ring system. That this rationale is correct
was demonstrated by subjecting α-acyloxy ketones **36–38**, each of which bears
at most one hydrogen in the α′-position of the ester substituent, to dimsyl-
sodium. As illustrated in Table I, 3(2H)-furanones **39–41** were obtained as
the sole product in good to excellent yield. Noteworthy here is the facile,
highly efficient synthesis of bullatenone **40**, a natural 3(2H)-furanone isolated
from blistered leaf myrtle (*Myrtus bullata*), a shrub endemic to New Zealand.[23]

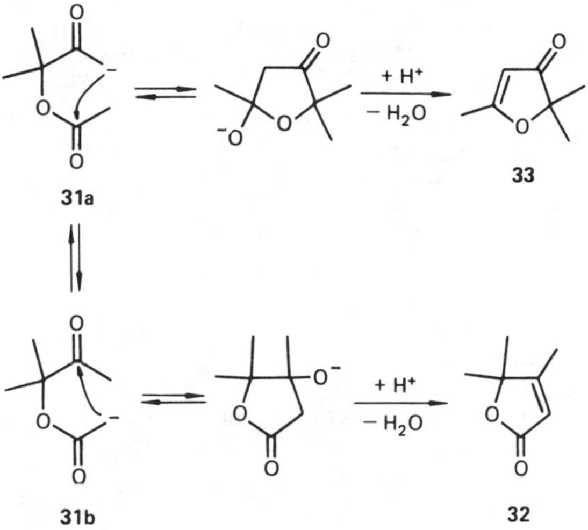

31a

33

31b

32

SCHEME 1

Alternatively, if the α′-position of the ester substituent possesses more than one hydrogen or does so through the principle of vinylogy (i.e., **31** and **42**, respectively), the corresponding butenolides **32** and **43** result. Thus, although we had succeeded in devising an efficient intramolecular cylization–dehydration strategy for the synthesis of selected 3(2H)-furanones, the lack of generality of this approach left much to be desired.

Convinced that the most efficient general approach to the 3(2H)-furanone ring system would involve the acid-catalyzed cyclization–dehydration

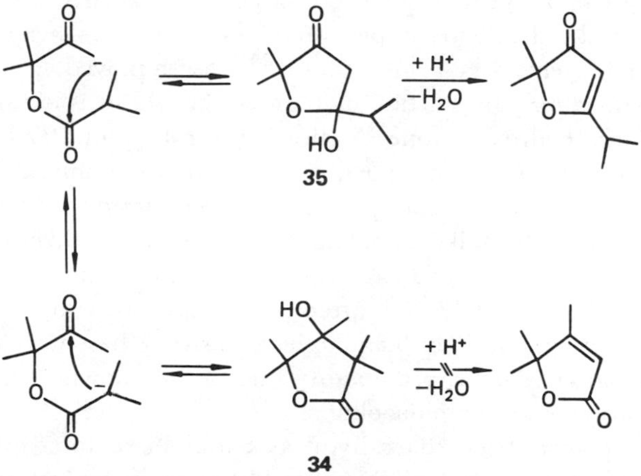

35

34

SCHEME 2

TABLE I
Synthesis of 3(2H)-Furanones from α-Acyloxy Ketones

Entry	α-Acyloxy ketone R			Product (% yield)		
a	$CHMe_2$	36	39	(64)		—
b	Ph	37	40	(88)		—
c	tBu	38	41	(79)		—
d	Me	31		—	32	(20)
e	$CH=CMe_2$	42		—	43	(75)

of an appropriately substituted α'-hydroxy-1,3-diketone (**44**), we reexamined the intermolecular acylation of hydroxy ketone **45**, employing kinetic deprotonation under aprotic conditions.

44

Initially, treatment of the dianion of **45**, generated at $-78°$C with 2.3 equivalents of LDA in THF, with a variety of acylating agents, such as EtOAc, Ac_2O, and AcCl, led only to complex mixtures. However, when the imidazole acyl transfer reagent, acetyl imidazolide,[24] was employed, 3(2H)-furanone **33**, accompanied only by starting hydroxy ketone **45**, was obtained in 70% yield. Similarly, hydroxy ketone **46** afforded spirofuranone **47** in 76% yield. Although the yields here were good, conversions were only in the range of 35–45%. These low conversions presumably result from the acidic nature of the initially derived 1,3-diketone. That is, the initial 1,3-diketone, a carbon acid, quenches the dianion of **45** at a rate faster than it can react with the acyl imidazolide. Unfortunately, all attempts to increase the conversion by employing reverse addition or 3 equivalents of strong base were counter productive; the latter result is presumably due to the known[24] base-catalyzed self-condensation of acetylimidazolide.

We wish to point out here that α-hydroxy ketone **46**, required for elaboration of furanone **47**, was readily available via addition of the acyl anion equivalent, lithium ethyl dithiane[25] to cyclopentenone, followed by hydrolysis of the dithiane functionality.* This approach to α-hydroxy ketone **46** proceeded in

* Hydrolysis of the dithiane functionality under acidic conditions leads efficiently to rearranged allylic alcohol *i*.[25] Advantage of this rearrangement is currently being taken in our approach to jatropholones A and B,[26] two diterpenes isolated from *Jatropha gossypiifolia* by Connally at the University of Glasgow. Our first generation strategy for construction of the jatropholones is outlined below. Interestingly, a viable chiral synthesis of enone *ii* also emanated from the jatrophone program; see p. 250.

Jatropholones A and B

good overall yield; furthermore, from the point of view of the jatrophone program, this approach to α-hydroxy ketones appeared quite viable for future elaboration of **48** and **49** required in phases II and III, respectively.

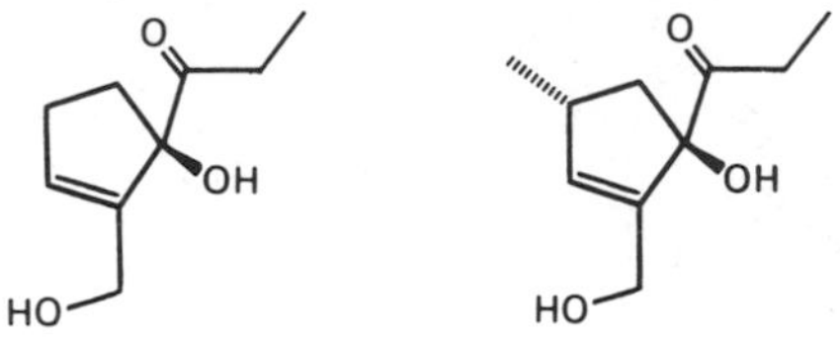

48   **49**

V. A Successful Conclusion to Phase I: Synthesis of 3(2H)-Furanone 21

Although our approach to the 3(2H)-furanone ring system proceeded only in modest yield, the reaction appeared applicable to the jatrophone problem, in that the sequence was quite clean, affording only the desired 3(2H)-furanone and recovered starting α-hydroxy ketone, which could be easily separated and recycled. With these considerations in mind, the decision was made to proceed with the synthetic adventure employing the then available 3(2H)-furanone protocol.

For construction of 3(2H)-furanone **21**, imidazolide **50** was required. Its preparation, illustrated below, was accomplished by nucleophilic addition of the dianion of acetylenic acid **51** to propionaldehyde. Protection of the secondary hydroxyl group with *tert*-butyldimethylsilyl chloride (TBDMSCl),[27] followed by conversion to the imidazolide, employing carbonyldiimidazole,[24] provided **50** in good overall yield.

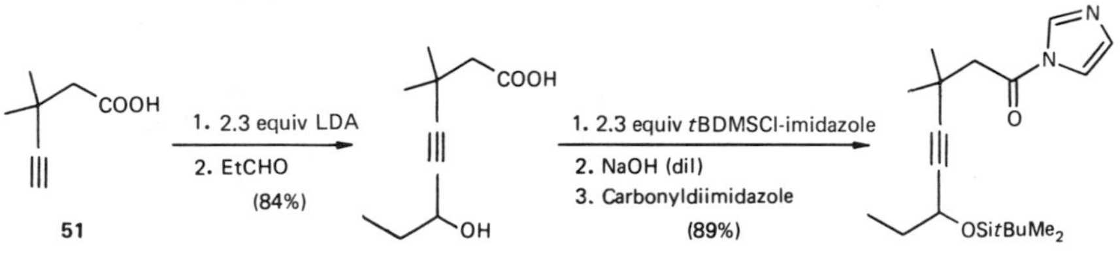

Next, α-hydroxy ketone **46** was treated with 2.3 equivalents of LDA in THF, followed by addition of imidazolide **50**; acid workup gave furanone **53**. The yield in this case, however, was only 19–23%. As anticipated, furanone **53** was isolated as a mixture of diastereomers, which could be separated by TLC. Hydrolysis of the diastereomeric silyl ethers, either individually or as the mixture, followed by oxidation with pyridinium chlorochromate[28] afforded ketone **21**, thereby demonstrating the validity of our overall approach.

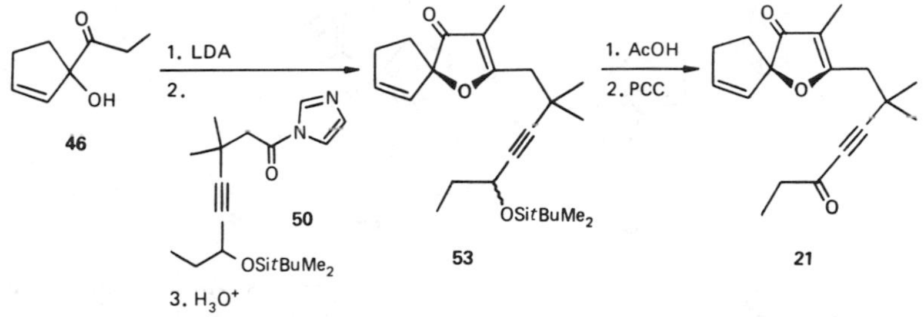

With a successful conclusion to phase I, except of course for the problem of yield in the 3(2H)-furanone synthetic protocol, we turned to phase II. It is important to emphasize here, as illustrated below, that imidazolide **50** is common to all three phases of the jatrophone venture. That is, the synthetic targets of phases II and III were designed such as to require only elaboration of modified α-hydroxy ketones (i.e., **48** and **49**).

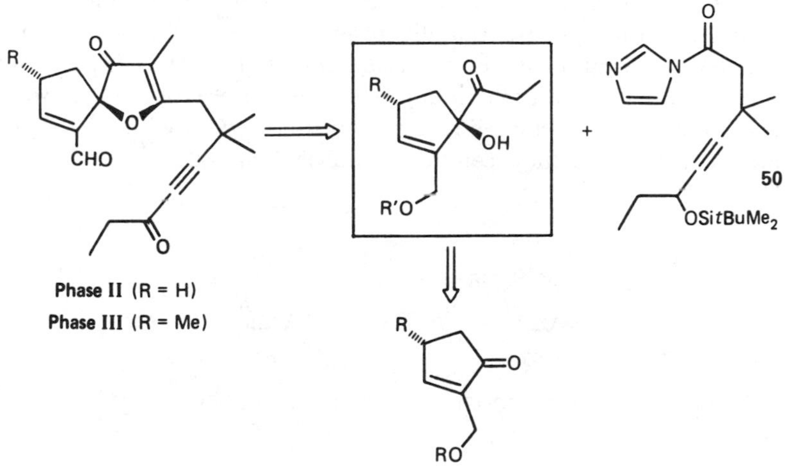

Phase II (R = H)
Phase III (R = Me)

VI. Synthesis of α-Hydroxymethylcyclopentenone: An Initial Solution

From the outset, the starting material for phase II of the jatrophone program was envisioned to be α-hydroxymethylcyclopentenone (**54**). Protection of the alcohol functionality (e.g., TBDMSCl) and treatment with the previously employed acyl anion equivalent, lithium ethyl dithiane, followed by hydrolysis of the dithiane functionality would then afford **55**, substrate for the 3(2*H*)-furanone protocol.

Although α-hydroxymethylcyclopentenone (**54**) is merely an olefinic positional isomer of the enolic form of 2-formylcyclopentanone, careful examination of the literature revealed, somewhat surprisingly, no previous reports for this compound. Indeed, at that time, α-(hydroxymethyl)-α,β-enones were conspicuous by their absence from the chemical literature.[29] Immediate attention therefore was turned toward the development of an efficient preparation of **54**.

Toward this end, 2-carbethoxy-2-cyclopentenone (**56a**),[30] available via the procedure of Reich and co-workers,[31] appeared to be an ideal starting material. Given the reported instability of **56a**,[30] we were somewhat surprised to find that ketalization under the usual conditions [(HOCH$_2$)$_2$, TsOH, benzene] afforded **57a**; the maximum yield, however, was only 45%. Marked improvement could be achieved by replacement of TsOH with the less acidic fumaric acid.[32] Under these conditions the yield was 82%.

Selective reduction of the ester functionality of **57a** proved more difficult. Indeed, initial attempts with such metal hydrides as LiAlH$_4$, etc.[33,34] proved unsuccessful, leading to rather complex mixtures. Diisobutylaluminum hydride (DIBAL), a reagent known[35] to be highly selective toward 1,2-reduction of α,β-unsaturated carbonyl systems, proved more useful. In particular, addition of 2 equivalents of DIBAL in toluene at −78°C, followed by quenching the reaction with methanol, led to the desired ketal alcohol **57b** as the major component. Hydrolysis, followed by purification, afforded **54** in

65% yield from **57a**. The overall yield from 2-carbethoxycyclopentanone was 40%.

VII. An Alternate Approach to α-Hydroxymethylcyclopentenone: Development of a Versatile Latent α-Ketovinyl Anion Equivalent

Although the above procedure was capable of providing usable quantities (~1–2 g) of α-hydroxymethylcyclopentenone (**54**), the protocol was not easily amenable to large-scale preparations (~10–20 g). This, coupled with the expense of phenylselenenyl chloride and the sometimes capricious nature of the ketalization and reduction steps, necessitated development of an alternate approach.

An ideal transformation in this regard would be the construction of α-substituted enones directly from the parent enone without intervention of the thermodynamic enolate. At the time, a general solution to this recurring synthetic problem was unavailable, although Corey,[36] Fuchs,[37] and Stork[38] had independently devised a reverse polarity (umpolung) strategy for the α-alkylation and α-arylation of α,β-unsaturated ketones. Central to their approach was the generation of an effective latent equivalent for α-ketovinyl cation **58**.[39] Such a strategy, however, is limited in that it depends critically on the availability of the requisite alkyl or aryl organometallic reagent.

a-Ketovinyl cation
58

a-Ketovinyl anion
59

A more versatile and possibly more direct approach would be the generation of α-ketovinyl anion **59**, followed by addition of an appropriate electrophile. While such an anion per se is not feasible, recent studies by Ficini,[40] House,[41] and Swenton[42] suggested that α-bromo ketal **60** could serve as a viable latent equivalent for **59**, provided the following transformations proceed efficiently: (a) metalation of **60**, (b) electrophilic capture of the resultant anion **61**, and (c) hydrolysis of ketal **62**. Assuming that each step proceeds without event, the entire transformation could possibly be effected in "one pot."

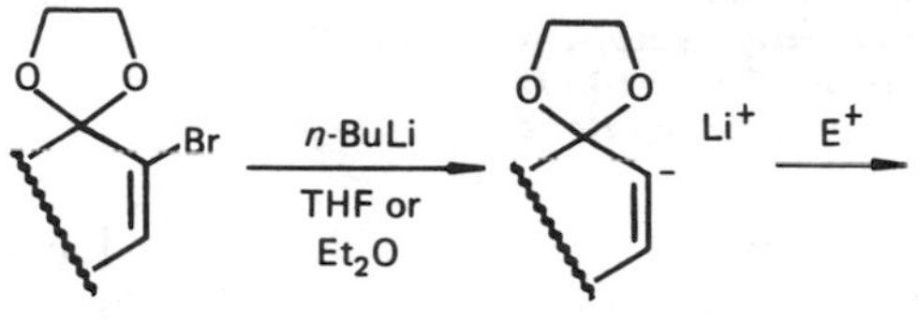

60   **61**

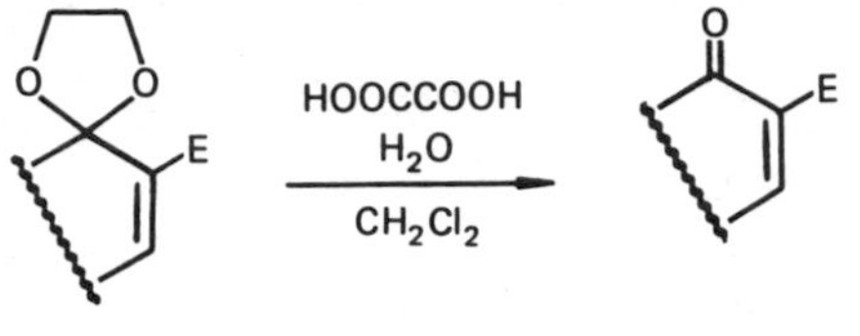

62   **63**

With these considerations in mind, we explored the metalation step.[43] Treatment of bromo ketal **64a** with 1.3 equivalents of *n*-butyllithium in THF at −78°C led smoothly to the corresponding anion, as determined by quenching with deuterium oxide.

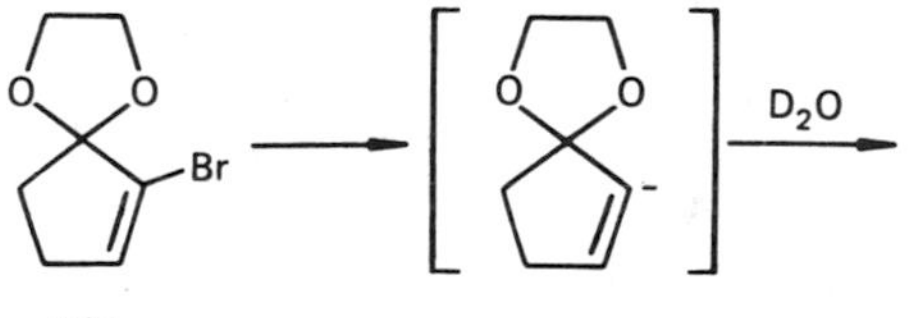

64a

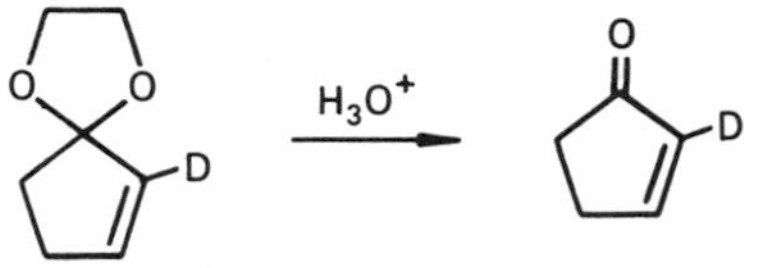

65a

Repetition of the above experiment, employing bromo ketals **64a–c** and a variety of electrophiles, afforded the respective α-substituted enone derivatives **65a–j** in good to excellent yield (see Table II).

The requisite bromo ketals **64a–c** were readily prepared from the corre-

TABLE II
SYNTHESIS OF α-SUBSTITUTED α,β-UNSATURATED KETONES

Entry	Substrate	Electrophile (E^+)	Solvent	Overall yield from 29a–c (%)	Product
64a	(structure)	D_2O	THF	79	65a
		MeI	THF–HMPA	53	65b
		n-C_5H_{11}I	Et_2O–HMPA	71	65c
		HCHO	THF	84	57b
		Me_2CO	THF–HMPA	62	65d
		ClCOOEt	THF	53	57a
		TMSCl	THF	56	65e
		MeSSMe	THF	73	65f
64b	(structure)	n-C_5H_{11}I	Et_2O–HMPA	60	65g
		ClCOOEt	THF	62	65h
64c	(structure)	n-C_5H_{11}I	Et_2O–HMPA	69	65i
		ClCOOEt	THF	81	65j

sponding parent enone in 50–75% yield via a three-step protocol: (a) bromination (Br_2–CCl_4), (b) dehydrobromination,[44] and (c) ketalization.

With monomeric formaldehyde as the electrophilic species, application of the reaction sequence afforded α-hydroxymethylcyclopentenone (54), the overall yield from cyclopentenone being 61%. Vis-a-vis the projected synthesis of normethyljatrophone, this alternative route to 54 was particularly attractive in that it represented a 20% increase in overall yield, could be carried out on a large scale (20–30 g), required about half the time to execute, and was less expensive in that phenylselenenyl chloride was not required.

VIII. Synthesis of the Pentenomycins, Their Epimers and Dehydropentenomycins: A Major Diversion

During our search for an effective preparation of α-hydroxymethylcyclo-pentenone (**54**), we became aware of a novel group of antibiotics that embody the α-hydroxymethylcyclopentenone skeleton. These antibiotics, known as the pentenomycins I–III (**66a–c**)[45] and dehydropentenomycins (**68**),[46] were first isolated in the 1970s. While beyond the scope of the present chapter,

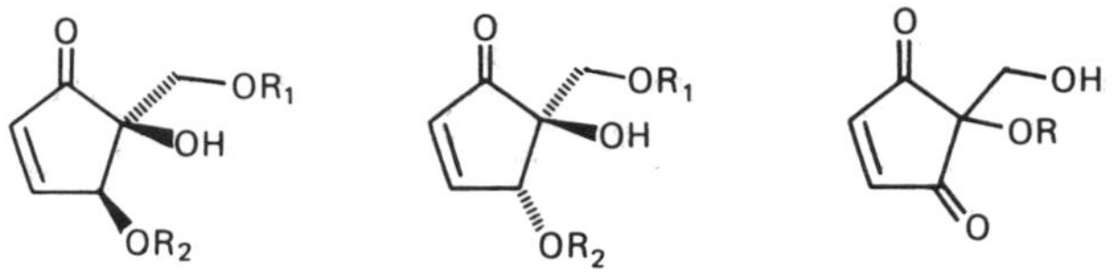

66   **67**   **68**

a: $R_1 = R_2 = H$
b: $R_1 = Ac$; $R_2 = H$
c: $R_1 = H$; $R_2 = Ac$

suffice it to say here that our general interest in the cyclopentanoid class of antibiotics, in conjunction with the ready availability of large quantities of **54**, made irresistible the temptation to undertake the synthesis of the pentenomycins (**66a–c**), their epimers (**67a–c**), and dehydropentenomycin (**68**).

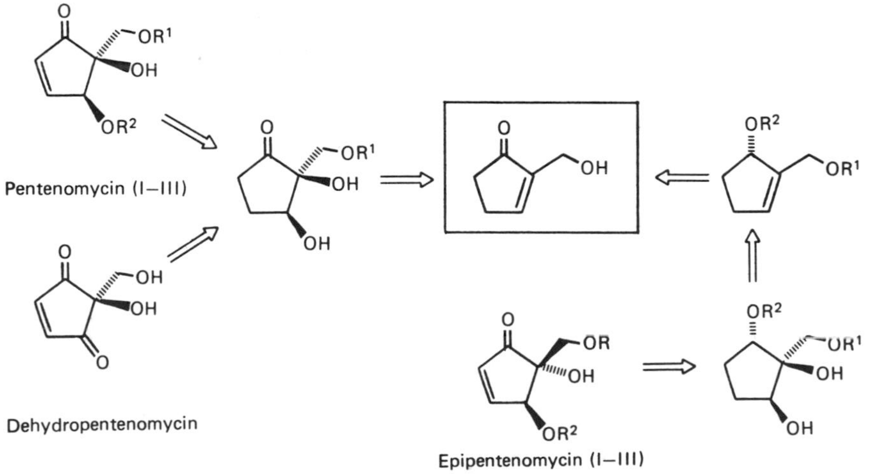

Pentenomycin (I–III)

Dehydropentenomycin

Epipentenomycin (I–III)

The successful conclusion to this venture, employing the synthetic analysis illustrated on page 242, has recently been recorded in detail.[3]

IX.   Return to the Jatrophone Problem: A Successful Failure

With a viable, large-scale preparation of **54** available, the stage was now set for a concerted effort toward completion of Phase II, synthesis of nor-methyljatrophone. The requisite α-hydroxy ketone (**55**) for phase II was prepared by treatment of the *t*BDMS ether of **54** with lithium ethyl dithiane, followed by hydrolysis; the yield was 59%. Subsequent generation of the dianion (2 equiv of LDA–THF), followed by addition of the same imidazolide (**50**) employed in phase I and acidic workup, led, *albeit in very low yield* (∼10%), to the desired 3(2*H*)-furanone (**69**), again as a mixture of epimers. Hydrolysis of the silyl ether functionality and oxidation with pyridinium chlorochromate afforded ketoaldehyde **22** as a beautifully crystalline solid (mp 91.5–92.5°C).

Thus, although we had formally achieved the first major objective in Phase II, namely the synthesis of **22**, the yield for the three-step sequence (**55** → **22**) was completely unacceptable. The only recourse at this point was to reconsider the approach to the 3(2*H*)-furanone ring system.

X. Development of An Alternative
3(2*H*)-Furanone Protocol:
A New 1,3-Diketone Synthesis

From the outset, our strategy for construction of the 3(2*H*)-furanone system involved reaction of a ketone enolate with an electrophile in the oxidation state of a carboxylic acid. There appeared, however, to be no reason that an aldehyde could not be employed for the initial carbon–carbon bond formation step, since at the time it was well known that low temperature aldol condensations proceeded in high yield.[48] Subsequent oxidation of the resultant β-hydroxy ketone **70** would then afford the desired 1,3-diketone **71**.[49] Furthermore, if the oxidation were to be carried out under acidic conditions, it might be possible in "one pot" to effect direct cyclization to the desired 3(2*H*)-furanone.

To explore the validity of this scenario, ketone **45** was treated with 1.1 equivalents of LDA, followed by condensation at low temperature with a variety of

45	**70**	**71**

aldehydes (**72a–h**). The result was β-hydroxy ketones **70a–h** (see Table III). Yields, in general, were good to excellent, except in the case of easily enolizable aldehydes (**72e–h**), where conversions were only modest (∼58%). Marked improvement here was realized by exploiting the zinc enolate prepared according to the method of House.[50,51]

Subsequent oxidation of the aldol products (**70a–i**) with Jones reagent[52] resulted both in formation of the diketone and in cyclization–dehydration to afford the 3(2*H*)-furanone. Best results, however, were obtained when oxidations were effected using Collins reagent,[53] followed in a subsequent step by treatment of the derived 1,3-diketone with mild aqueous acid to effect cyclization–dehydration. *This two-step protocol for the elaboration of 1,3-diketones holds considerable promise, we believe, as a general synthetic method:* Indeed, we have recently prepared no less than twenty-two 1,3-dicarbonyl derivatives via this method.[6b]

TABLE III
SYNTHESIS OF SIMPLE 3(2H)-Furanones

Entry	Ketone	Aldehyde RCHO 72	Aldol 70 (% yield)	1,3-Diketone 71 (% yield)	3(2H)-Furanone (% yield)
a	Me₃SiO **45**	CH₂=CHCHO	89[a]	57	50
b	45	CH₂=C(Me)CHO	92[a]	60	60
c	45	tBuMe₂SiO...CHO H	98[a]	90	43
d	45	PhCHO	100[a]	95	78
e	45	EtCHO	81[a]	86	66
f	45	n-BuCHO	77[b]	79	45
g	45	Me(CH₂)₅CHO	93[b]	80	62
h	45	PhCH₂CHO	86[b]	67	50
i	Me₃SiO **45a**	PhCH₂CHO	82[a]	93	20

[a] Lithium enolate employed.
[b] Zinc enolate employed.

Several additional comments specific to the 3(2H)-furanone protocol are in order. First, the generality of this approach to furanones disubstituted in the 2-position is illustrated by the variety of systems prepared. Even the very unstable 3(2H)-furadienones (Table III; entries a–c) could be prepared in useful, albeit modest, yields. Furthermore, the method depends only on the availability of the appropriate aldehyde.

XI. Exploitation of the Improved 3(2H)-Furanone Protocol

Having devised an alternative 3(2H)-furanone synthesis, we returned to the jatrophone problem. For preparation of normethyljatrophone, the disilylated derivative of hydroxy acid **52** (i.e., **73**), prepared in connection with our synthesis of imidazolide **50**, was reduced with LiAlH$_4$ and, in turn, oxidized with Collins reagent[53] to yield the desired aldehyde **74**.

Condensation with protected hydroxy ketone **75** then afforded aldol **76** in nearly quantitative yield. Initial attempts, however, to effect the oxidation, acid-catalyzed deprotection, and dehydration–cyclization in one step (i.e.,

76 → 22), not unexpectedly, failed. On the other hand, careful oxidation with the prescribed amount of Collins reagent (i.e., 6 equiv), followed by acid-catalyzed dehydration–cyclization, led to **78**; the yield was still only modest. An extremely important observation was made however. In particular, both the β-hydroxyl group and the allylic silyl ether had undergone oxidation. This observation suggested that utilization of 12 equivalents of Collins reagent, instead of the usual 6 equivalents, followed by acid-catalyzed cyclization–dehydration, might lead to an improvement. Such proved to be the case! In particular, a 74% isolated yield of hydroxy aldehyde **78** was obtained. Removal of the silyl group in **77** and oxidation of the secondary acetylenic alcohol afforded ketoaldehyde **22**, identical in all respects to that prepared previously. Significantly, the overall yield from **74** was 56% after purification.

XII. Macrocyclization: A Difficult Step

With both an efficient and *convergent* approach to ketoaldehyde **22** available, the stage was set to explore the all-important aldol cyclization of **22** to **23**. Subsequent stereocontrolled reduction of the acetylenic linkage to a trans olefin would then afford normethyljatrophone (**2**).

| 22 | 23 | Normethyljatrophone 2 |

Unfortunately, all attempts to effect *direct* aldol cyclization of **22**, employing a wide variety of acidic or basic protocols (∼35 different conditions!), proved fruitless. Several observations are particularly noteworthy. First, ketoaldehyde **22** was found to be extremely stable to acid treatment; even concentrated H_2SO_4 at 0°C for several hours resulted in a nearly quantitative recovery of **22**. On the other hand, when **22** was subjected to a variety of strong and/or weak bases under either protic or aprotic conditions, only destruction of **22** resulted, with no discernible products being formed.

At this point let us consider the events required for the successful cyclization of **22**. First, chemospecific generation of the enol or enolate of the C-7 ketone, and not the aldehyde, is required. Second, the enolate (enol) must undergo *irreversible* addition to the aldehyde.

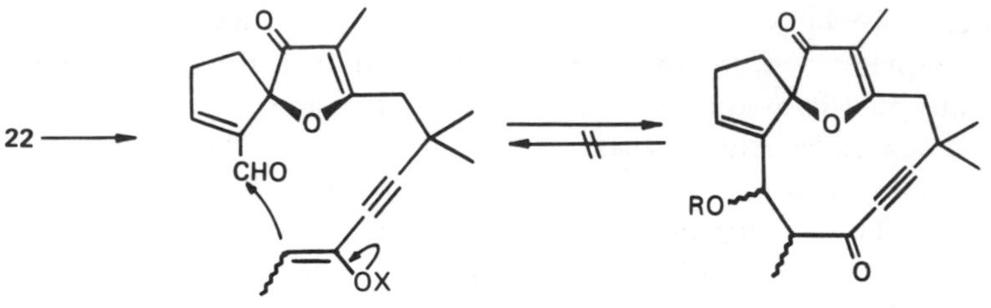

22 ⟶

To control chemoselectively the generation of the ketone enolate, in the presence of the aldehyde functionality, we initially selected the "Reformatsky-like" aldol cyclization conditions introduced by Yamamoto.[54] Toward this end, bromoaldehyde **80** was prepared from hydroxy aldehyde **78**. The optimal bromination agent proved to be 2-carboxytriphenylphosphonium perbromide, introduced by Ramage for selective α-bromination of ketones.[55] In our case, to avoid extensive acid-catalyzed decomposition, it was necessary to employ $CaCO_3$ to remove the HBr produced in the reaction. When this precaution was taken, a 64% yield of **80** was obtained.

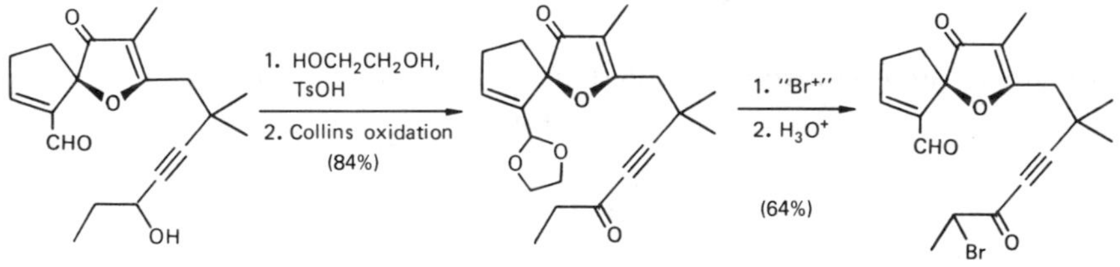

1. $HOCH_2CH_2OH$, TsOH

2. Collins oxidation
(84%)

1. "Br⁺"

2. H_3O^+

(64%)

78   **79**   **80**

Subjection of bromo aldehyde **80** to the Yamamoto conditions of Et_2AlCl and activated Zn led to enolate formation, followed by closure of the macrocyclic ring. The yield of this transformation (**80** → **81**), although not maximized, was 23%.

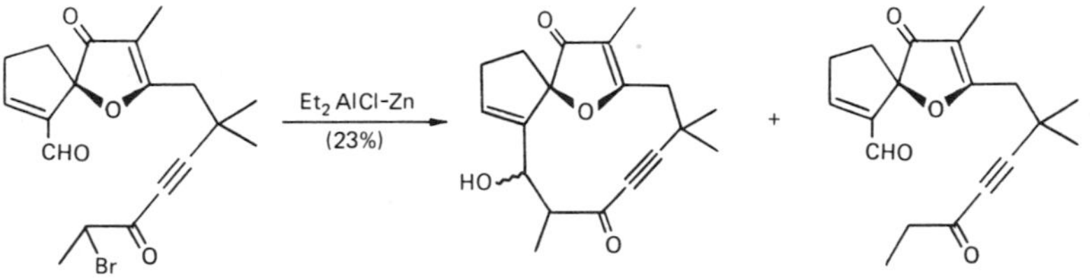

Et_2AlCl-Zn
(23%)

+

80   **81**   **22**

While successful, the above protocol did not eliminate the possibility of reversal of the aldol reaction; the latter, of course, would contribute to the observed low yield. In fact, ketoaldehyde **22** was isolated as the major side product. This observation suggested that a reverse aldol process was, in fact, occurring under the Yamamoto conditions. To prevent such a reversal, we reasoned that the free β-hydroxy group of the aldol product must in some way be captured. Ideal in this regard would be the availability of an internal protecting or trapping agent. Such a stringent set of requirements appeared to be fulfilled by the TiCl$_4$-catalyzed condensation of acetals with enol silyl ethers, a reaction introduced by Mukaiyama.[56] Interestingly, only one

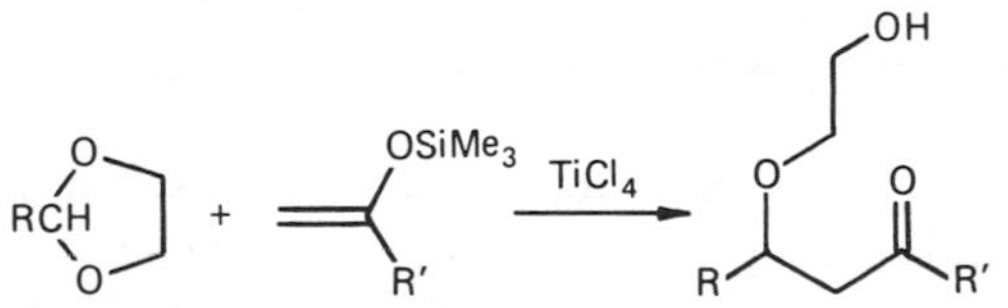

example of the intramolecular version of this crossed-aldol strategy was on record at that time; the example was that of Posner shown below.[57,58]

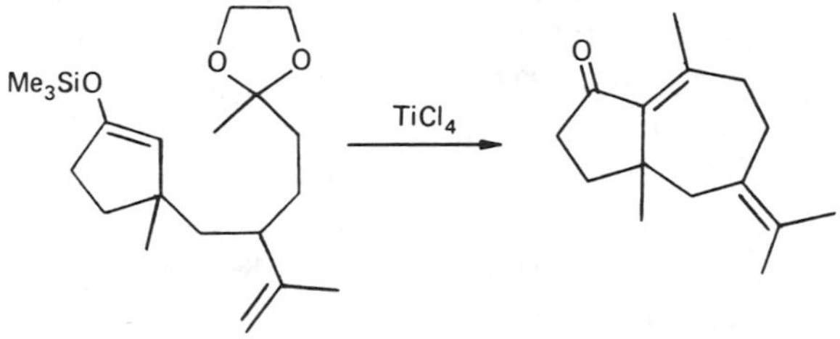

In our case, treatment of enol silyl ether **82**, prepared from ketoacetal **79**, with TiCl$_4$ led to two diasteromeric products **83a** and **83b** (47%, 2:1);

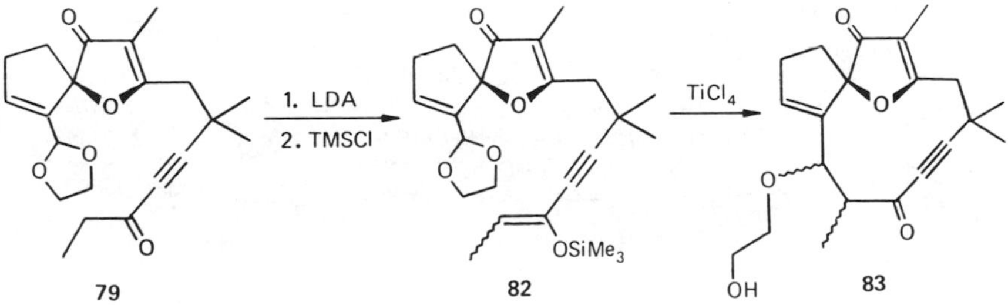

79

82

83

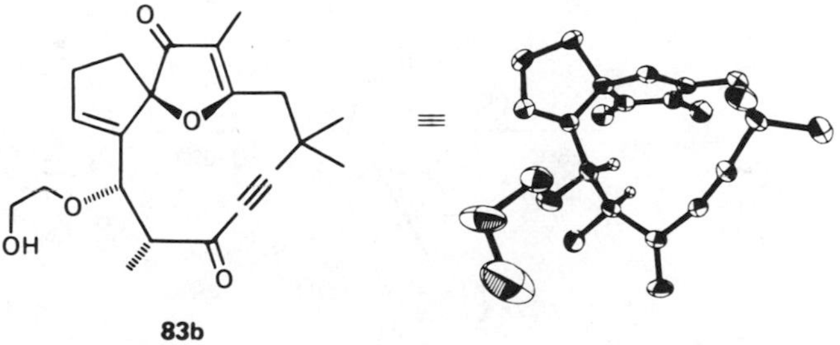

83b

the major product (**83b**) proved to be crystalline (mp 156–157°C).* To demonstrate beyond doubt that the macrocyclic ring was intact, we completed a single-crystal X-ray analysis; the result of that study is illustrated on page 249.[60]

XIII. Completion of Phase II: Synthesis of Normethyljatrophone

At this point, all that remained to complete a synthesis of normethyl-jatrophone (**2**), and thereby phase II in our jatrophone program, was elimination of the elements of ethylene glycol and conversion of the acetylene to a trans olefin. Toward this end, both the major and minor isomers (**83a** and **83b**) were subjected to acid-catalyzed elimination of ethylene glycol; *p*-TsOH in benzene proved to be the method of choice. Both cases afforded the same crystalline solid (mp 185–186°C) in 75–80% yield. *There was, however, a problem.* The high field (250 MHz) ^{1}H-NMR spectrum of the elimination product revealed a substantial downfield shift for the C-5 vinyl proton. This result was clearly *inconsistent* with the chemical shift expected on the basis of analogy to jatrophone (δ 5.80 versus δ 7.20 in $CDCl_3$). In fact, it suggested that we had generated the less stable *E* isomer (**84**). X-Ray crystallographic analysis demonstrated this to be the case.[60] *Significant for our purposes, furanone*

---

* In connection with a pending synthesis of jatropholones A and B (see p. 235), we have exploited the intramolecular Mukaiyama strategy for preparation of bicyclic enone **A**. The synthesis of **A** is outlined below.[59]

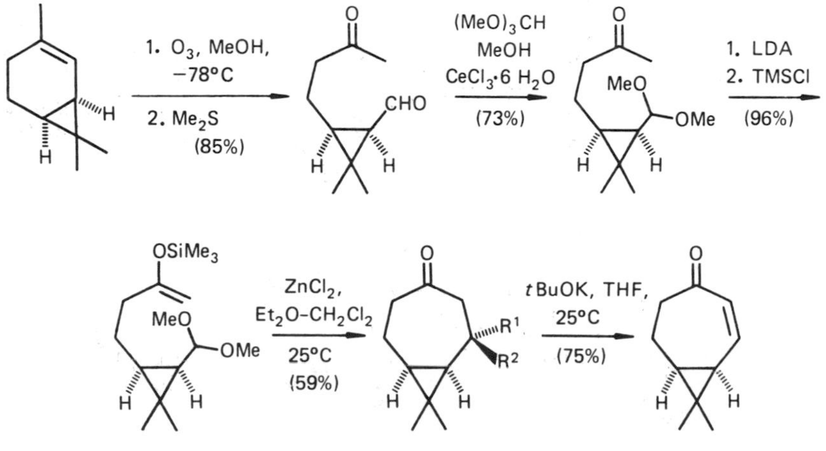

R¹ = H, R² = OMe
R¹ = OMe, R² = H

A

(**84**) *was found to undergo slow isomerization to the more stable* Z *isomer* (**23**) *upon prolonged exposure to the above elimination conditions* (*2 weeks, 88%*).

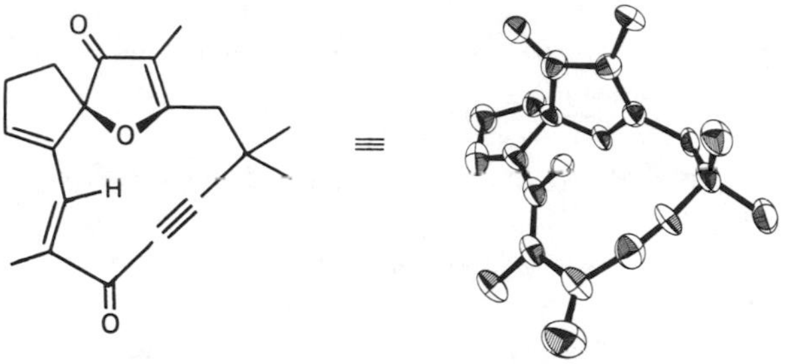

84

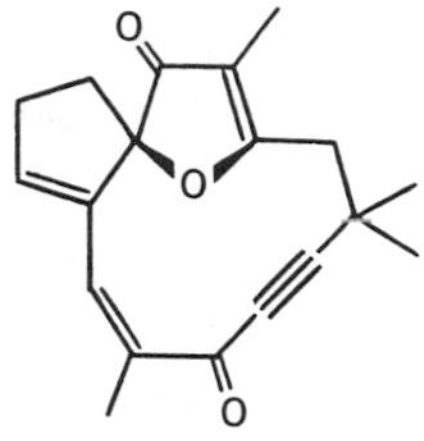

23

At this point, we had arrived at the final transformation necessary to complete the synthesis of normethyljatrophone, namely, reduction of the acetylene functionality to a trans olefin. From the outset we had planned to exploit either LiAlH$_4$ or CrSO$_4$ to effect the desired conversion. Both strategies, however, were thwarted. In the case of the LiAlH$_4$ reduction, the 3(2H)-furanone carbonyl proved to be a major problem. Simultaneous reduction of the latter *per se* would not invalidate the strategy, in that oxidation would in the end regenerate the 3(2H)-furanone functionality. We found, however, that the resultant allylic alcohol system could not be prevented from undergoing a 1,3-hydroxyl rearrangement. Such a result was devastating to the

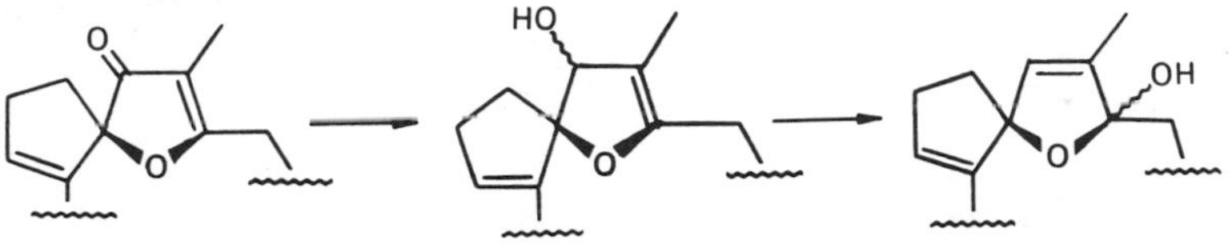

jatrophone strategy. The alternative protocol, that of treatment of **23** with chromous sulfate, led to transannular bond formation. It should be noted here that we had by this time explored the chromous sulfate reduction of furanone **85**, prepared in connection with Phase I. The results of that study, illustrated below, indicated that **86** was by far the major product. This observation

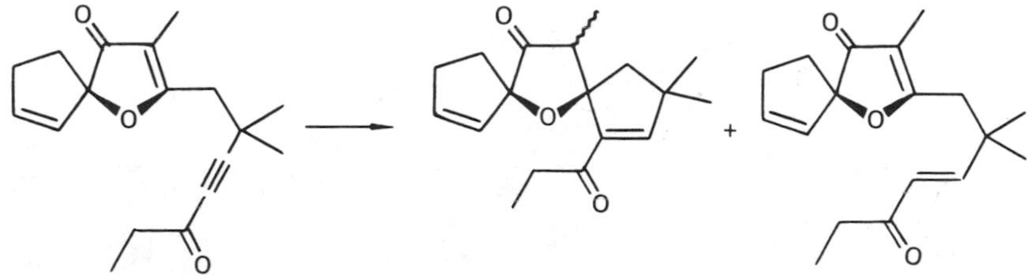

85

86

87

strongly suggested that in systems wherein the furanone ring and acetylene functionality were confined to a macrocyclic ring (as in **23**) only transannular bond formation would occur.*

Fortunately for our purposes, **23** could be selectively cis hydrogenated (PdSO$_4$–pyridine)[61] to the cis-Z isomer **88** and, in turn, isomerized through aegis of KI–AcOH at room temperature to **2**, a beautifully crystalline solid (mp 135–136°C). That normethyljatrophone (**2**) was in hand derived from careful comparison of the 250-MHz ^{1}H-NMR spectral data with that of an authentic sample of jatrophone as well as by completion of a single-crystal X-ray analysis. The result of the latter is illustrated on page 253.[60]

To summarize at this point, the total synthesis of normethyljatrophone (**2**), and thereby completion of phase II as outlined in Scheme 3, had been achieved. The overall yield for the 16 steps, beginning with cyclopentenone, was 5.6%.

---

* Three additional examples of this interesting cyclization process have been uncovered recently with substrates that were readily available in our laboratory. To date, the yields indicated below have not been maximized.

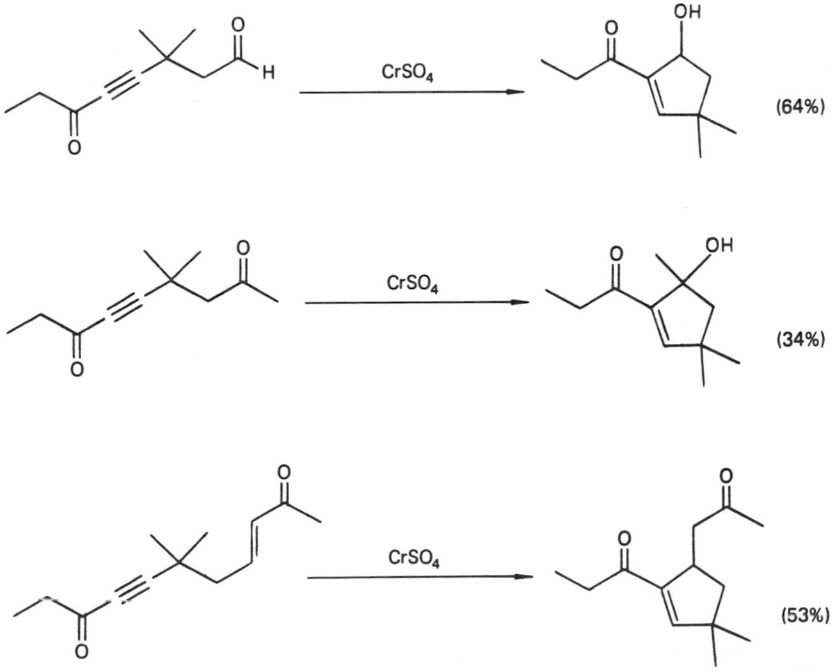

(64%)

(34%)

(53%)

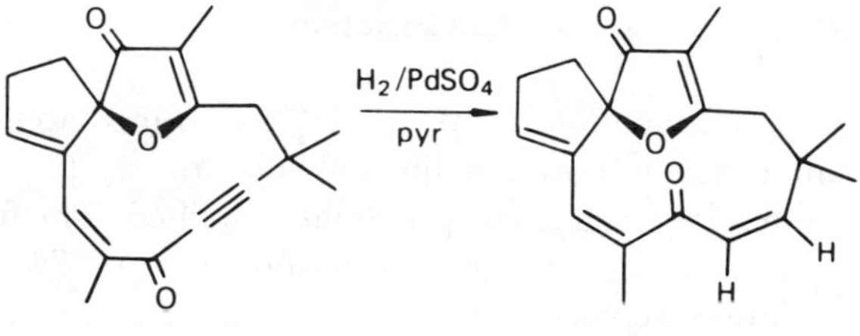

23 88

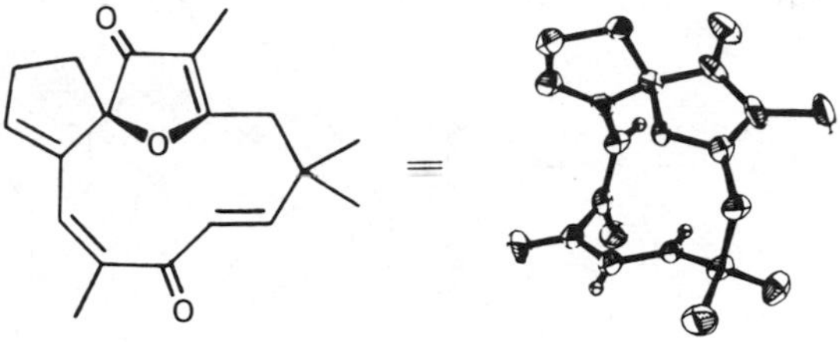

2

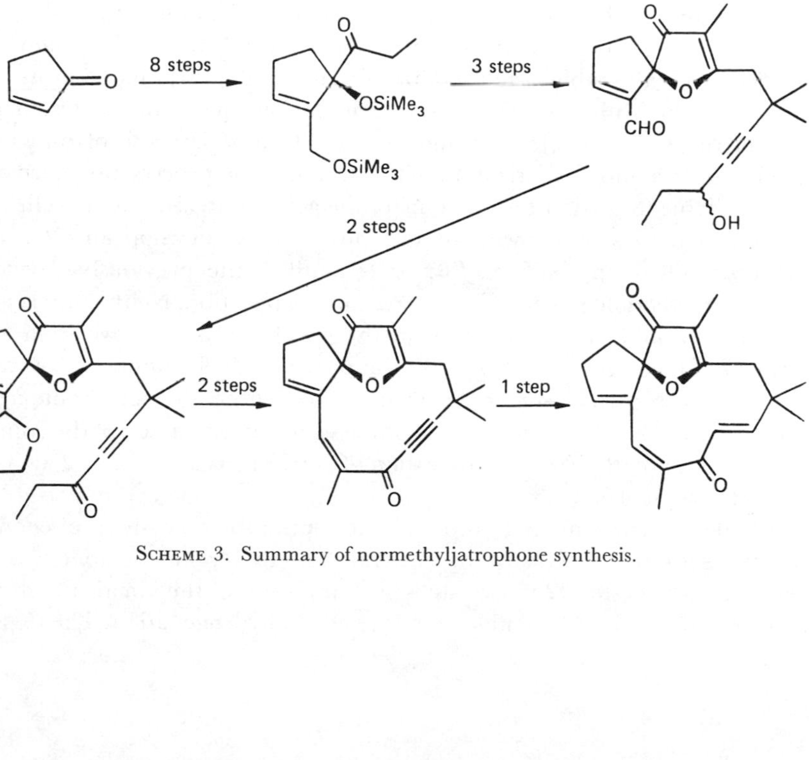

SCHEME 3. Summary of normethyljatrophone synthesis.

XIV. Phase III: The Stereochemical Problem

With a successful conclusion to phase II, we immediately turned our
attention to exploiting the normethyljatrophone strategy for the total synthesis of jatrophone. Two significant problems remained. The first related to
an effective source of α-hydroxymethylcyclopentenone **89**. The second
concerned the relative stereochemistry of the C-2 secondary methyl group
vis-a-vis the furanone oxygen of jatrophone (**1**). From the outset we envisioned
that addition of lithium ethyl dithiane to enone **89** would afford a diastereomeric mixture of **90** and **91**. Furthermore, we could anticipate that **91** would
predominate. That is, molecular-model studies suggested that attack of the

dithiane would occur *anti* to the β-methyl group to yield the undesired isomer
(i.e., **91**). The preference however was anticipated to be small.

A further complicating feature in the application of the normethyl strategy
to jatrophone concerned the possibility that epimerization of the C-2 methyl
substituent might occur at an advanced stage in the synthesis. Two steps were
particularly vulnerable: the acetalization of the aldehyde function in preparation for the Mukaiyama cyclization and the final acid-catalyzed isomerization of the C-5–C-6 olefin. While we could anticipate use of quite mild
conditions to avoid epimerization in the acetalization process, no information
was available to predict the outcome of the acid-catalyzed macrocyclization.

From a tactical sense we had two options: (A) develop an alternative
stereocontrolled approach to **90**; or (B) utilize the previously developed
dithiane methodology and effect the required separation. Neither alternative,
of course, would have any effect on the possibility of epimerization at an
advanced stage of the synthesis. On the positive side for option B, the availability of the incorrect isomer (**91**) would allow synthesis of epijatrophone (**3**).
Furthermore, from an operational point of view, intermediates of the epimeric
series would greatly aid in the detection of any isomerization at C-2 along the
synthetic sequence.

The decision was made to proceed with option B. A synthesis of enone **89**
was therefore required. Toward this end, 3-methylcyclopentenone was converted to α-ketoester **92** in two steps: (a) acylation of the kinetically derived
enolate with N-carbethoxyimidazolide; and (b) hydrogenation. The required

precedent for generation and alkylation of the kinetic enolate of 3-substituted cyclopentenone derivatives derived from previous work of our laboratory.[62]

Conversion of **92** to its selenide, followed by Sharpless–Reich oxidative elimination,[63] provided enone **93** in 82% overall yield, which, in turn, was subjected to ketalization with ethylene glycol. Here conventional protocols, employing as catalysts fumaric or adipic acid, resulted in partial migration of the double bond to the α',β'-position. To circumvent this rather annoying problem, we employed the aprotic ketalization conditions introduced by Noyori.[64] In particular, stoichiometric quantities of the bistrimethylsilyl ether of ethylene glycol and a catalytic amount of trimethylsilyltrifluoromethyl sulfonate led in high yield (73%) to the desired ketal **94**.

Reduction of **94** with 2 equivalents of DIBAL, followed by hydrolysis of the ethylene ketal protecting group with aqueous adipic acid in CH_2Cl_2, afforded hydroxymethylcyclopentenone **89**. The overall yield of **89** from 3-methyl-cyclopentenone was excellent (21% for 8 steps).

More recently, we have devised a greatly improved synthesis of **89** from **96**, the latter readily available from 1,3-cyclopentane dione **97**.[65a] In this case, alkylation of the kinetically derived enolate of **96** with methyl iodide, followed by $LiAlH_4$ promoted 1,2-reduction of the vinylogous ester functionality and subsequent acid-catalyzed hydrolysis, led to **89** in 45% overall yield.[65b]

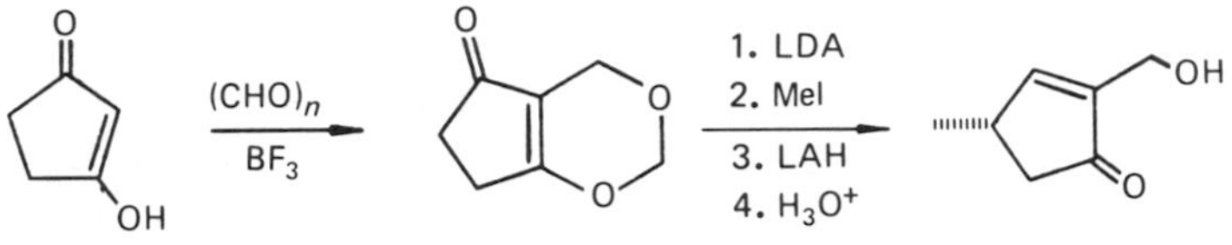

With ample quantities of **89** in hand, the derived *t*BDMS ether (**95**) was treated with lithioethyl dithiane, followed by alkylative hydrolysis, to afford a diastereomeric mixture of adducts (**98** and **99**) in 54% yield. The stereochemistry of these adducts, however, could not be conclusively defined via

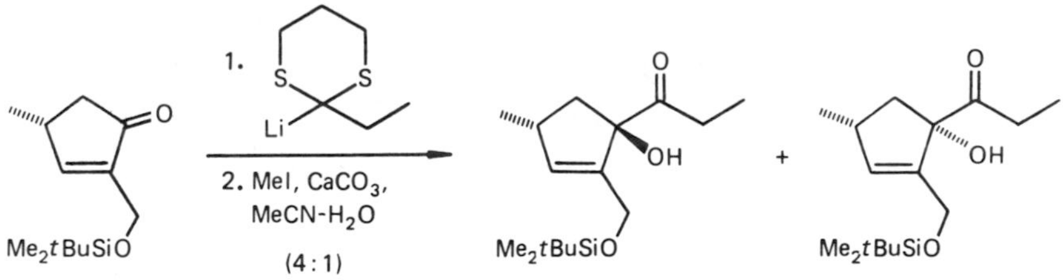

conventional spectroscopic techniques. Instead, the major isomer was converted to diol **100**, the primary hydroxyl group derivatized with *p*-bromo-benzoyl chloride, and the resultant crystalline benzoate (**101**) subjected to X-ray analysis; the result of that study is illustrated below.[60]

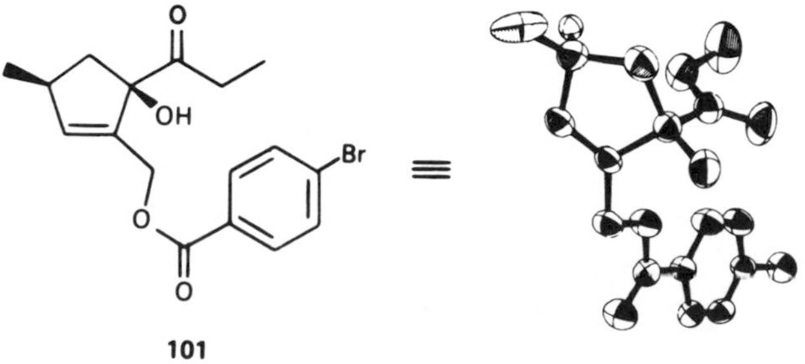

As anticipated, the preferred mode of addition was trans to the secondary methyl substituent; that is, to yield the incorrect diastereomer (**99**) vis-a-vis jatrophone. *While not surprised by the preferred mode of addition, the high degree of selectivity was clearly not anticipated!* The immediate consequence of this fact would be to impose a severe material bottleneck on the further progress of the jatrophone synthesis.

To circumvent this stereochemical problem, we required an effective method to invert either C-2 or C-15 of the major isomer (i.e., **99**). While at first sight such a transformation does not appear feasible, further reflection suggested the series of reactions illustrated below. In particular, treatment of

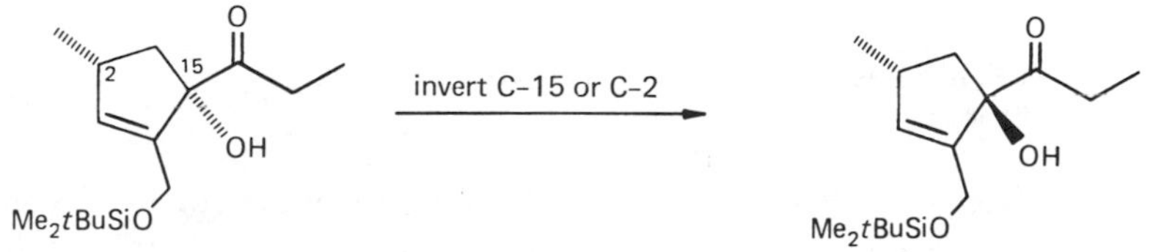

hydroxy ketone **99** with phenylsulfenyl chloride was anticipated to react initially with the alcohol substituent, followed immediately by a [2,3]-sigmatropic rearrangement to afford sulfoxide **102** (i.e., Mislow–Evans allyl sulfenate–sulfoxide rearrangement).[66] Taking advantage of the fact that the phenyl sulfoxide and methyl groups are now vicinal on a cyclopentene ring, base-catalyzed equilibration was expected to epimerize the center α to the sulfoxide group, with the trans isomer (i.e., **103**) greatly predominating over the cis isomer.[67] Subsequent thermal reversal of the allyl sulfoxide–sulfenate equilibration in the presence of a nucleophile with a high preference for sulfur, [(i.e., $(EtO)_3P$)], would then lead to the desired isomer.[66] It is important to note that this series of reactions need not be carried out on the pure diastereomer (i.e., **99**) but could in principle be applied to the diastereomeric mixture.

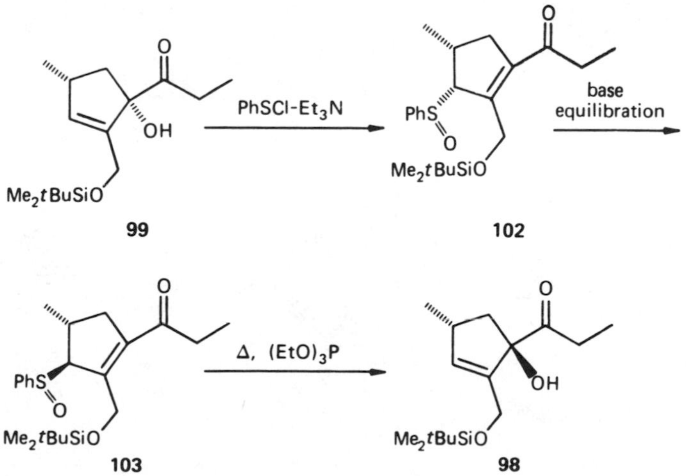

In the event, treatment of the mixture of diastereomers **99** and **98** (3:1, respectively) with phenylsulfenyl chloride (triethylamine, ether, −10°C) led to a mixture of sulfoxides (**102** and **103**). Without purification, equilibration with either potassium carbonate or diazabicyclo[5.4.0]undecene, followed by addition of triphenyl phosphite, led to a diastereomeric mixture in which the ratio was now 1:3. Needless to say, the improved ratio of **99** to **98** was a very welcome result.

XV. Total Synthesis of Jatrophone and Its Epimer: The Final Goal

Having devised an effective stereocontrolled synthesis of both diastereomers (**98** and **99**), we were now in position to complete the synthesis of jatrophone as well as its epimer. All that was required, assuming the integrity of the C-2 methyl substituent, was to repeat the normethyljatrophone sequence with both diastereomers. Chronologically, epijatrophone (**3**) was prepared first, due to the initial availability of **99**. Both sequences proceeded essentially as experienced in the normethyljatrophone series. Significantly, the stereochemical integrity of the C-2 center was maintained throughout both the synthesis of epijatrophone and jatrophone. For the sake of brevity here, the overall sequence requires 16 steps and proceeds in 1.2% overall yield from

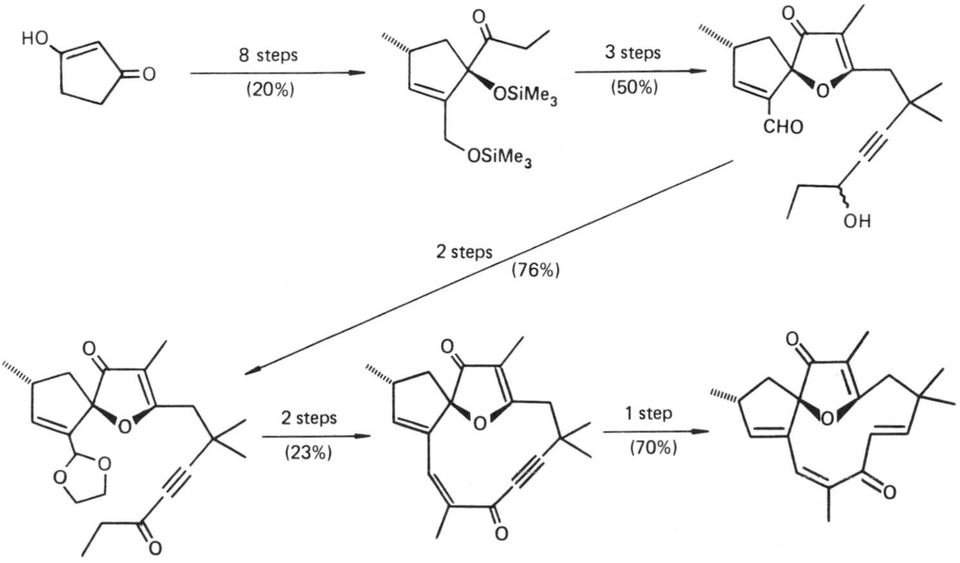

SCHEME 4

cyclopentane-1,3-dione (see Scheme 4). Synthetic ($\pm$)-jatrophone, prepared in this manner, was shown to be identical to an authentic sample of ($+$)-jatrophone via all the usual spectroscopic techniques. For the record, the high field ^{1}H-NMR (250 MHz) of natural and synthetic jatrophone are reproduced below (Fig. 1).

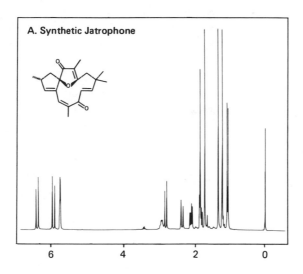

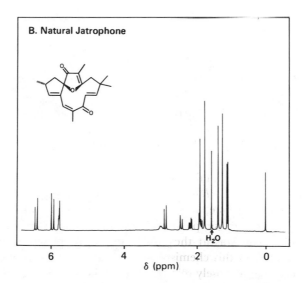

FIG. 1. Synthetic (A) and natural (B) jatrophone.

XVI. Further Diversions during the Jatrophone Synthetic Program

During the evolution of our jatrophone strategy, we had occasion to explore the chemistry of a number of simple $3(2H)$-furanones[6] as well as prepare, for the first time, geiparvarin (**5**),[4] its configurational isomer (isogeiparvarin **121**), and four synthetic analogs of the jatrophone molecule [i.e., cis- and trans-normethyljatropholactone **7** and **8**,[5] normethyljatrophone (**2**),[2] and epijatrophone (**3**)].[2] In addition, on the basis of a lead obtained by the Kupchan

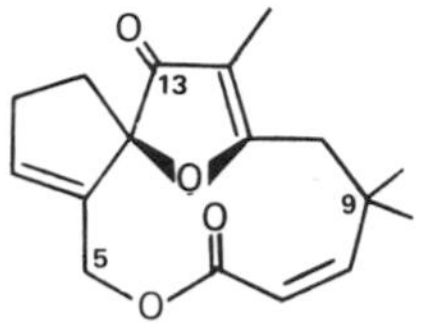

7

cis-Normethyljatropholactone

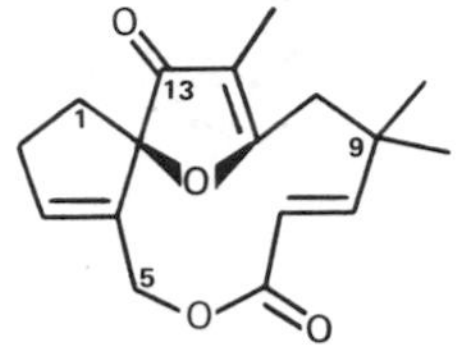

8

trans-Normethyljatropholactone

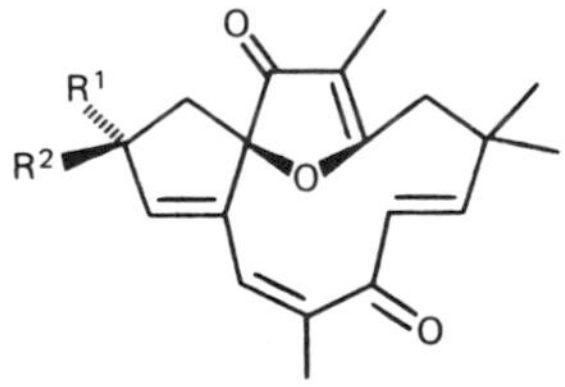

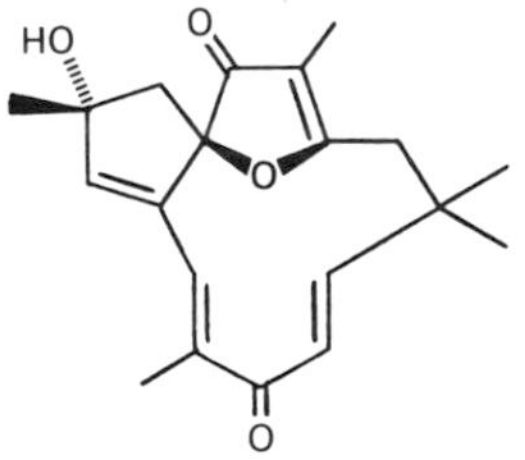

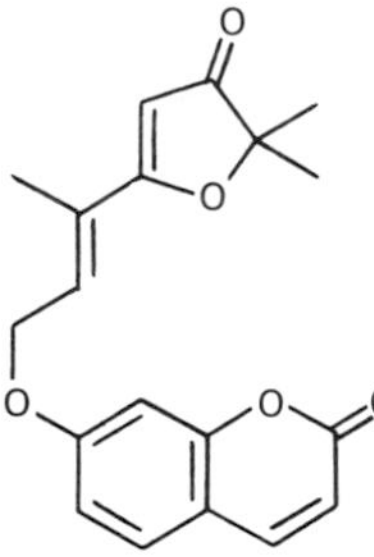

9: R^1 = OH, R^2 = CH$_3$
10: R^1 = CH$_3$, R^2 = OH

11

5

laboratory during the original jatrophone isolation studies, we in collaboration with Professor Cordell at the University of Illinois-Chicago Circle Campus, reexamined *Jatropha gossypiifolia* for related natural products. This effort led to the identification of three new jatrophone derivatives, which we termed hydroxyjatrophones A, B, and C (**9–11**).[8]

Our interest in simple $3(2H)$-furanones, as well as closely related derivatives of the jatrophone molecule, was twofold. First, as mentioned in the introduction of this chapter, simple transformational chemistry of the $3(2H)$-furanone ring system was at the outset almost completely unexplored. A detailed knowledge of this chemistry was mandatory, given our interest in jatrophone and several closely related synthetic targets (i.e., geiparvarin and the eremantholides). Second, it appeared that such systems would be ideal substrates to model the mode of biological action. In particular, both jatrophone and the eremantholides were known to undergo facile conjugate addition of propanethiol to yield monoadducts **104** and **105**;[1,10] this reactivity has been proposed by Kupchan and Le Quesne, respectively, to be responsible for the observed antileukemic and cytotoxic activity of jatrophone and the eremantholides.[1,10] Details of similar reactions with simple $3(2H)$-furanones,

104   **105**

however, had not been investigated. Furthermore, the recent discovery in our laboratory of bis-β,β'-conjugate addition,[68] suggested that $3(2H)$-furanones, such as budlein A[69] or goyazensolide[70] each of which possesses a potential leaving group in the δ'-position, could undergo bis-δ,δ'-conjugate addition with an appropriate bionucleophile.[71] That is, conjugate addition of pro-

Budlein A   Goyazensolide

panethiol, followed by loss of H_2O, would lead to a new furanone (i.e., **107**), which in a subsequent step could accept a second equivalent of propanethiol to afford **108** via 1,6-conjugate addition. Such a transformation *in vivo* would permit cross-linking, a process that has been suggested to enhance activity of many antitumor agents.[72]

106   107   108

In the remainder of this chapter we will describe several of the diversions that relate to our efforts to define the possible mode of biological action of the antitumor agents that possess the $3(2H)$-furanone ring system.

XVII. Reactions of Simple $3(2H)$-Furanones and Furadienones with Propanethiol

To define the relative reactivity of the $3(2H)$-furanone ring system with potential bionucleophiles, we subjected model systems **109–112** to reaction with excess propanethiol; acidic, neutral, and basic reaction conditions were explored. In addition to defining the degree of reactivity of the $3(2H)$-furanones, we were particularly interested in establishing the feasibility of bis-δ,δ'-conjugate addition with a furadienone, such as **111**. Our results are illustrated in Table IV.

Several comments concerning this study are in order. First, simple $3(2H)$-furanones and furadienones, such as **109** and **110**, respectively, undergo 1,4- and 1,6-conjugate addition of propanethiol under acidic and basic conditions, *but not* under neutral conditions. The reactivity of furadienone **110** with propanethiol supports the suggestion of Le Quesne[10] that the γ position of eremantholide A serves as the electrophilic site for bionucleophiles. *Of considerable significance in this regard is the fact that our model $3(2H)$-furadienone* **110** *was shown by the NCI to possess presumptive antitumor activity.*[73]

The reactivity of furadienone **111**, a candidate for bis-δ,δ'-conjugate addition, was found to be quite dependent on the reaction conditions: protocols employing basic conditions afforded only monoadduct **115**, whereas reactions effected under neutral or acidic conditions led only to bisadduct **116**. (Sulfoxide **117** was shown to arise via air oxidation of **116**). Presumably, under basic conditions loss of water is inhibited, thereby preventing addition of a second equivalent of propanethiol. Alternatively, under acidic or neutral conditions loss of water is not precluded, the result being generation of a new furadienone system (i.e., **120**) capable of accepting a second equivalent of

TABLE IV

REACTIONS OF SIMPLE 3(2H)-FURANONES AND FURANDIENONES WITH PROPANETHIOL

Entry	Substrate	Reaction conditions	Product	Yield (%)
1 (a)	**109**	Benzene–iPr$_2$NH	**113**	54
(b)		Benzene		NR
(c)		Benzene–TsOH		48
2 (a)	**110**	Benzene–iPr$_2$NH	**114**	42
(b)		Benzene		NR
(c)		Benzene–TsOH		70
3 (a)	**111**	Benzene–Et$_3$N or borate buffer (pH 9)	**115**	62
				57
(b)		Benzene	**116**	32
			117	22
(c)		Benzene–TsOH	**116**	60
4 (a)	**112**	Benzene	**118**	32
(b)		Benzene–TsOH	**116**	58
(c)		Benzene–Et$_3$N	**116**	30
5 (a)	**5**	Benzene–iPr$_2$NH	**119**	40
(b)		Benzene–TsOH		47

propanethiol. We wish to emphasize that the latter observation demonstrates the feasibility of the bis-δ,δ'-conjugate addition process.

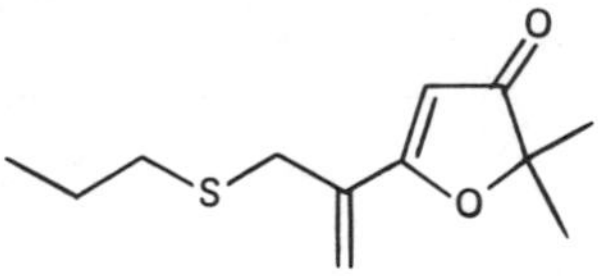

120

Finally, we examined the possibility that a free-radical mechanism may play a significant role in these reactions. To explore this possibility, we examined the addition of propanethiol to **111** and **112** in the presence of the free radical inhibitor 2,5-di-*tert*-butylhydroquinone.[74] No change in product

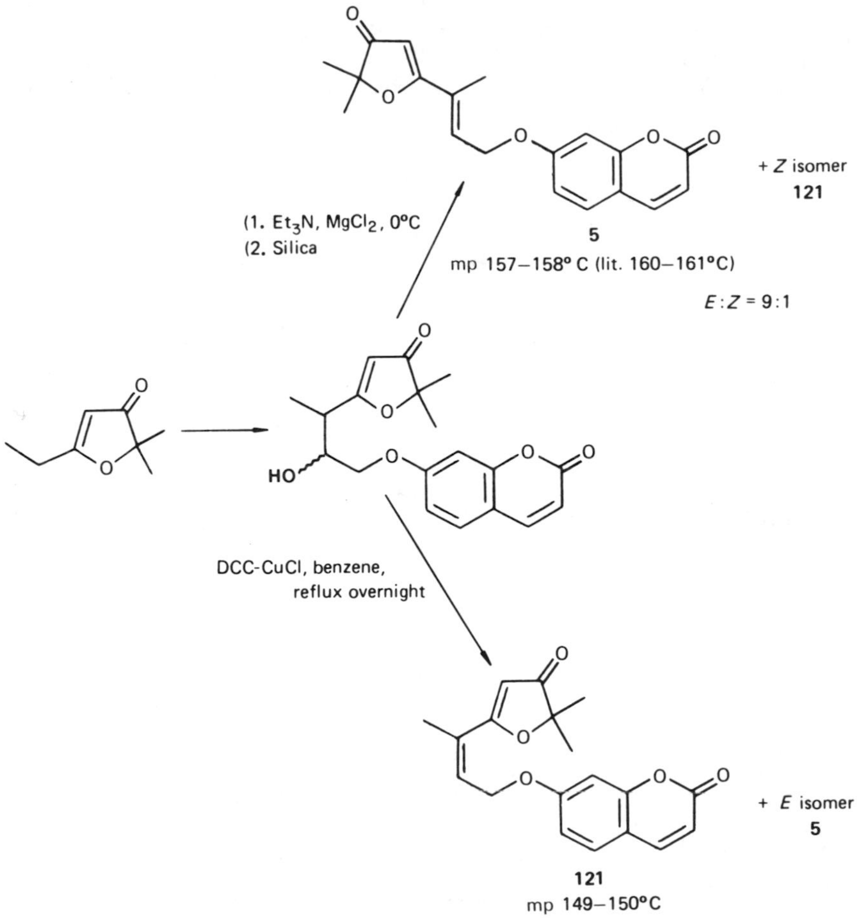

(1. Et₃N, MgCl₂, 0°C
(2. Silica

5
mp 157–158° C (lit. 160–161°C)

+ *Z* isomer
121

E : *Z* = 9 : 1

DCC-CuCl, benzene,
reflux overnight

+ *E* isomer
5

121
mp 149–150°C

formation occurred under the acidic and basic conditions, whereas conjugate addition was completely suppressed under the neutral conditions.

Before concluding this section, we wish to note that geiparvarin (**5**) (Table IV) has been shown to display significant activity in the NCI P-388 screen as well as *in vitro* activity against human sarcoma of the nasopharynx.[75] The fact that geiparvarin also reacts with propanethiol via 1,6-conjugate addition is noteworthy vis-a-vis the possible mode of action.

For the sake of completeness, we note here that the first total synthesis of both geiparvarin (**5**) and its configurational isomer, isogeiparvarin (**121**), were carried out in our laboratory during the jatrophone era.[4] An overview of this synthesis is illustrated on the previous page.

XVIII. Jatrophone Analogs: *cis*- and *trans*-Normethyljatropholactones

Having explored the reactivity of propanethiol with several *simple* 3(2*H*)-furanones as well as geiparvarin (**5**), we next turned to analogs of jatrophone. Our interest in *cis*- and *trans*-normethyljatropholactones (**7** and **8**) as possible substrates for such a study derived from the original Kupchan observation[1] that the principle reactive site in the jatrophone molecule was C-9.

From the point of view of structural activity relationships, it did not appear unreasonable to replace the α,β-unsaturated ketone system at C-5—C-6 with a saturated ester linkage in that such a modification would not preclude conjugate addition at C-9. Furthermore, the availability of both the cis and trans isomers would provide for the first time a measure of the importance of the ground-state conformation of the macrocyclic ring system on the course and/or rate of the conjugate addition process (*vide infra*).

Having established *cis*- and *trans*-normethyljatropholactones (**7** and **8**) as synthetic targets, acetylenic macrolide **122** appeared to be an appropriate advanced intermediate from which both isomers could be derived, assuming, of course, the availability of a stereocontrolled reduction protocol for conversion of the acetylenic linkage to the cis or trans olefin. Central to this strategy was the prospect of exploiting our 3(2*H*)-furanone synthetic protocol for construction of spirofuranone **123a**; subsequent macrocyclic lactonization would then afford **122**. Suffice it to say here that macrolide **122** could be obtained in 23% overall yield from **75** and **124**.[5] Closure of the macrocyclic ring (**123a** → **122**) was accomplished via the Mukaiyama[76] procedure (i.e., 1-methyl-2-chloropyridinium iodide–Et$_3$N–CH$_3$CN, high dilution). Semihydrogenation of the C-7—C-8 acetylenic linkage then afforded *cis*-normethyljatropholactone (**7**) as a beautifully crystalline solid. That in fact **7** was in hand derived from careful comparison of the high field ^{1}H-NMR

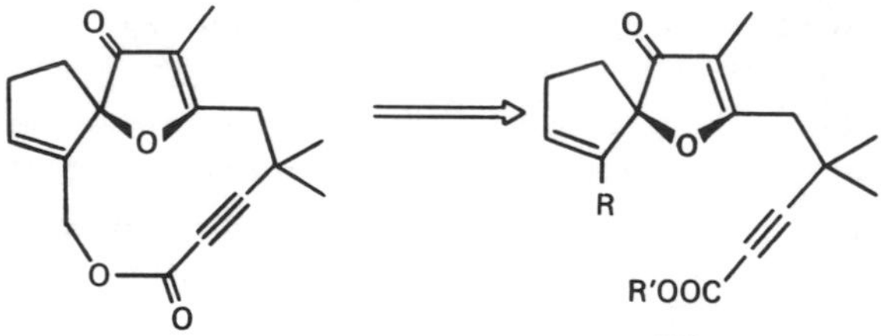

122

123

a: R = CH$_2$OH, R' = H
b: R = CH$_2$OH, R' = Et
c: R = CHO, R' = Et

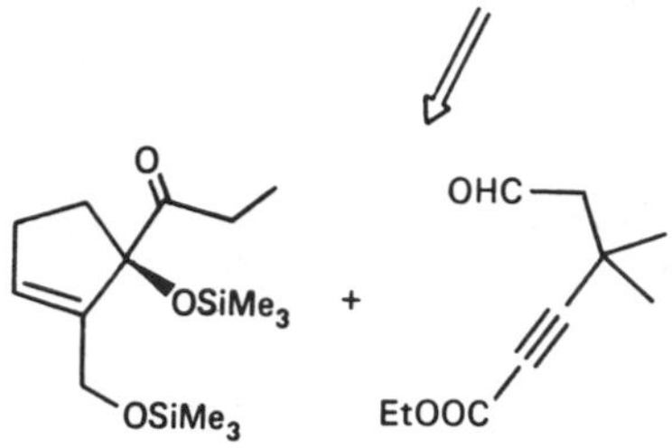

75   +   124

spectra of **7** with that of **88**, prepared in connection with our normethyl-jatrophone synthesis.

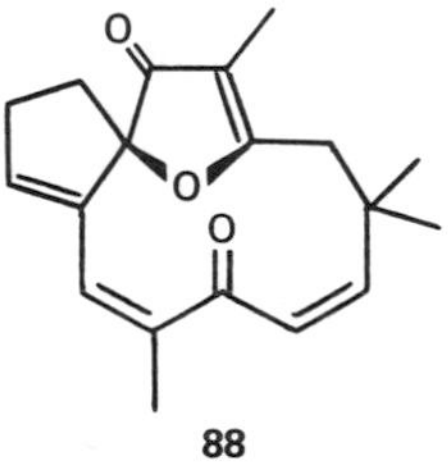

88 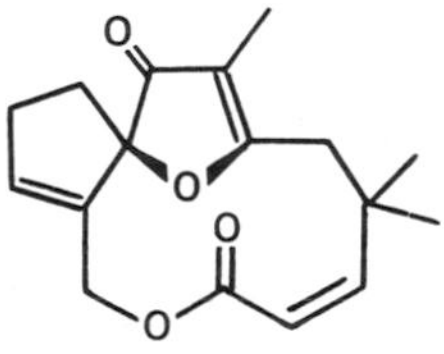   7

Turning next to the conversion of macrolide **122** to *trans*-normethyl-jatropholactone **8**, we had anticipated from the beginning that isomerization of *cis*-normethyljatropholactone would afford the desired trans isomer. Such a strategy was, in fact, exploited to great advantage in our jatrophone synthesis. Unfortunately, all attempts to effect the required isomerization of **7** by employing numerous acids, light (with and without I$_2$), and/or transition metal catalysts under a wide variety of time, temperature, and solvent regimes proved fruitless.

The only alternative at this point, aside from abandoning the goal of *trans*-normethyljatropholactone (**8**), was to prepare the requisite trans alde-hyde ester (**125**) and then to subject this intermediate to the 3(2*H*)-furanone synthetic protocol. Although the latter appeared straightforward, there was one major concern. It related to the propensity with which the obvious precursor of aldehyde **125** (i.e., **126a**) might undergo intramolecular cyclization and thereby preclude its utilization in the preparation of **125**. Such a concern was not without precedent. Indeed, in the early stages of development of the jatrophone strategy we had observed that the corresponding ethyl ketone (**126b**) could not be prevented from undergoing cyclization to **127b**.

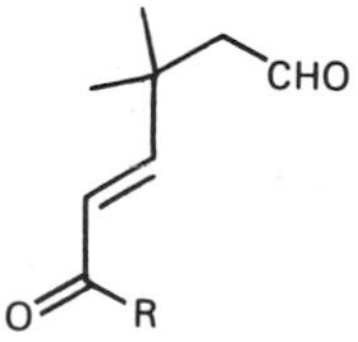

125: R = OEt 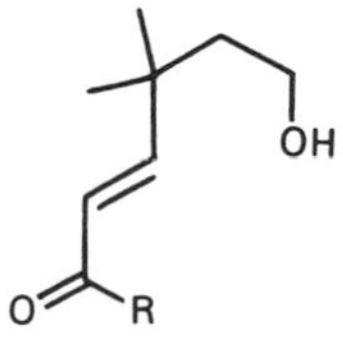   **126**   **127**

a: R = OEt
b: R = Et

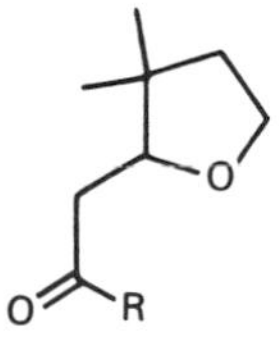

The latter, of course, is an example of an allowed 5-exo-trigonal cyclization.[77] Fortunately for our purposes, the corresponding trans hydroxy ester **126a** could be prepared and converted to the desired aldehyde with only a minor amount ($\sim 10\%$) of furan formation (i.e., **127a**).

With the preparation of aldehyde **125** secure, execution of the $3(2H)$-furanone synthesis, followed by macrolide cyclization, led to *trans*-normethyljatropholactone **8**. The overall yield from **75** was 19%. Confirmation that *trans*-normethyljatropholactone was in hand derived from completion of a single-crystal X-ray analysis.[60]

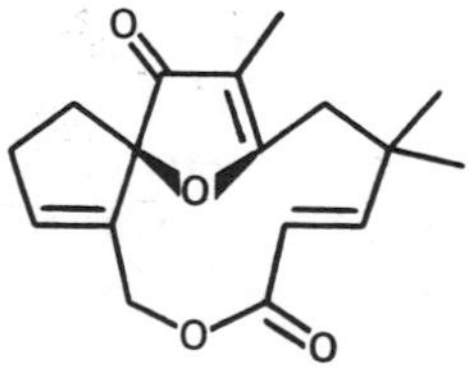

8 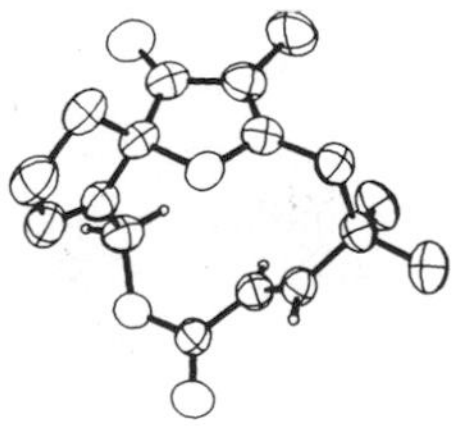   **8**

XIX. Isolation of Three New Natural Jatrophone Derivatives

As mentioned in the introduction, we, in collaboration with Professor Cordell, re-examined *Jatropha gossypiifolia* for related natural products. This effort led to the isolation and identification of hydroxyjatrophones A, B, and

C (**9–11**).[8] Structural assignments were based on high-field (5.89 T) NMR studies carried out at the University of Pennsylvania. In particular, extensive double resonance experiments, including two dimensional ^{1}H J-resolved experiments, were performed, which allowed virtually complete assignment of the ^{1}H- and ^{13}C-NMR spectra. The reader is referred to our full account of this work for the specific details of the NMR assignments. Suffice it to say here that the structures of hydroxyjatrophones A and B proved to be **9** and **10**, whereas hydroxyjatrophone C (**11**) possessed the E configuration of the C-5—C-6 trisubstituted olefin in contrast to Z, as found in jatrophone and hydroxyjatrophones A and B. More recently the structure and stereochemistry of hydroxyjatrophone C has been confirmed via an X-ray crystal analysis carried out in our laboratory.[78]

9: R^1 = OH, R^2 = Me
10: R^1 = Me, R^2 = OH

11

XX. Interplay of Structure and Conformation of the Jatrophone Nucleus vis-a-vis Antitumor Activity

The availability of the hydroxyjatrophones A–C (**9–11**) as well as synthetic samples of normethyljatrophone (**2**), jatrophone (**1**), and the two normethyl-jatropholactones, in conjunction with our previous study of the reactions of n-propanethiol with simple 3(2H)-furanones, provided us with the opportunity to explore the reactivity of the jatrophone nucleus with this model biological nucleophile. This work, which expands the earlier study of Kupchan on the reaction of jatrophone itself with propanethiol[1] was initiated on the premise that the range of observed cytotoxicity (see Table V) may in some way be related to the ground state solution conformation of the macrocyclic ring system. Furthermore, we postulated that the solution conformation would have an effect on the nature of the derived products. Toward this end, the conformations of a number of jatrophone systems were determined in the solid state by X-ray crystallography and in solution via NOE studies. For a description of the crystal and solution conformation the reader is referred to our full account.[8] For our purposes here, no differences were noted between the crystal and solution conformations. Reactions were then carried out with *n*-propanethiol, products were isolated, and structures assigned.

In general, jatrophone, normethyljatrophone, and hydroxyjatrophones A and B were found to afford three products, although in different amounts, upon treatment with *n*-propanethiol under basic conditions (pH 9.2). Two of these products were monoadducts (**128** and **129**) differing only in stereochemistry at the site of addition (i.e., C-9), whereas the third was a bisadduct having structure **130**. Interestingly, monoadducts of type **128** were readily converted to bisadduct **130** upon further treatment with propanethiol, whereas adducts **129** were completely inert to further addition.

128   129   130

In the case of hydroxyjatrophone C (**11**), two adducts were obtained, however, in this case both proved to be bisadducts having the structure **131**, wherein the stereochemistry at C-5 and C-6 remains unassigned.

Somewhat different results were obtained when the synthetic analogs of jatrophone [i.e., *cis*- and *trans*-normethyljatropholactones (**7**) and (**8**)] were subjected to the same reaction protocol. In the case of *trans*-normethyljatropholactone (**8**), a single monoadduct (**132**) was formed, which did not involve transannular bond formation. *cis*-Normethyljatropholactone (**7**), on the other hand, was completely inert. The lack of reactivity observed for the cis isomer (**7**) is presumably due to the rigidity of the macrolide ring system, which prohibits effective overlap of the α,β-unsaturated carbonyl system.

131   132

Concerning the question of thiol reactivity versus cytotoxicity, the susceptibility of the jatrophone derivatives and analogs to thiol addition is, in the most fundamental sense, a significant indicator of cytotoxic activity. Each derivative

TABLE V

CYTOTOXIC ACTIVITY OF JATROPHONE, ITS DERIVATIVES AND ANALOGS[a]

| | P388 | | | | |
| | In vivo | | In vitro | Hela 14, | |
Compound	TLc	ED_{50} (mg/kg)	ED_{50} ($\mu g/nK$)	I C_{50} ($\mu g/ml$)	KB, ED_{50} ($\mu g/ml$)
Jatrophone (1)	145	12.0	0.01	1.56	8.7×10^{-5}
2-α-Hydroxyjatrophone (9)	125	6.0	0.03	NT	0.16
2-β-Hydroxyjatrophone (10)	132	10.0	0.06	NT	0.07
2-Normethyljatrophone (2)	120	10.0	—	NT	NT
2-α-Hydroxy-5,6-isojatrophone (11)	Inactive		2.2	NT	0.03
trans-Normethyljatropholactone (8)	NT		—	25	NT
cis-Normethyljatropholactone (7)	NT		—	100	NT

[a] All tests were carried out under the auspices of the National Cancer Institutes, using standard protocols except for cytotoxicity against Hela cells, which were carried out by T. Takita of the Microbial Chemistry Research Foundation, Institutes of Microbial Chemistry, Tokyo, Japan.

that formed adducts with n-propanethiol exhibited some level of cytotoxicity. The one derivative, cis-normethyljatropholactone (7), that did not form an adduct also did not exhibit cytotoxic activity. However, none of the results obtained appeared to have any significance in predicting the degree of activity. (Table V).

The observation that the principle product of thiol addition to jatrophone (1) and the hydroxyjatrophones A–C (9–11), when allowed to go to completion, is a bisaddition product is, we believe, of considerable significance in terms of the chemical mechanism of the cytotoxic effect. That is, cross-linking of proteins may be the principal mode of cell inactivation.

ACKNOWLEDGMENTS

I am greatly indebted to the many students and postdoctoral colleagues whose dedication and enthusiasm for chemistry is best illustrated by the results recorded in this chapter. Foremost here were Steven R. Schow, Michael D. Taylor, Michael A. Guaciaro, Peter M. Wovkulich, Tsi (Eric) Hall, Michael Malamas, Stephen J. Branca, and Bruce H. Toder. Other members of the Smith group during the jatrophone era and who contributed greatly to our overall research program included Robert M. Scarborough, Jr., Nancy N. Pilla, Diane Boschelli, Patricia Levenberg, and Paula Jerris.

I would also like to acknowledge here the generous and continuing support of the National Institutes of Health (National Cancer Institute) and the National Science Foundation (Chemical Instrumentation Program). Finally, I wish to thank Regina Zibuck for her generous help in the preparation of this manuscript.

REFERENCES

1. (a) S. M. Kupchan, C. W. Sigel, M. J. Matz, J. A. S. Renauld, R. C. Haltiwanger, and R. F. Bryan, *J. Am. Chem. Soc.* **92,** 4476 (1970); (b) S. M. Kupchan, C. W. Sigel, M. J. Matz, C. J. Gilmore, and R. F. Bryan, *ibid.* **98,** 2295 (1976).

2. (a) A. B. Smith, III, M. A. Guaciaro, S. R. Schow, P. M. Wovkulich, B. H. Toder, and T. Hall, *J. Am. Chem. Soc.* **103,** 219 (1981); (b) A. B. Smith, III, S. R. Schow, M. A. Guaciaro, P. M. Wovkulich, B. H. Toder, T. W. Hall, and M. Malamas, in preparation.

3. (a) S. J. Branca and A. B. Smith, III, *J. Am. Chem. Soc.* **100,** 7767 (1978); (b) A. B. Smith, III and N. N. Pilla, *Tetrahedron Lett.* **21,** 4691 (1980); (c) A. B. Smith, III, S. J. Branca, N. N. Pilla, and M. A. Guaciaro, *J. Org. Chem.* **47,** 1855 (1982).

4. (a) B. H. Toder, D. Boschelli, and A. B. Smith, III, *J. Cryst. Mol. Struct.* **9,** 189 (1979); (b) A. B. Smith, III and P. J. Jerris, *Tetrahedron Lett.* p. 711 (1980); (c) P. J. Jerris, and A. B. Smith, III, *J. Org. Chem.* **46,** 577 (1981).

5. A. B. Smith, III and M. Malamas, *J. Org. Chem.* **47,** 3442 (1982).

6. (a) A. B. Smith, III, P. A. Levenberg, P. J. Jerris, R. M. Scarborough, Jr., P. M. Wovkulich, and T. W. Hall, *J. Am. Chem. Soc.* **103,** 1501 (1981). (b) A. B. Smith, III and P. A. Levenberg, *Synthesis* p. 567 (1981).

7. A. B. Smith, III, M. A. Guaciaro, and P. M. Wovkulich, *Tetrahedron Lett.* p. 4193 (1978); A. B. Smith, III, S. J. Branca, M. A. Guaciaro, P. M. Wovkulich, and A. Korn, *Org. Synth.* **61,** 65 (1983).

8. A. B. Smith, III, M. D. Taylor, G. T. Furst, S. P. Gunasekura, C. A. Bevelle, G. A. Cordell, N. R. Farnsworth, S. M. Kupchan, H. Uchida, A. R. Branfman, R. G. Dailey, Jr., and A. T. Snedden, *J. Am. Chem. Soc.* **105,** 3177 (1983).

9. M. D. Taylor, A. B. Smith, III, and M. S. Malamas, *J. Org. Chem.* **48,** 4257 (1983).

10. (a) R. F. Raffauf, P.-K. Huang, P. W. Le Quesne, S. B. Levery, and T. F. Brennen, *J. Am. Chem. Soc.* **97,** 6884 (1975); (b) P. W. Le Quesne, S. B. Levery, M. D. Menachery, T. F. Brennen, and R. F. Raffauf, *J. Chem. Soc., Perkin Trans. I* p. 1572 (1978).

11. (a) F. N. Lahey and J. K. MacLeod, *Aust. J. Chem.* **20,** 1943 (1967); R. M. Carman, F. N. Lahey, and J. K. MacLeod, *ibid.* p. 1957; (c) D. L. Dreyer and A. Lee, *Phytochemistry* **11,** 763 (1972); (d) K. Padmawinata, *Acta Pharm.* **4,** 1 (1973); *Chem. Abstr.* **79,** 75897n (1973).

12. (a) S. M. Kupchan, T. J. Giacobbe, and I. S. Krull, *Tetrahedron Lett.* p. 2859 (1970); (b) J. R. Lillehaug, K. Kleppe, C. W. Sigel, and S. M. Kupchan, *Biochim. Biophys. Acta* **327,** 92 (1973).

13. (a) J. Attenburrow, A. F. B. Cameron, J. H. Chapman, R. M. Evans, B. A. Hems, A. B. A. Jansen, and T. Walker, *J. Chem. Soc.* p. 1094 (1952); (b) E. B. Bates, E. R. H. Jones, and M. C. Whiting, *ibid.* p. 1854 (1954); (c) also see K. R. Bharucha and B. C. L. Weedon, *ibid.* p. 1584 (1953).

14. G. Stork and M. Tomasz, *J. Am. Chem. Soc.* **86,** 471 (1964).

15. R. Ratcliffe and R. Rodehorst, *J. Org. Chem.* **35,** 4000 (1970).

16. K. E. Wilson, R. T. Seidner, and S. Masamune, *J. Chem. Soc., Chem. Commun.* p. 213 (1970); also 9-BBN is very effective in the reduction of α, β-unsaturated ketones to allylic alcohols, see ref. 25.

17. J.-L. Luche, *J. Am. Chem. Soc.* **100,** 2226 (1978); also see J.-L. Luche, L. Rodriques-Hahn, and P. Crabbé, *J. Chem. Soc., Chem. Commun.,* p. 601 (1978).

18. (a) C. E. Castro and R. D. Stephens, *J. Am. Chem. Soc.* **86,** 4358 (1964); (b) J. Inanaga, T. Katsaki, S. Takimoto, S. Ouchida, K. Inoue, A. Nakano, N. Okukado, and M. Yamaguchi, *Chem. Lett.* p. 102 (1979). For preparation of this reagent, see C. E. Castro, *J. Am. Chem. Soc.* **83,** 3262 (1961).

19. P. Margaretha, *Tetrahedron Lett.* p. 4891 (1971).

20. H. G. Lehmann, *Angew. Chem., Int. Ed. Engl.* **4,** 783 (1965). For additional examples of this reaction, see G. W. Moersch, D. E. Evans, and G. S. Lewis, *J. Med. Chem.* **10,** 254 (1967); T. L. Popper, J. N. Gardner, and R. Neri, and H. L. Herzog, *ibid.* **12,** 393 (1969); N. H. Dyson, J. A. Edwards, and J. H. Fried, *Tetrahedron Lett.* p. 1841 (1966); S. J. Halkes, J. Hartog, L. Morsink, and A. M. de Wachter, *J. Med. Chem.* **15,** 1288 (1972); G. Teutsch, L. Weber, G. Page, E. L. Shapiro, H. L. Herzog, R. Neri, and E. J. Collins, *ibid.* **16,** 1370 (1973); J. R. Bull and A. Tuinman, *Tetrahedron Lett.* p. 4349 (1973); H. A. C. M. Keus, *Tetrahedron* **32,** 1541 (1976).
21. H. O. House, "Modern Synthetic Reactions," 2nd ed., p. 494. Benjamin, Menlo Park, California 1972.
22. Consider the ease with which ketones versus esters react with HCN (i.e., cyanohydrin formation); see, for example, D. T. Mowry, *Chem. Rev.* **42,** 189 (1948).
23. (a) C. W. Brandt, W. I. Taylor, and B. R. Thomas, *J. Chem. Soc.* p. 3245 (1954); (b) W. Parker, R. A. Raphael, and D. I. Wilkinson, *J. Chem. Soc.* p. 3871 (1958).
24. H. A. Staab, *Angew. Chem, Int. Ed. Engl.* **1,** 351 (1962).
25. E. J. Corey, D. J. Seebach, *J. Org. Chem.* **40,** 231 (1975).
26. K. K. Purushothaman, S. Chandrasekharan, A. F. Cameron, J. D. Connolly, C. Labbe, A. Maltz, and D. S. Rycroft, *Tetrahedron Lett.* p. 979 (1979).
27. E. J. Corey and A. Venkateswarlu, *J. Am. Chem. Soc.* **94,** 6190 (1972).
28. E. J. Corey and G. Schmidt, *Tetrahedron Lett.* p. 399 (1979).
29. A number of natural products possessing the α-(hydroxymethyl)-α,β-unsaturation having antitumor properties are known. For example (a) glycalase I inhibitor 2-(crotonyloxy)-methyl]-4,5,6-trihydroxycyclohex-1-en-3-one [T. Takeuchi, H. Chimura, M. Hamada, H. Umezawa, O. Yoshioka, N. Oguci, Y. Takahashi, and A. Matsuda, *J. Antibiot.* **28,** 737 (1975)]; (b) the sesquiterpenoid lactone alliacol A [W. Steglich, *Pure Appl. Chem.* **53,** 1233 (1981), and references cited therein], and (c) the diterpenoids.
30. J. N. Marx, J. H. Cox, and L. R. Norman, *J. Org. Chem.* **37,** 4489 (1972).
31. H. J. Reich, J. M. Renga, and I. L. Reich, *J. Am. Chem. Soc.* **97,** 5434 (1975).
32. J. W. De Leeuw, E. R. De Waard, T. Beetz, and H. O. Huisman, *Recl. Trav. Chim. Pays-Bas* **92,** 1047 (1973).
33. Unsuccessful attempts included addition of **57a** to $LiAlH_4$ in Et_2O, inverse addition of a THF solution of $LiAlH_4$ to **57a** in THF at 0 and $-10°C$ [G. W. K. Cavill and F. B. Whitfield, *Aust. J. Chem.* **17,** 1260 (1964); F. A. Hochstein and W. G. Brown, *J. Am. Chem. Soc.* **70,** 3484 (1948)], and treatment of **23a** in Et_2O with $LiAlH_4$ and 1 equiv of EtOH [R. S. Davidson, W. H. H. Gunter, S. M. Waddington-Feather, and B. Lythgoe, *J. Chem. Soc., London* p. 4907 (1964); also see for a review, J. Malek and M. Cerny, *Synthesis* p. 217 (1972)]. In contrast, $LiAlH_4$ in THF added inversely to methyl cyclohexenecarboxylate in THF at 0°C was observed to give a nearly quantitative conversion to the corresponding allylic alcohol.
34. (a) W. Brown, *Org. React.* **6,** 469 (1951); (b) J. Dilts and E. C. Ashby, *Inorg. Chem.* **9,** 855 (1970); P. Lansbury and R. MacLeavy, *J. Am. Chem. Soc.* **87,** 831 (1965).
35. K. E. Wilson, R. T. Seidner, and S. Masamune, *J. Chem. Soc., Chem. Commun.* p. 213 (1970).
36. E. J. Corey, L. S. Melvin, Jr., and M. F. Haslanger, *Tetrahedron Lett.* p. 3117 (1975).
37. P. L. Fuchs, *J. Org. Chem.* **41,** 2935 (1976).
38. G. Stork and A. A. Ponaras, *J. Org. Chem.* **41,** 2937 (1976).
39. A latent equivalent of a β-ketovinyl anion was recently reported by Okamura [see M. L. Hammond, A. Mourino, and W. H. Okamura, *J. Am. Chem. Soc.* **100,** 4907 (1978)].
40. J. Ficini and J.-C. Depezay, *Tetrahedron Lett.* p. 4797 (1969).
41. H. O. House and W. C. McDaniel, *J. Org. Chem.* **42,** 2155 (1977).

42. P. W. Raynolds, M. J. Manning, and J. S. Swenton, *J. Chem. Soc., Chem. Commun.* p. 499 (1977); C. Shih, E. L. Fritzen, and J. S. Swenton, *J. Org. Chem.* **45,** 4467 (1980).
43. For examples of the metalation of vinyl halides, see D. Seebach and H. Neumann, *Chem. Ber.* **107,** 847 (1974); P. Radlick and H. T. Crawford, *J. Chem. Soc., Chem. Commun.* p. 127 (1974); and E. Negishi, A. Abramovitch and R. E. Merrill, *ibid.* p. 138 (1975).
44. K. Sato, S. Inoue, S. Kuranami, and M. Ohashi, *J. Chem. Soc., Perkin Trans. I* p. 1666 (1977); also see G. L. Dunn, V. J. DiPasquo, and J. R. E. *J. Org. Chem.* **33,** 1454 (1968); Ref. *38,* last reference.
45. (a) K. Umino, T. Furumai, N. Matsuzawa, Y. Awataguchi, Y. Ito, and T. Okuda, *J. Antibiot.* **26,** 506 (1973); (b) K. Umino, N. Takeda, Y. Ito, and T. Okuda, *Chem. Pharm. Bull.* **22,** 1233 (1974); (c) T. Date, K. Aoe, K. Kotera, and K. Umino, *ibid.* p. 1963; T. Shomura, J. Hoshida, Y. Kondo, H. Watanabe, S. Omoto, S. Inouye, and T. Niida, *Kenkyu Nempo— Nihon Daigaku Bunrigakubu (Mishima), Shizen Kagaku Hen* **16,** 1 (1976); T. Shoumura *et al.* (Meiji Seika), Japanese Kokai 19276–82 (1976); K. Hatano, T. Hasegawa, M. Izawa, M. Asai, and H. Kwasaki (Takeda Chemical Industries, Ltd.), Japanese Kokai 750, 597 (1975); *Chem. Abstr.* **84,** 3287 (1976); K. Hatano, M. Izawa, T. Hasegawa, S. Tonida, M. Asai, H. Iwasaki, and T. Yamano, *J. Takeda Res. Lab.* **38,** 22 (1979).
46. M. Noble, D. Noble, and R. A. Fletton, *J. Antibiot.* **31,** 15 (1978).
47. For a review of our interest in the cyclopentanoid class of antibiotics, see A. B. Smith, III and D. Boschelli, manuscript in preparation.
48. See, for example, W. A. Keschick, C. T. Buse, and C. H. Heathcock, *J. Am. Chem. Soc.* **99,** 247 (1977).
49. The oxidation of β-hydroxy ketones to 1,3-diketones, although not unprecedented, has not been widely exploited, presumably because of the enolic nature of the derived diketone and the potential of oxidative cleavage. Studies in our laboratory with simple β-hydroxy ketones and esters indicate that the latter is not a serious problem when Collins's or Swern's oxidation conditions are employed (unpublished results of P. Levenberg).
50. H. O. House, D. S. Crumrine, A. Y. Teranishi, and H. D. Olmstead, *J. Am. Chem. Soc.* **95,** 3310 (1973).
51. It is of interest that the lithium enolate derived from ketone **45a,** unlike that of **45,** afforded an excellent conversion to aldol **70i.** Presumably, the added methyl group decreases the basicity of the lithium enolate such that proton transfer in this case is not a serious problem.
52. K. Bowden, I. M. Heilbron, E. R. H. Jones, and B. C. L. Weedon, *J. Chem. Soc.* p. 39 (1946).
53. J. C. Collins, W. W. Hess, and F. J. Frank, *Tetrahedron Lett.* p. 3363 (1968); see also Ref. *15.*
54. K. Maruoka, S. Hashimoto, Y. Kitagawa, H. Yamamoto, and H. Nozaki, *J. Am. Chem. Soc.* **99,** 7705 (1977); J. Tsuji and T. Mandai, *Tetrahedron Lett.* p. 1817 (1978).
55. V. W. Armstrong, N. H. Chishti, and R. Ramage, *Tetrahedron Lett.* p. 373 (1975).
56. T. Mukaiyama and S. Hayashi, *Chem. Lett.* p. 15 (1974); K. Banno and T. Mukaiyama, *ibid.* p. 741 (1975); T. Mukaiyama and A. Ishida, *ibid.* p. 312.
57. A. Alexakis, M. J. Chapdelaine, G. H. Posner, A. W. Runquist, *Tetrahedron Lett.* p. 4205 (1978).
58. K. Isaac and P. Kocienski, *J. Chem. Soc., Chem. Commun.* p. 460 (1982).
59. M. D. Taylor, G. Minaskanian, K. N. Winzenberg, P. Santone, and A. B. Smith, III, *J. Org. Chem.* **47,** 3960 (1982).
60. Unpublished results of B. H. Toder, a graduate student, in our laboratory; a detailed account will be published in due course.
61. L. Fieser and M. Fieser, "Reagents for Organic Synthesis," p. 566. Wiley, New York, 1967.
62. B. H. Toder, S. J. Branca, R. K. Dieter, and A. B. Smith, III, *Synth. Commun.* **5,** 435 (1975).

63. K. B. Sharpless and R. F. Lauer, *J. Am. Chem. Soc.* **95,** 2697 (1973); H. J. Reich, I. L. Reich, and J. M. Renga, *ibid.* p. 5813.

64. T. Tsunada, M. Suzuki, and R. Noyori, *Tetrahedron Lett.* p. 1357 (1980).

65. (a) For precedent for this transformation, see M. L. Bolter, W. D. Crow, and S. Yoshida, *Aust. J. Chem.* **35,** 1411 (1982). (b) Unpublished results of Mr. M. Malamas, a graduate student in our laboratory.

66. D. A. Evans and G. C. Andrews, *Acc. Chem. Res.* **7,** 147 (1974), and references cited therein.

67. J. E. Baldwin, R. E. Hackler, and D. P. Kelly, *J. Chem. Soc., Chem. Commun.* p. 538 (1968).

68. A. B. Smith, III, B. A. Wexler, and J. S. Slade, *Tetrahedron Lett.* **21,** 3237 (1980).

69. A. R. de Vivar, C. Guerrero, E. Diaz, and A. Ortega, *Tetrahedron* **26,** 1657 (1970).

70. W. Vichnewski, S. J. Sarti, B. Gilbert, W. Herz, *Phytochemistry* **15,** 191 (1976).

71. Conjugate addition at the α-methylene functionality would also be expected.

72. S. M. Kupchan, D. C. Fessler, M. A. Eakin, and T. J. Giacobbe, *Science* **168,** 376 (1970).

73. Furandienone **110** displayed presumptive activity against P-388 lymphocytic leukemia (T/C = 128 at 100 mg/kg).

74. K. Griesbaum, *Angew. Chem., Int. Ed. Engl.* **9,** 273 (1970).

75. Geiparvarin (NSC 142227) displayed activity against P-388 lymphocytic leukemia (T/C = 130 at 400 mg/kg and 123 at 600 mg/kg) and in the KB screen (ED_{50} = 3 mg/mL). It proved inactive in the B-15 melanoma, L-1210 leukemia, and Lewis lung carcinoma screens.

76. T. Mukaiyama, M. Usui, and K. Saigo, *Chem. Lett.* p. 49 (1976).

77. J. E. Baldwin, *J. Chem. Soc., Chem. Commun.* p. 734 (1976); J. E. Baldwin, J. Cutting, W. Dupont, L. Kruse, L. Silberman, and R. C. Thomas, *ibid.* p. 736; J. E. Baldwin, *ibid.* 738; J. E. Baldwin, R. C. Thomas, L. I. Kruse, and L. Silverman, *J. Org. Chem.* **42,** 3846 (1977).

78. Unpublished results of P. Carroll and V. Santo Pietro in our laboratory; a detailed account will be published in due course.

Chapter 10

ON THE STEREOCHEMISTRY OF NUCLEOPHILIC ADDITIONS TO TETRAHYDROPYRIDINIUM SALTS: A POWERFUL HEURISTIC PRINCIPLE FOR THE STEREORATIONALE DESIGN OF ALKALOID SYNTHESES

Robert V. Stevens

Department of Chemistry and Biochemistry
University of California
Los Angeles, California

Because of the continued advent and acquisition of ever more powerful physical methods for the isolation, purification, and proof of structure of natural products, chemists are now equipped as never before to explore the biosphere in search of new substances. It is interesting to note that these technological advances have permitted the isolation and characterization of minute (sometimes even microscopic!) quantities of material, thus prohibiting further evaluation of their chemical and/or biological properties. This is no small price to pay and under these circumstances the synthetic organic chemist can be called upon even further to provide such knowledge. It is also interesting to note that this revolution has even caused changes in some of the basic concepts of natural products chemistry. For example, the classic definition of an alkaloid as being a nitrogenous *plant* product with basic properties is clearly obsolete now that, as we shall see, compounds of this type have been found in the insect and animal kingdoms as well. For some time now my co-workers and I have been interested in the total synthesis of these naturally occurring substances, but before embarking on one such adventure I would first like to summarize some of the criteria that we have grown to value highly in the development of a synthetic plan.

At the outset, we place a very high premium on efficiency of bond construction. Clearly those reactions in which more than one new bond is formed are favored over those that require step-by-step methodology. For example, the acid-catalyzed rearrangement of various cyclopropylimines to the corresponding Δ^2-pyrrolines (cf. **1** to **2**) provides an illustration wherein two new bonds (one σ and one π bond) are produced as does the annulation of these and other endocyclic enamines (e.g., **4**) to afford exclusively cis-fused hydroindolones (**3**) or hydroquinolones (**5**).[1] Indeed, these two methods, either individually or in tandem, have been extensively exploited for the total synthesis of a variety of alkaloids. These examples also illustrate a second major goal of our research, namely, the development of *general methodology*, methodology that can be applied to a whole host of structurally diverse targets (see Scheme 1). A third criterion, which is universally acknowledged, is the challenge of stereochemical control, in both the relative and absolute senses. Many otherwise promising synthetic plans suffer from either lack of stereochemical control or require wasteful steps to adjust the offending

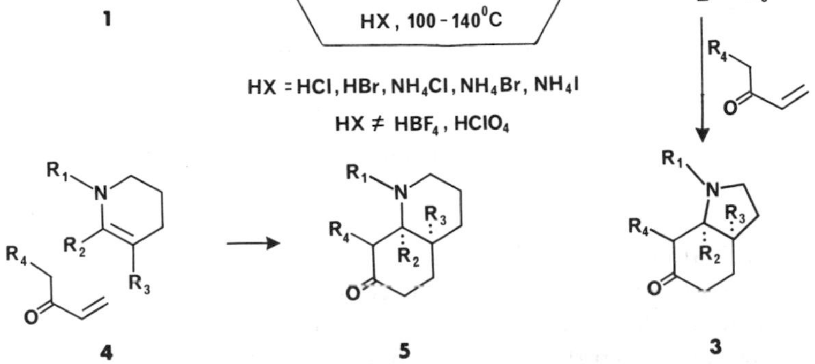

$$HX = HCl, HBr, NH_4Cl, NH_4Br, NH_4I$$

$$HX \neq HBF_4, HClO_4$$

SCHEME 1

stereochemical centers. A fourth goal that guides our planning is concerned with protection–deprotection sequences. Unless such operations can be somehow merged with another useful task, they are inherently wasteful, and we endeavor to eliminate or at the very least minimize such operations. In this regard we have taken a lesson from nature itself wherein such operations are rare, although not unknown.

Webster defines *heuristic* as "serving to discover or to stimulate investigation; of methods of demonstration which tend to lead a person to investigate further." It is the goal of this account to outline the genesis of a powerful heuristic principle and to retrace the course of events from its inception to the present time, and finally even to project a little into the future. This story began when the structures of a family of novel tricyclic alkaloids first emerged. Small amounts of these substances were isolated from the defensive secretions of various species of ladybugs in pioneering investigations headed by Professors B. Tursch at l'Université Libre Bruxelle and W. A. Ayer at the University of Alberta.[2]

In contemplating the synthesis of these conformationally rigid systems, it was noted that in each case the methyl group occupies the thermodynamically more stable equatorial position and accordingly should offer no particular cause for concern stereochemically. Furthermore, we noted that myrrhine (**7**) with its all-trans ensemble of quinolizidine rings is clearly thermodynamically more stable than either precoccinelline (**6**) or hippodamine (**8**) and that the latter two bases should be of comparable energy. Therefore, it is clear at the outset that one could take advantage of the thermodynamic stability of myrrhine (**7**) in synthetic planning but that any such plans for either coccinelline (**6**) or hippodamine (**8**) would probably have to rely on some sort of kinetically controlled process (see Scheme 2). Actually, in less time than it took to write this paragraph we had developed on paper an attractive approach that relies on one of the oldest reactions known in alkaloid synthesis, namely, the classic Robinson–Schöpf condensation. Clearly, if one can mix together the three reagents shown in Scheme 3 and obtain the tropane alkaloid skeleton, then one could predict with confidence that a mechanistically similar condensation between the amine dialdehyde **9** and acetonedicarboxylic ester should afford the parent ring system found in this new class of alkaloids. For this reason, this idea was very nearly not pursued insofar as there superficially appeared to be very little in the way of new chemistry involved, which, of course, is a fifth fundamental goal of our research. However, in contemplating other possible routes, we quickly became aware of the fact that from the standpoint of efficiency of bond construction, stressed above, this classic reaction is truly remarkable (and difficult to beat). Note that a total of four new bonds are formed in a single

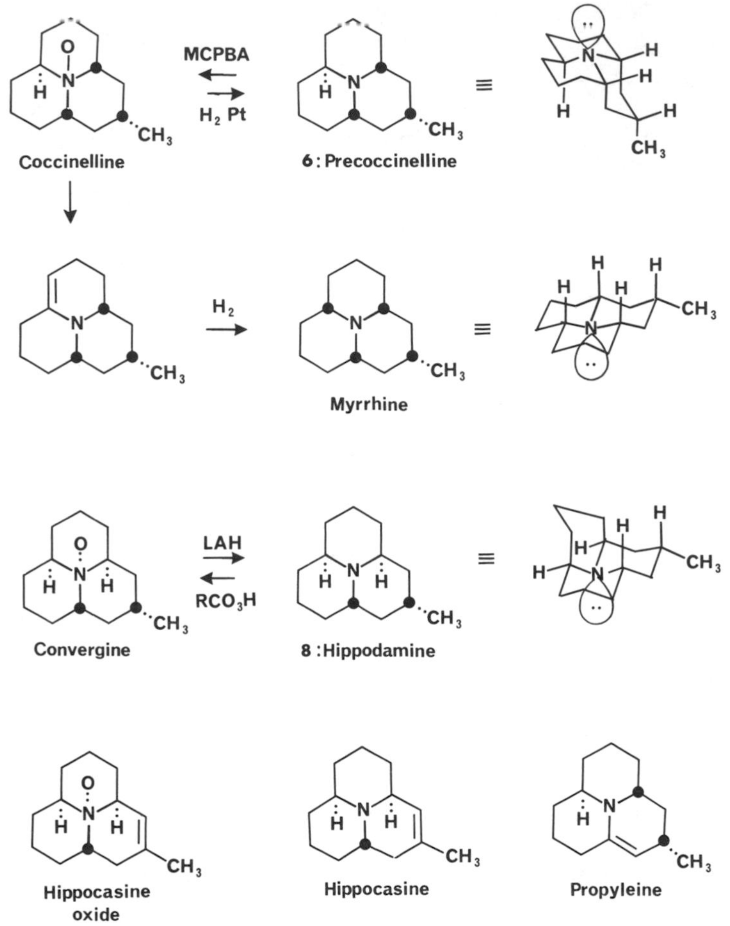

SCHEME 2

vessel! Upon closer scrutiny we began to realize that this reaction should not only provide the desired tricyclic skeleton from two simple acyclic precursors, but that these achiral precursors also generate a total of five chiral centers! The question then arose—what is the stereochemistry of this remarkable old reaction? The literature precedent sheds no light on this issue insofar as the tropane skeleton can arise only in the manner indicated. It was to clarify these stereochemical questions that we decided to initiate the research outlined below. It should be noted at this point that we had no idea at the time that the results of this investigation would provide us with a new and powerful heuristic principle upon which much of our subsequent research has been

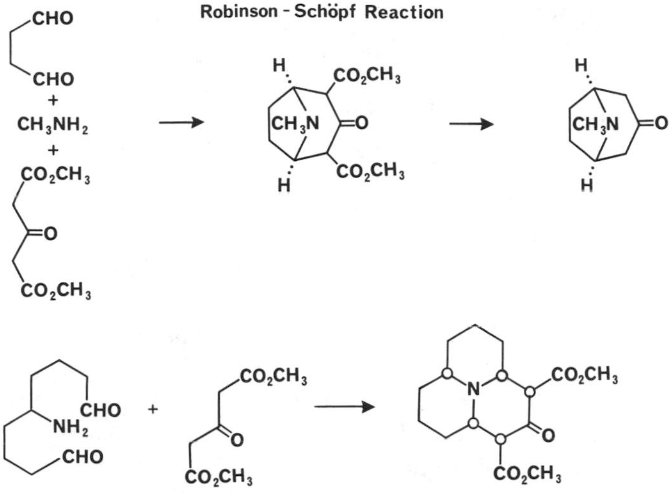

Robinson - Schöpf Reaction

SCHEME 3

based, but more about that later. What was clear at the outset was the fact that selection of the proper conditions for the condensation reaction would be crucial insofar as all of the chiral centers could, in principle, be scrambled either via retro-Mannich and/or retro-Michael processes, as illustrated in Scheme 4. These thoughts, then, set the stage for the laboratory experiments that were placed in the very capable hands of Mr. Albert Lee.[3]

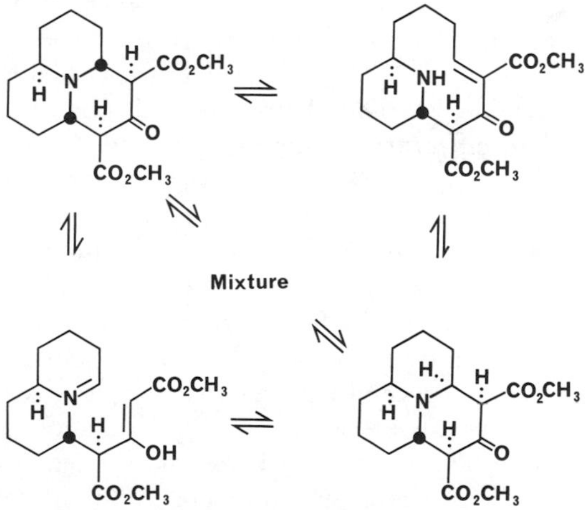

SCHEME 4

The requisite amino dialdehyde **9** was prepared (as its dimethyl acetal **10**) by the route shown in Scheme 5. Conjugate addition of dimethyl malonate to acrolein afforded **11**, which was transformed to the corresponding acetal (**12**) in the usual way. Decarbomethoxylation of **12** proceeded smoothly, under the neutral conditions indicated, to provide **13**. Claisen condensation of the latter intermediate with itself afforded β-ketoester **14**, which was similarly decarbo-methoxylated under neutral conditions to yield ketone **15**. Reductive amina-tion of **15** by the Borch method proved difficult until it was discovered that the addition of 3A molecular sieves improved the yield of amine **10** drama-tically. It is perhaps worthy of note that this relatively simple sequence of reactions employs only 2 equivalents each of malonic ester and acrolein as building blocks.

Intermediate **10** being now in hand, we turned our attention to the crucial condensation with acetonedicarboxylic ester (see Scheme 6). The presence of the basic nitrogen dictated the employment of strong mineral acid to effect hydrolysis of the two acetal functions. Accordingly, hydrochloric acid at pH 1 was employed. The pH of the solution was then adjusted to about 5.5 by the careful addition of base, followed by the addition of a citrate–phosphate buffer. Then, a solution of acetonedicarboxylic ester in the same buffer was added dropwise at room temperature to the stirred solution. Much to our delight, a single isomer of the correct empirical formula crystallized directly from the reaction mixture and in good yield. The structural and stereochemi-cal assignments followed rapidly from the IR and ^{13}C-NMR spectra as well as other data. If one ignores for the moment the two chiral centers alpha to the ester moieties, there remain three possible perhydro[9b]azaphenalene stereoisomers, which correspond to precoccinelline (**6**), myrrhine (**7**) and hippodamine (**8**). Of these, only the stereochemistry equivalent to that of myrrhine (**7**) is capable of satisfying the empirical stereoelectronic require-ments for absorption in the 2700–2800 cm^{-1} region of the IR spectrum (Bohlmann-Wenkert bands). The absence of such absorptions allowed us to rule out this stereochemistry in our intermediate. As noted above, this stereochemistry also corresponds to the thermodynamically most stable arrangement. Therefore we also knew on the basis of this single spectrum that we were probably dealing with a kinetically controlled process. A distinction between the other two possible perhydro[9b]azaphenalene stereoisomers could be made from the ^{13}C-NMR spectrum of our intermediate, which recorded only 9 lines, as is consistent with a meso compound, thus ruling out the hippodamine-like stereochemistry. In order to be a meso compound, the two ester groups must clearly be cis but, as inspection of perspective formula **20** indicates, the alternate possibility wherein both esters are cis but axial can be readily dismissed on simple steric grounds. A third decarbomethoxyla-tion under the neutral conditions employed previously afforded ketone **22**,

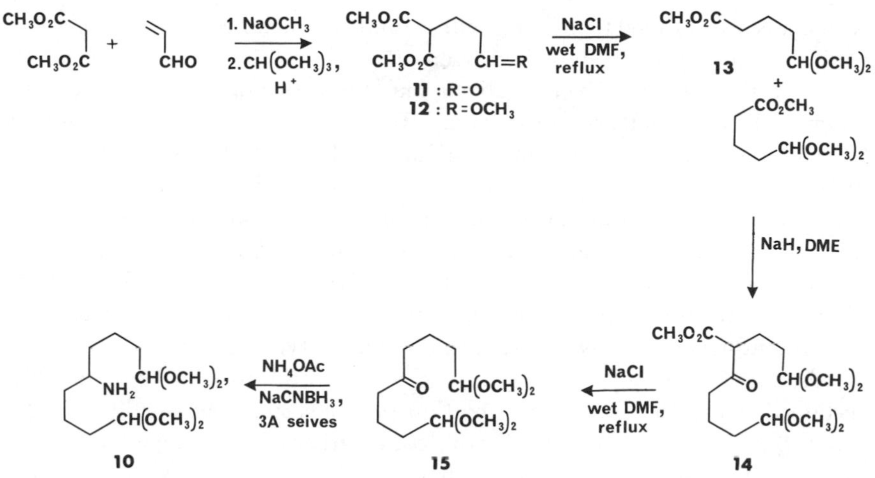

SCHEME 5

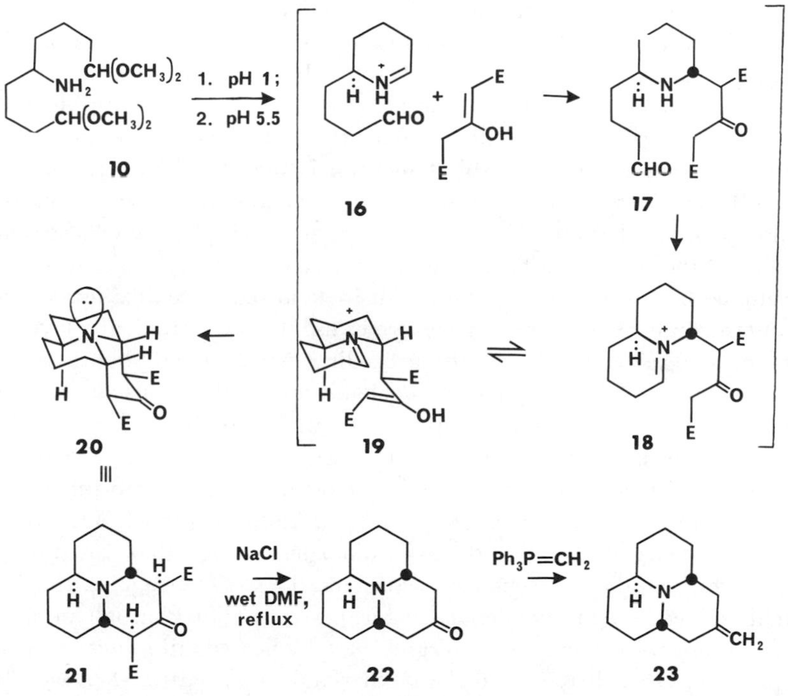

SCHEME 6

which was transformed to precoccinelline (and coccinelline) by Wittig methylenation, followed by catalytic hydrogenation.

At the time we intuitively realized that the remarkable stereoselectivity
observed in the condensation reaction (**10** to **20**) involved some heretofore
little appreciated and important principles of dynamic stereochemistry, but
little did we know that they would provide the basis for a powerful new
heuristic principle. Although there were no intermediates actually isolated
in this reaction, we made the reasonable assumption that the first step in
which relative stereochemistry is introduced involves the tetrahydropyridinium salt **16** and the enol of acetonedicarboxylic ester. Because we know the
stereochemistry of these two centers in the final tricyclic product and because
we also know that they correspond to kinetic control, we can be reasonably
certain that the **16** to **17** transformation proceeds stereochemically as indicated; but why? Intermediate **16** can exist in either of the two conformations
shown in Scheme 7. Accordingly, there are four possible transition states
wherein maximum orbital overlap is maintained between the approaching
nucleophile and the developing lone-electron pair on nitrogen or, to put it
another way, there are four possible transition states that lead to products
wherein the sp^3-hybridized orbitals generated are anti and coplanar. Two
of these, **25** and **26**, require boat-like transition states and are accordingly
kinetically disfavored. The remaining two possibilities both involve chair-like
transition states, but one of them (**27**) clearly suffers from an unfavorable
1,3-diaxial interaction between the ring side chain (R) and the incoming
(in this case sterically demanding) nucleophile. Therefore, of the four possibilities wherein maximum orbital overlap is maintained (stereoelectronic
control), **24** is the least objectionable and explains the observed product.
It also turns out that once these two centers are established, they also dictate
the fate of the remaining three centers. Thus, a second cyclization to afford
the annulated tetrahydropyridinium salt **18**, followed by enolization, provides
an intermediate (**19**) whose ring closure is also stereoelectronically favorable.

As pleased as we were with the preceding experimental results, we were
painfully aware of the fact that the mechanistic considerations advanced had
not one shred of experimental support! Perusal of the literature uncovered
a few more examples that were consistent with our hypothesis and therefore
inspired some degree of confidence, but most of these fell into the category
of being ambiguous for one reason or another. At this stage we began to search
for synthetic targets of chemical and/or biological interest that would provide
further tests of these mechanistic considerations. Once again, the insect
domain served as a point of departure for this study. The Pharoah ant (*Mono-
morium pharaonis* L.) can be a serious pest in heated buildings, especially
hospitals, in Great Britain and the Netherlands. F. J. Ritter *et al.* were able
to isolate and determine the structure of one of the trail pheremones of these

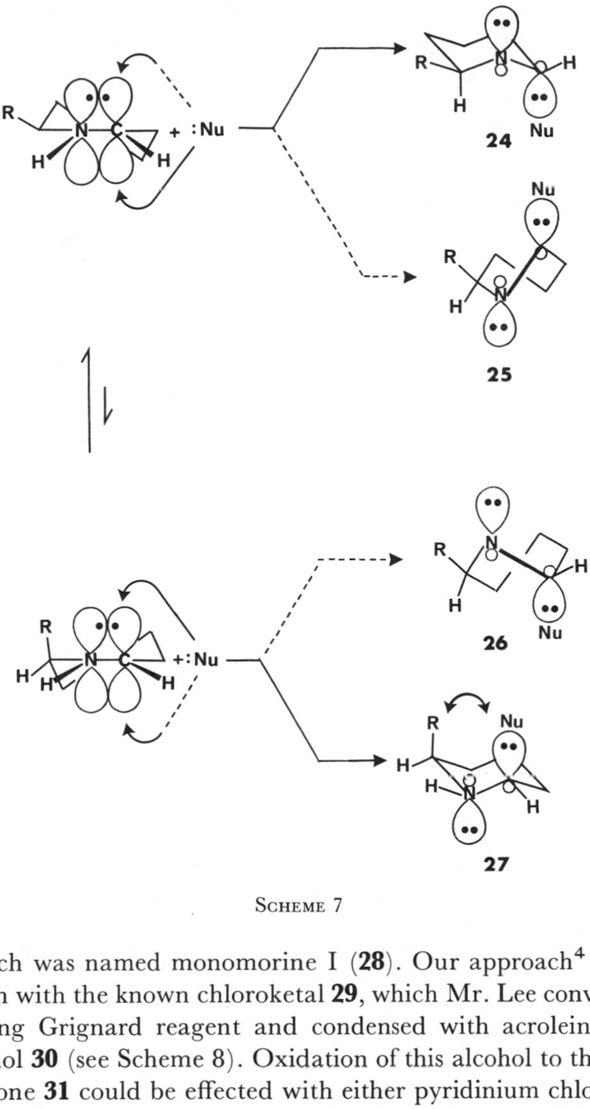

Scheme 7

insects, which was named monomorine I (**28**). Our approach[4] to this substance began with the known chloroketal **29**, which Mr. Lee converted to the
corresponding Grignard reagent and condensed with acrolein to provide
allylic alcohol **30** (see Scheme 8). Oxidation of this alcohol to the somewhat
unstable enone **31** could be effected with either pyridinium chlorochromate
or MnO_2, but use of the latter reagent simplified workup. Michael addition
of 1-nitropentane to the enone **31**, catalyzed with tetramethylguanidine,
afforded nitro ketone **32**. Reductive cyclization of **32** over Pd–C was rather
slow. However, it was found that the addition of anhydrous Na_2SO_4 had a
remarkable accelerating effect, presumably because it removed the water
formed in the formation of the pyrroline intermediate. Although no systematic
study on the stereochemistry of reduction of 2,5-disubstituted pyrrolines had

been made, we noted that all such reductions previously reported proceeded in a syn fashion to provide the thermodynamically favored cis product, which in this instance was isolated as the beautifully crystalline oxalate salt **33**. The latter intermediate in hand, we were now in a position to explore the key reaction, stereochemically speaking. Hydrolysis of the ketal in acid followed by a basic workup and extraction provided the rather unstable endocyclic enamine **34**, which was immediately exposed to NaCNBH$_3$ under acidic conditions. Racemic monomorine I (**28**) was the only volatile product produced in this sequence of reactions (68% overall yield from the salt **33**). It should also be noted that removal of the ketal protecting group was merged with another useful step, namely, formation of the enamine **34**, thus compensating for its employment.

The stereochemical outcome observed in the **35** and **28** transformation is consistent with our hypothesis. Thus there are once again four possible transition states for this reduction wherein maximum orbital overlap can be

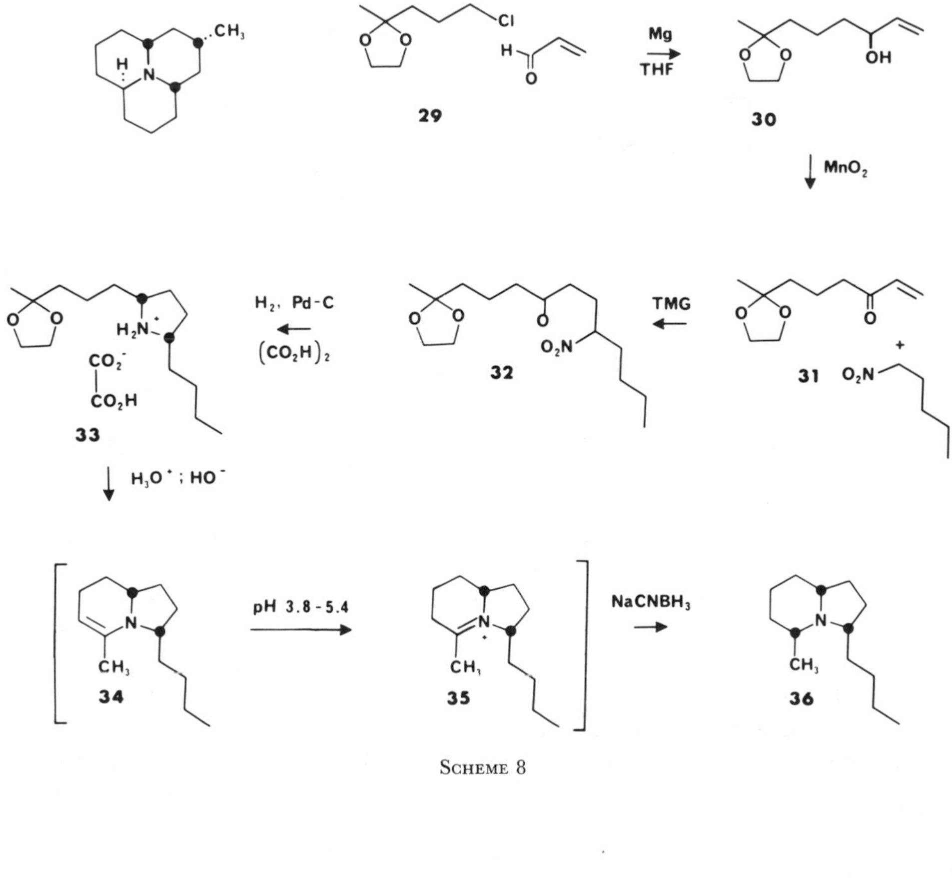

SCHEME 8

maintained with respect to the developing lone-electron pair on nitrogen and the attacking hydride reagent. Two of these (cf. dashed arrows in **36** and **37**) require boat-like transition states leading to **38** and **39** and are disfavored kinetically (see Scheme 9). Of the two chair-like transition states shown in the scheme (**40** and **41**), the latter suffers from a strong peri interaction with the proton at C-8 and is therefore likewise disfavored, leaving **40** as the preferred pathway, which is what was observed experimentally.

The latter synthesis not only shed a little light on stereoelectronic control in nucleophilic additions to tetrahydropyridinium salts but also served as a vehicle for the synthesis of a closely related alkaloid, namely, gephyrotoxin 223. This substance occurs in the skins of certain frogs (family *Dendrobatidae*) of pantropic distribution. Extracts from these frogs have been used by the natives of that region to poison the tips of darts and arrows. In recent years these frogs have proved to be a rich source of neurotoxin alkaloids. In one study headed by Dr. John W. Daly[5] at the National Institutes of Health, a crude extract was separated and analyzed by gas chromatographic–mass spectroscopic techniques. One of the minor alkaloids had a molecular weight

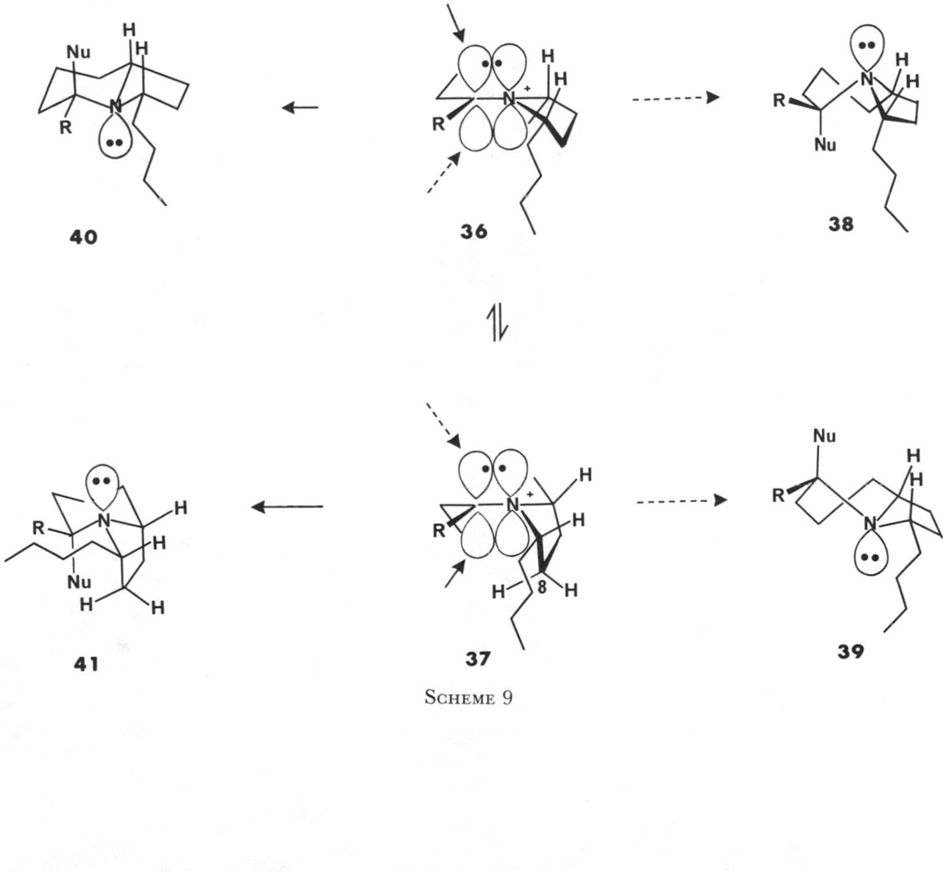

SCHEME 9

of 223 (hence gephyrotoxin 223) and a fragmentation pattern consistent with the gross structure shown in Scheme 10. Of course, the issue of stereochemistry remained unresolved and provided us with a unique opportunity to test further the stereoelectronic principles discussed above and to produce material for biological evaluation. It should be noted that in the structure elucidation no material was actually "isolated" in the practical sense of the word. The structure of monomorine I (**28**) and its C-1 epimer (**42**) is shown in Scheme 10 for comparison with the four possible diastereomers of gephyrotoxin 223. One of the diastereomers (**46**) had already been synthesized by Prof. Tim L. Macdonald[6] at Vanderbilt University and submitted to the NIH group for comparison with the crude extract. It was initially reprted that the fragmentation pattern in the mass spectrometer was virtually identical to that of the natural material, thus confirming the structural assignment. However, the retention time on the gas chromatograph appeared different, leaving the important issue of stereochemistry in doubt. Later, after our work was complete, it was reported that his result "could not be reproduced"—a statement that suggests that the relative stereochemistry of the natural material is, in fact, that shown in formula **46**.

The experimental work executed by Albert Lee is outlined in Scheme 11.[7] Thus alkylation of the dianion of β-keto sulfoxide **47** with bromo acetal **48** provided **49** in 70% yield. Pyrolysis of the latter intermediate in refluxing chloroform afforded the somewhat unstable enone **50**, which was exposed immediately to 1-nitropentane and tetramethylguanidine (TMG) to yield nitro ketone **51**. Reductive cyclization, achieved with a Raney nickel catalyst,

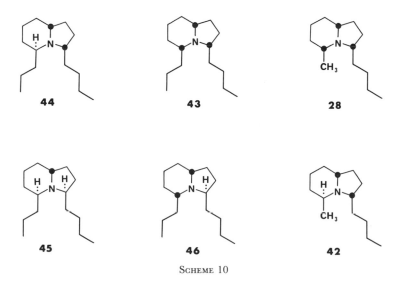

SCHEME 10

provided a single stereoisomer, which was isolated and purified as the corresponding oxalate salt **52** in 64% yield. Hydrolysis of **52** and cyclization afforded the tetrahydropyridinium salt **53**. All attempts to isolate **53** as the corresponding endocyclic enamine led to dimerization (presumably to give **54**). However, treatment of the acidic mixture with cyanide led to cyanoamine **55** in 96% yield. Once again, only one stereoisomer could be detected. In order to confirm the stereochemical assignments, **55** was reduced first to the corresponding aldehyde **56** with iBu$_2$AlH and thereafter by Wolff–Kishner reduction to the known epimer of monomorine I (**42**). The cyanoamine also serves as a latent synthon for iminium salt **53**. Thus, when treated with excess CH$_3$MgBr in ether at 0°C it was converted to **42** in 87% yield. No other stereoisomer was detected. Similarly, when exposed to n-PrMgBr, cyanoamine **55** led stereospecifically to gephyrotoxin 223 stereoisomer **44**.

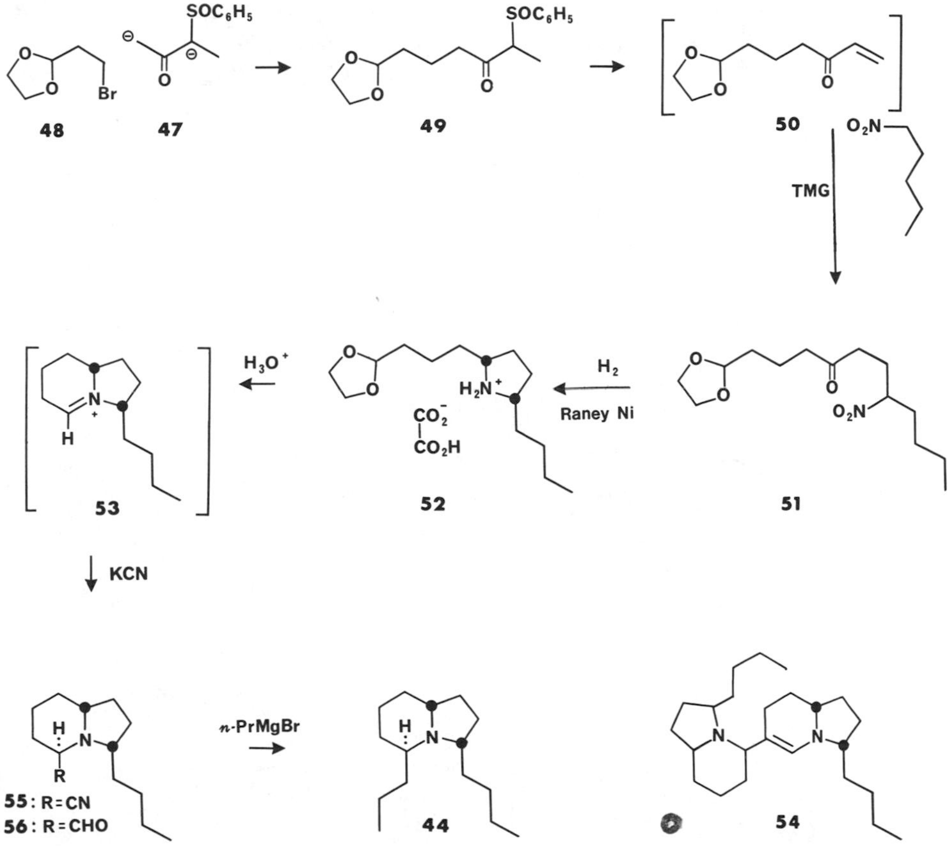

SCHEME 11

In an analogous fashion, Yumi Nakagawa[8] has synthesized stereospecific-ally a second stereoisomer of gephyrotoxin 223 by the route outlined in Scheme 12. The extraordinarily high degree of stereoselectivity observed in all of the preceding nucleophilic additions is in agreement with the stereoelectronic arguments advanced above. When one also considers the wide range of nucleophiles employed in these studies (enols, hydrides, cyanide, and Grig-nard reagents) and the corresponding spectrum of transition states, these results are all the more remarkable.

With the above results in hand, let me now focus attention on how these considerations have come to influence our synthetic planning. Scheme 13 shows the structures of aristoteline, makomakine, and hobartine. From the isolation studies, it was known that when exposed to acid, makomakine cyclizes, as expected, to aristoteline. Inspection of the preferred conformation of makomakine shows that the α-hydrogen is anti and coplanar to the lone-electron pair on nitrogen. Accordingly, stereoelectronically one would expect hydride reduction of the corresponding imine 57 to proceed stereospecifically to provide makomakine. However, nucleophilic attack at this position is also clearly obstructed by the axial methyl group. Which factor is more impor-

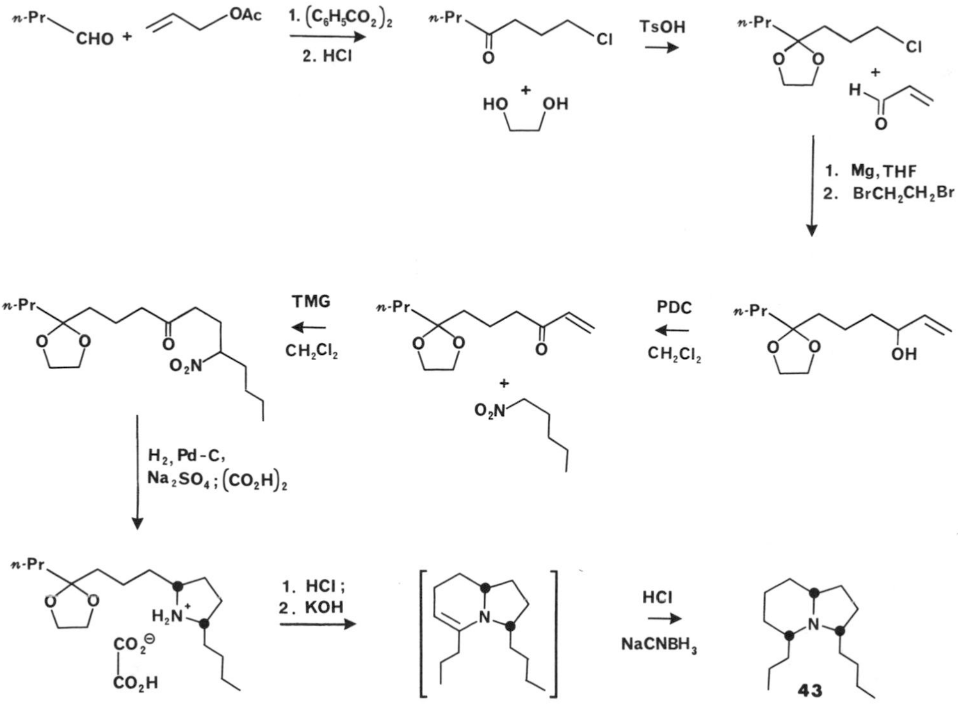

SCHEME 12

tant—stereoelectronic control or steric hindrance? This issue was settled convincingly in favor of stereoelectronic control by Paul M. Kenney,[9] who designed the exceedingly simple experiments shown in Scheme 13. Thus (+)-β-pinene (58) was added to a dichloromethane solution of 3-indolyl-acetonitrile and mercuric acetate at −30°C. Upon warming to 0°C, the resultant imine (57) was reduced with sodium borohydride in 3 N sodium

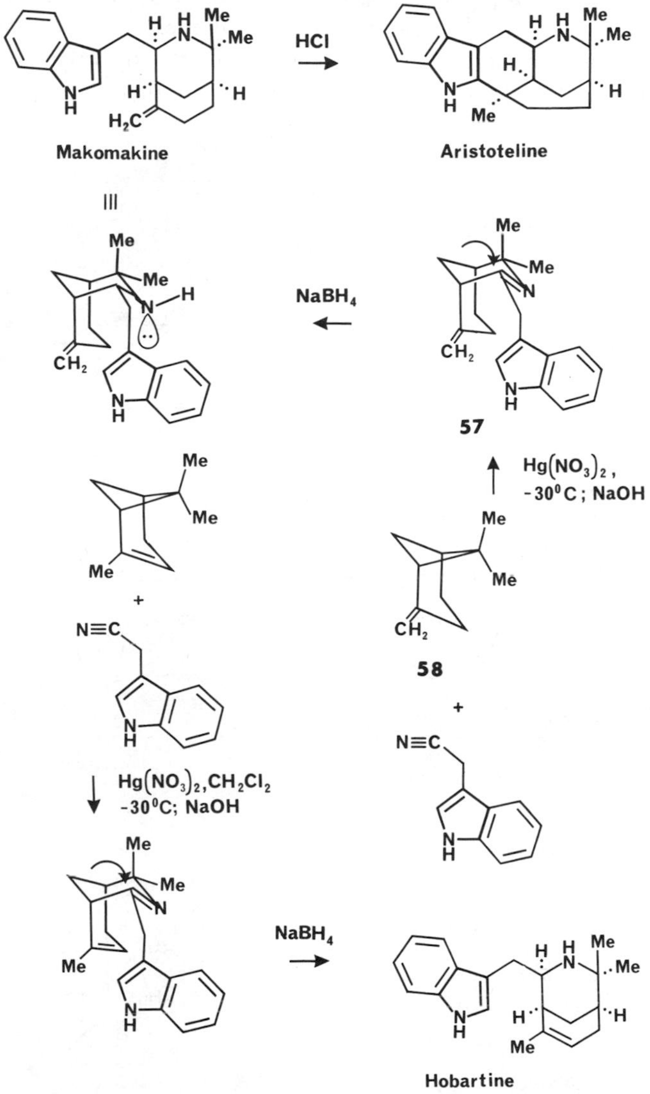

SCHEME 13

hydroxide–methanol, a reaction that led stereospecifically to (−)-mako-
makine—no other stereoisomer was detected. Treatment of makomakine
with concentrated hydrochloric acid afforded (+)-aristoteline. In a similar
fashion, racemic hobartine was prepared from (+)-α-pinene.

A mechanism that accounts for these remarkable cyclization reactions is
outlined in Scheme 14. Note that the enantiospecificity observed in the
cyclization of (−)-β-pinene to (+)-makomakine is not observed in the
(+)-α-pinene case, which led to racemic hobartine. In the former case, the
intermediate allylic mercurial derivative (**59**, or its equivalent) can cyclize
only in the fashion shown, thus retaining the chirality of its precursor. In

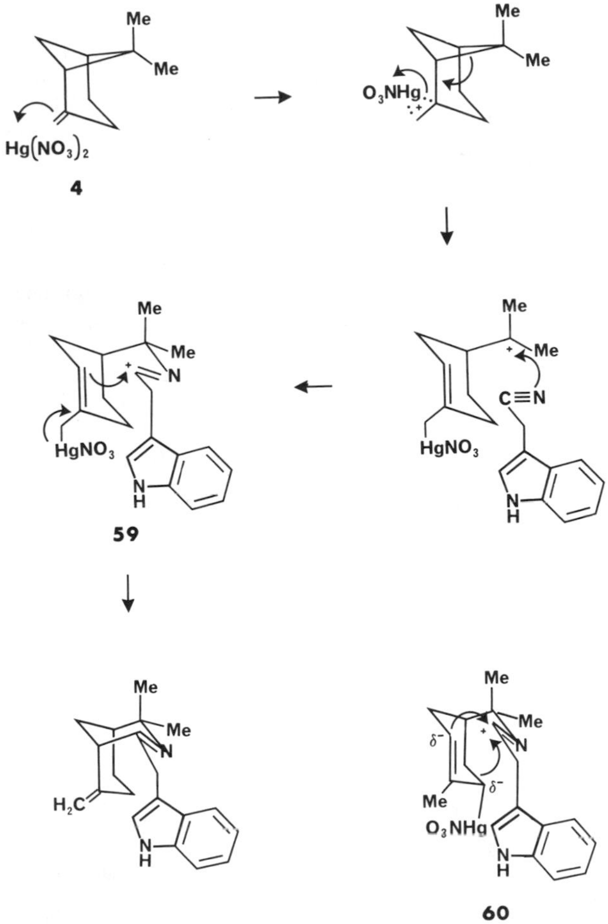

SCHEME 14

contrast, the equivalent intermediate (**60**), derived from α-pinene, can cyclize at either of the two enantiomeric sites shown, leading to a racemic product.

As noted earlier, annulation of Δ^2-pyrrolines or Δ^2-tetrahydropyridines with methyl vinyl ketone, or derivatives thereof, has found wide application in the total synthesis of alkaloids. Invariably, annulation of these endocyclic enamines afford exclusively cis-fused hydroindolones or hydroquinolones (e.g., **61** to **62**) (Scheme 15). The latter case is particularly intriguing. Because it is well established that hydroquinolones prefer thermodynamically to exist as the corresponding trans isomer, it is clear that this process is kinetically controlled. What we did not realize at the time is that this can be rationalized

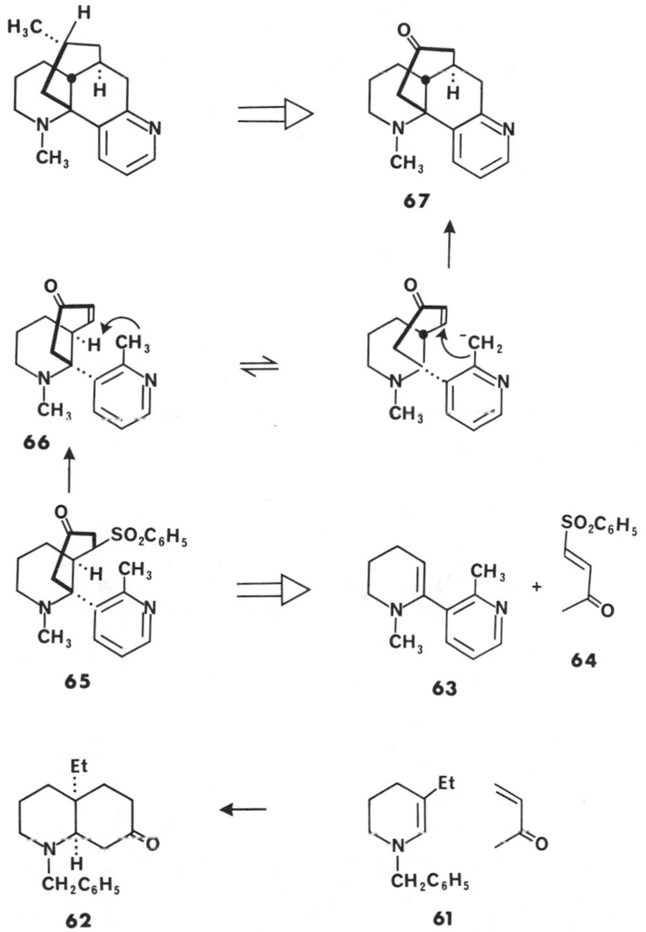

SCHEME 15

stereoelectronically and that failure to understand and appreciate this can have a decisive impact on the outcome of an otherwise reasonable synthetic plan. Nick Hrib and I were to learn this in a most brutal fashion in connection with a projected synthesis of N-methyllycodine.[10] Thus, by analogy with successful annulations, such as the **61** to **62** transformation, we anticipated that Δ^2-tetrahydropyridine **63** would react with enone **64** to afford the cis-fused hydroquinolone **65**. Elimination of benzenesulfinic acid from the latter intermediate would then provide enone **66**, which could be epimerized and cyclized in the presence of a suitable base to provide the tetracyclic skeleton **67**.

The desired endocyclic enamine **63** was obtained via the very efficient route shown in Scheme 16 and we were pleased to observe a smooth reaction with

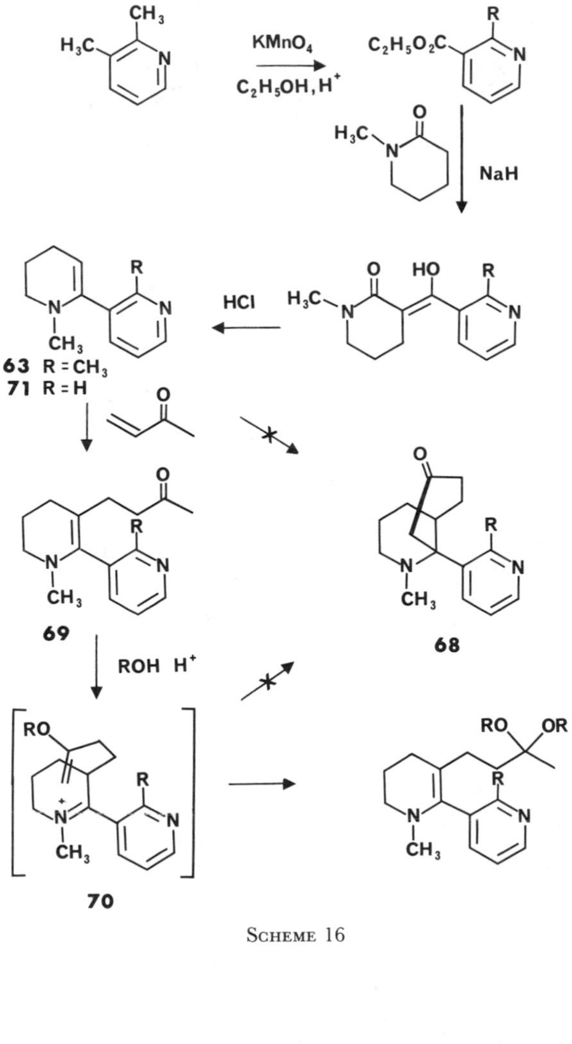

SCHEME 16

methyl vinyl ketone (and derivatives thereof). However, we were surprised to learn that the product of this reaction was not the desired hydroquinolone **68** but rather the alkylated but uncyclized enamine **69**. This unexpected result was followed by a series of unsuccessful experiments aimed at forcing the cyclization to occur under acidic conditions via enol ether intermediates such as **70**. At first we attributed these failures to the methyl group on the pyridine ring, which, after all, would provide steric hindrance to the nucleophilic addition. However, this possibility was ruled out when we prepared the corresponding desmethyl derivative **71** and observed identical and disheartening behavior.

As attempt after attempt to bring about these annulations failed, we began to realize that some heretofore unknown and obviously serious limitation must be operating against us. We now believe we understand thoroughly the fundamental reasons for these observations and, as we shall see, that this knowledge can be exploited. It is instructive to consider first the successful annulation sequence **61** to **62**. When analyzed closely, three-dimensionally, a number of subtle, yet decisive, considerations unfold. Consider first the initial Michael addition step. In principle, this step can occur via a number of stereoisomeric transition states. However, if we make the reasonable assumption that there will be a steric bias favoring staggered transition states in the emerging carbon–carbon bond, then the number is reduced to 12 stereoisomeric transition states. The products of six of these, in which the geometry of the enolate is fixed as being Z, are illustrated in Scheme 17. The remaining six would be of E geometry but do not alter significantly the subsequent analysis. We submit that the transition state leading to the zwitterion shown in the box will be strongly favored for a combination of steric and electronic considerations. At the outset, the bottom two entries can be ruled out sterically. As inspection of the perspective formula at the bottom of Scheme 17 indicates, a severe 1,3-diaxial interaction between the enolate (R_3) and the C-5 axial proton would disfavor this mode of attack. The remaining four possibilities are all similar sterically (two gauche interactions each) but only the one enclosed in the box brings the developing positively and negatively charged regions of the zwitterion into close proximity. The other three possibilities, although sterically feasible, require much greater separation of charge and accordingly are disfavored.

Scheme 18 shows an alternate view of the initial Michael addition and illustrates a second but potentially very important consideration. From X-ray and other data on other enamine systems, it is known that the hybridization of nitrogen is somewhere between sp^3 and sp^2. The important point here is that if the lone-electron pair on nitrogen is anything other than sp^2-hybridized, it can only overlap with the adjacent π-cloud if it adopts a quasi-axial conformation. As such, electrophilic addition can occur, in principle,

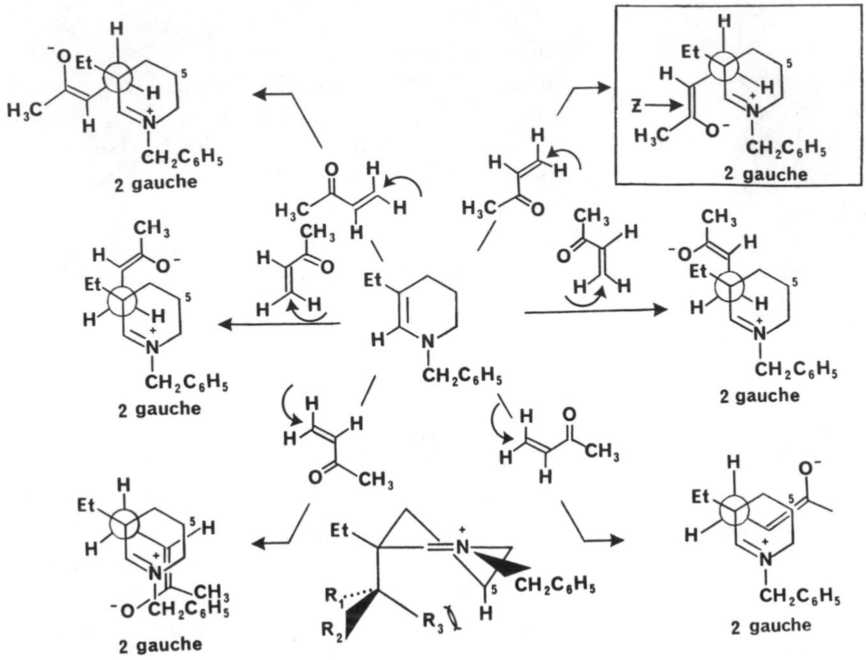

SCHEME 17

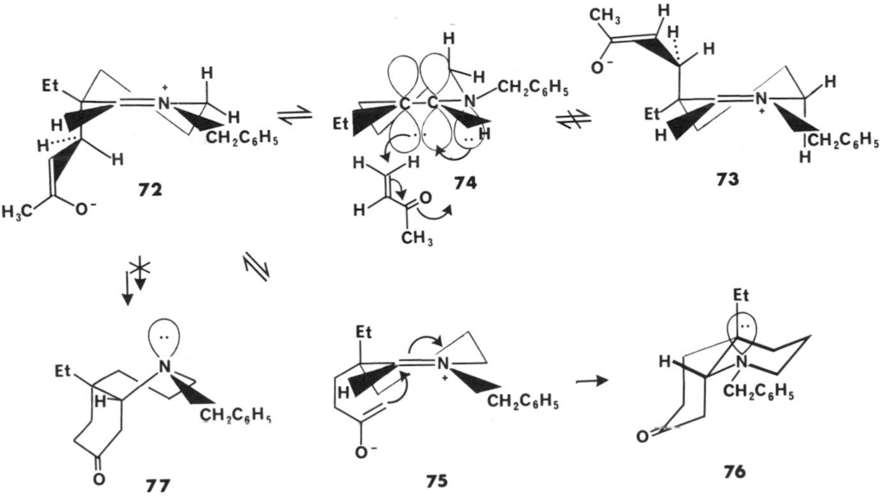

SCHEME 18

on either the same face of the molecule as the lone pair to afford **72** or on the opposite face to provide **73**. We assert that the latter mode of attack is highly unlikely. The problem involved can be seen most readily by following the fate of the axial proton α to the nitrogen. Note that in the product (**73**) this proton becomes equatorial. Therefore, a boat-like transition state is required in order to maintain maximum orbital overlap between the participating centers. In contrast, attack on the α face to produce **72** can proceed smoothly through a half-chair-like transition state. Note that the axial proton in **74** remains axial in **72**. For these reasons, we assert that in the absence of over-riding steric or electronic considerations, electrophilic additions will prefer-entially occur on the same face of the molecule as the lone-electron pair on nitrogen. The full implications of these considerations on other systems that are conformationally rigid or biased is presently being investigated. Having considered the initial Michael addition, let us turn our attention to the sub-sequent ring closure. After isomerization of the enolate to the terminal position, this closure can occur from either the conformation in which the enolate moiety is axial, as in **72** or, by conformational inversion, to one in which it is equatorial (**75**). As pointed out by Prof. Ziegler[11] of Yale Uni-versity, the latter pathway is preferred because maximum orbital overlap can be achieved via a chair-like transition state with respect to both rings (cf. **75** to **76**). Maximum orbital overlap in the alternate conformation (**72** to **77**) can be achieved only at the expense of a boat-like transition state with respect to both rings and is accordingly disfavored.

The preceding considerations provide an explanation for our inability to effect annulation of **63** to **68**. As shown in Scheme 19, the initially formed zwitterion **78** cannot cyclize readily because it would require a double boat-like transition state. In order to complete the annulation process, **78** must first undergo conformational inversion to **79**. However, in this case there is a substituent of substantial size at C-2 (Ar = 2-picolyl or 3-pyridyl). As such, this conformation would undergo substantial 1,2-allylic strain. In the face of these high energy requirements for ring closure, the intermediate zwitterion **78** is simply deprotonated at the site indicated to afford the alkylated but uncyclized enamine **69**. Thus these results provide a new perspective on, and respect for, the potential importance of stereoelectronic control in determin-ing the fate of reactions of this type.

The preceding considerations have allowed us to design a number of stereorational syntheses that incorporate the fundamental principles raised by the experimental work outlined above. Several of these are presently under active investigation; however, I would like to end this account with one such study just completed, which illustrates the type of analysis now possible. The problem was to design an efficient synthesis of karachine (**80**)

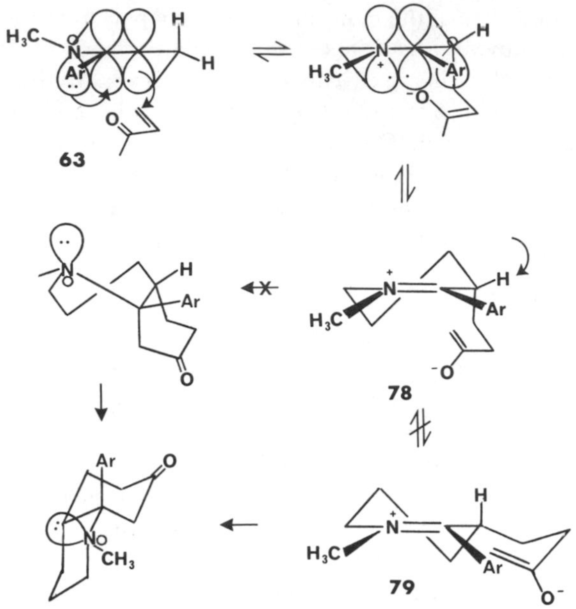

SCHEME 19

(Scheme 20), a recently isolated alkaloid of a new structural type. At first we were discouraged by the fact that the ketone moiety is not anti and coplanar to the lone-electron pair on nitrogen and therefore would require a boat-like transition state for its formation from zwitterion **81**. However, closer inspection showed that this bond and the lone-electron pair are precisely syn to one another, and although anti additions are normally favored, examples of syn additions to olefins are known. Still, we were disturbed by the energy requirements for a boat-like transition state until we realized with some degree of embarrassment that zwitterion **81** is already in a boat conformation! Thus the down payment in terms of energy would have been made in advance in the formation of this bicyclic system. Working backwards, we quickly realized that this bicyclic system could, in principle, be assembled from an intramolecular annulation of endocyclic enamine **82**, which, in turn, could be derived, in situ, from γ-alkylation of the known siloxydiene **83** with commercially available berberine. The only other potential problem envisaged concerned the stereochemistry of the Michael-addition step (**82** to **84**), but we soon realized that the stereoisomeric transition state would require much greater charge separation and accordingly should not be favored. It was with these thoughts in mind that Jim Pruitt reduced to practice[12] this simple two-step synthesis with most gratifying results.

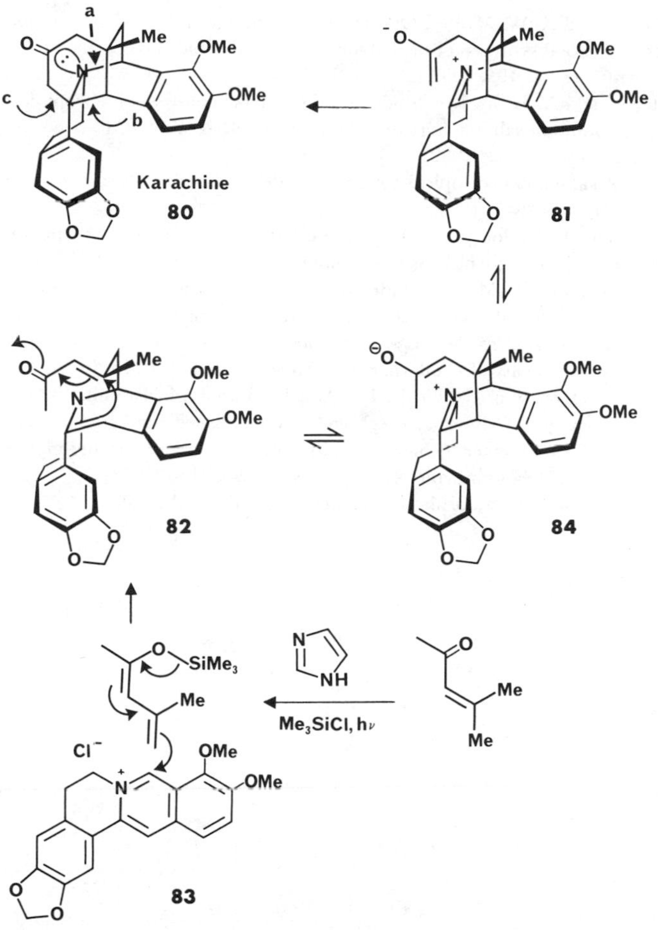

SCHEME 20

ACKNOWLEDGMENTS

I would like to take this opportunity to express my gratitude to those graduate students cited whose efforts have borne so much fruit. The support of the National Science Foundation (NSF CHE 81-15444) and the National Institutes of Health (NIH GM 28122) is also gratefully acknowledged.

REFERENCES

1. R. V. Stevens, General methods of alkaloid synthesis. *Acc. Chem. Res.* **10,** 193 (1977).
2. For a stimulating earlier overview, see W. A. Ayer and L. M. Browne, *Heterocycles* **7,** 685 (1977).

3. R. V. Stevens and A. W. M. Lee, On the stereochemistry of the Robinson Schöpf reaction A stereospecific total synthesis of the ladybug defense alkaloids precoccinelline and coccinelline. *J. Am. Chem. Soc.* **101,** 7032 (1979).

4. R. V. Stevens and A. W. M. Lee, Studies on the stereochemistry of nucleophilic additions to tetrahydropyridinium salts. A stereospecific total synthesis of (±)-Monomorine I. *J. Chem. Soc., Chem. Commun.* **2,** 102 (1982).

5. J. W. Daly, Alkaloids of neotropical poison frogs (dendrobatidae). *Fortsch. Chem. Org. Naturst.* **41,** 205 (1982).

6. T. L. Macdonald, "Indolizidine alkaloid synthesis. Preparation of the pharaoh ant trail pheremone and gephyrotoxin 223 stereoisomers. *J. Org. Chem.* **45,** 193 (1980).

7. R. V. Stevens and A. W. M. Lee, Studies on the stereochemistry of nucleophilic additions to tetrahydropyridinium salts. A stereospecific total synthesis of one of the stereoisomers of Gephyrotoxin 223. *J. Chem. Soc., Chem. Commun.* **2,** 103 (1982).

8. Y. Nakagawa, University of California, Los Angeles (unpublished results).

9. M. Kenney, University of California, Los Angeles (unpublished results).

10. N. Hrib, University of California, Los Angeles (unpublished results).

11. F. Ziegler and E. B. Spitzner, A biogenetically modeled synthesis *via* an indole acrylic ester. A total synthesis of (±)-minovine. *J. Am. Chem. Soc.* **95,** 7146 (1973).

12. J. Pruitt, University of California, Los Angeles (unpublished results).

Chapter 11

A NONBIOMIMETIC APPROACH TO THE TOTAL SYNTHESIS OF STEROIDS: THE TRANSITION METAL-CATALYZED CYCLIZATION OF ALKENES AND ALKYNES

K. Peter C. Vollhardt

Department of Chemistry
University of California, Berkeley
Berkeley, California

ISBN 0-12-450280-6

I. Introduction

A. BACKGROUND

This story begins in 1973 at the California Institute of Technology, Pasadena, where I was employed as a postdoctoral research associate under the tutelage of Prof. R. G. Bergman (who has since moved to U.C. Berkeley). Having been trained as a physical organic chemist with Dr. P. J. Garratt, University College London, I decided to broaden my horizon and work in the area of gas phase rearrangements involving strained-ring alkynes. At that time I was also eager to gain academic employment and find a new research area of my own. In this connection I was intrigued by the possibilities that were offered by transition metals, both as catalysts and as reagents, in synthesis.

Just at that time there appeared an interesting report by Macomber[1] in which a highly substituted 1,5-hexadiyne was brought to reaction with $CpCo(CO)_2$ to give an unusual strained-ring fused cyclopentadienone (Fig. 1).

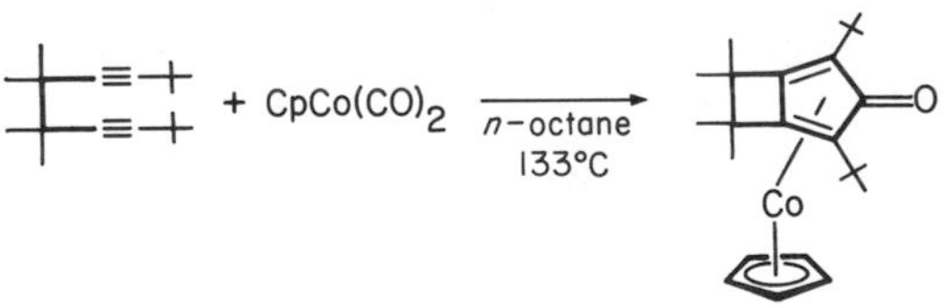

Fig. 1

B. THE INITIAL DISCOVERY

It appeared to me that it should be possible to use the same reaction to prepare the unsubstituted (parent) derivative, and I went immediately to the laboratory to treat 1,5-hexadiyne with the same cobalt complex. This was simple, since both compounds were commercially available. To my amazement, on purification of the product mixture by column chromatography, I recovered a large amount of $CpCo(CO)_2$ as the only colored material (the expected product would have been red) and a colorless, crystalline substance, which, according to mass spectrometry, turned out to be a trimer of the organic starting material. The ^1H-NMR spectrum revealed one aromatic multiplet and two aliphatic absorptions.

After some guesswork, I decided that the transformation shown in Fig. 2 must have taken place. This was remarkable because two strained rings were formed in the coordination sphere of a transition metal, in the presence of which they were obviously stable.[2] Moreover, the transition metal had

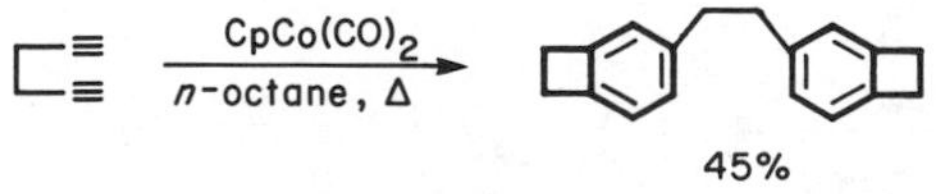

FIG. 2

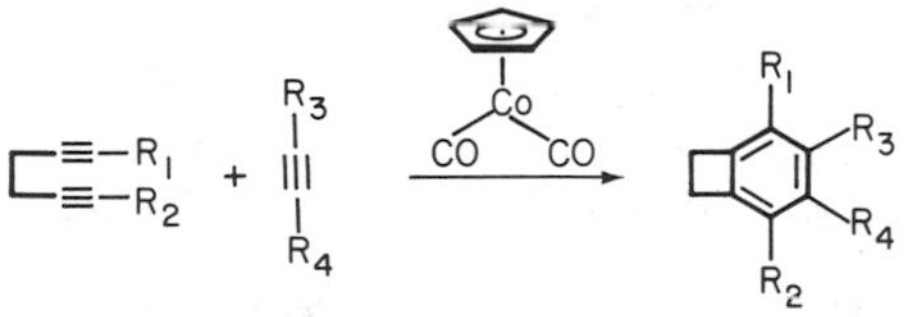

FIG. 3

apparently acted as a catalyst and had effected the formal $[2 + 2 + 2]$ cycloaddition of the alkyne units to give a fairly complex benzene derivative.

The cyclization of alkynes to benzene was well known at the time[3] but had seen only limited applications in organic synthesis.[4] Having the blessing of my postdoctoral mentor, I quickly established that 1,5-hexadiyne could be cocyclized with a variety of monoalkynes to provide a general synthesis of benzocyclobutenes (Fig. 3).[5]

C. LONG-TERM PLANS

Although the yields were only moderate, it appeared to me that this synthesis might open up useful synthetic routes to theoretically interesting strained molecules.[6] In addition, it was well known that the four-membered ring in benzocyclobutenes could be reversibly opened, on heating (Fig. 4), to the so-called o-xylylenes. These are excellent dienes in the Diels–Alder reaction, and therefore there seemed to be additional synthetic possibilities.[7] It is to the credit of Prof. R. H. Schlessinger of the University of Rochester, who was passing through Caltech at the time, to have pointed out the potential of this route in natural product syntheses, particularly steroids. An especially powerful exploitation of this method had just appeared in the literature, involving the intramolecular trapping of o-xylylene intermediates to give polycycles. Oppolzer, in this pioneering work, suggested that such an approach could be used in the construction of natural products, e.g., (±)-chelidonine (Fig. 5).[8] This strategy was later utilized by others, including us, in a variety of total syntheses.[9]

FIG. 4

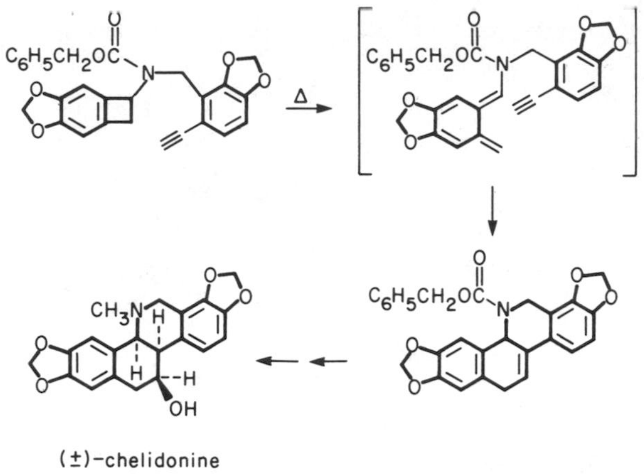

(±)-chelidonine

FIG. 5

Armed with this knowledge and some good advice from Prof. D. Seebach of the ETH Zürich, who was spending a sabbatical at Caltech, I migrated north to start an assistant professorship at the University of California, Berkeley. One of my research projects was the study of CpCo(CO)$_2$ as a potential catalyst in alkyne cyclizations en route to natural products. I was encouraged by the alacrity with which my new colleagues accepted the feasibility of what I regarded as rather farfetched ideas. They were particularly intrigued by my suggested approaches to the female sex hormone estrone, in which ring B might be derived by an intramolecular o-xylylene Diels–Alder reaction.

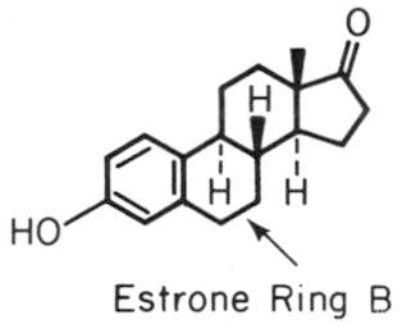

Estrone Ring B

Estrone has been the target of numerous synthetic strategies,[10] it has obvious importance as a synthetic relay point to contraceptive drugs,[11] and it appeared a reasonable challenge on which to test the usefulness of our novel synthetic methodology. But before a molecule of such complexity could be tackled, the scope and limitations of the basic cyclization reaction had to be established. It was at this point that I was joined by two superb graduate students, R. L. Funk and R. L. Hillard III, who undertook to explore this chemistry as part of their efforts to obtain a Ph.D.

II.   Model Studies in Cobalt-Mediated
Alkyne Cyclizations

A.   IMPROVING YIELDS: THE USE OF TRIMETHYLSILYL ALKYNES

The most immediate problem in this project was to find means by which
the yields of the cocyclization reaction leading to benzocyclobutenes could
be improved. We argued that in order to increase chemoselectivity, a sterically
bulky monoalkyne should be used, expected to be incapable of autocyclization, yet unhindered enough so that it would undergo cocyclization. To
investigate whether this notion was at all feasible, di-*tert*-butylacetylene was
treated with 1,5-hexadiyne in the presence of $CpCo(CO)_2$. This, however,
only furnished the diyne trimer (as in Fig. 2), indicating that the monoalkyne
was now too bulky to enter into the cyclization.

It occurred to us that attenuated bulkiness would be achieved by going
down the periodic table from carbon to silicon and switching from the *tert*-
butyl groups to trimethylsilyl substituents. This expectation was based on the
fact that the carbon–silicon bond is longer than the carbon–carbon bond. In
addition, cocyclization of a silylated alkyne would lead to a silylbenzene in
which the silicon unit might be replaced by electrophiles.[12] Our choice,
bis(trimethylsilyl)acetylene (BTMSA), was commercially available and
could also be readily made from ethyne dianion and chlorotrimethylsilane.
We quickly established that BTMSA was indeed unreactive on its own, at
least on the time scale of the ordinary cyclization experiment,[13] but that it
cocylized with 1,5-hexadiyne with increasing efficiency as its concentration
was raised. Ultimately, it was found that using BTMSA as solvent in the
presence of the catalyst and adding the diyne slowly, using syringe-pump
techniques, gave excellent yields of the cocyclized material, 4,5-bis(trimethylsilyl)benzocyclobutene (Fig. 6). This compound turned out to be a

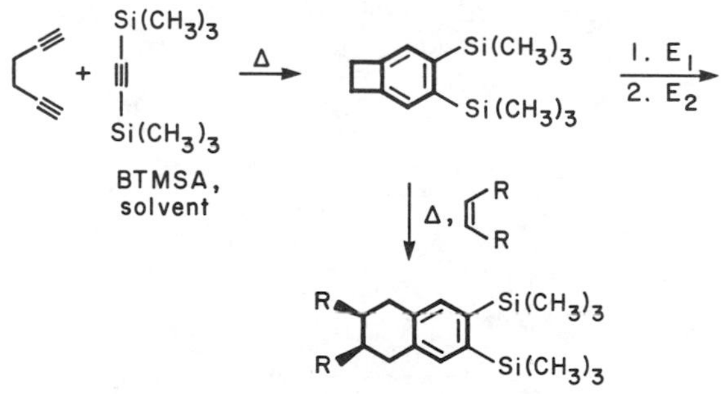

FIG. 6

very useful synthetic intermediate because the two trimethylsilyl groups could be substituted by electrophiles, selectively and stepwise (Fig. 6). Moreover, a variety of tetralin derivatives became available through its use as an o-xylylene precursor.[14]

In addition to BTMSA, we discovered that almost any trimethylsilylated alkyne could be fairly efficiently cooligomerized, not only with 1,5-hexadiyne, but also with other diynes.[14,15]

B. Mechanisms

Mechanistically, it is likely that these cyclization reactions proceed through defined organometallic intermediates; in other words, they are not concerted. A reasonable mechanistic pathway is outlined in Fig. 7. It appears that initial dissociation of one molecule of carbon monoxide is rate determining in the catalysis reaction of $CpCo(CO)_2$. Since BTMSA is used as a solvent in most cases, we assume that it functions as the first new ligand. Subsequently, the second molecule of carbon monoxide is expelled and replaced by one end of the diyne to give the bisalkyne complex **A**. The two alkyne ligands in **A** then seem to undergo so-called oxidative coupling to give the metallacycle **B**. The appended third alkyne unit in **B** is then probably inserted into the vinyl cobalt bond to give the metallacycloheptatriene **C**. Ring closure and extrusion of CpCo then gives the final product. An alternative possibility of

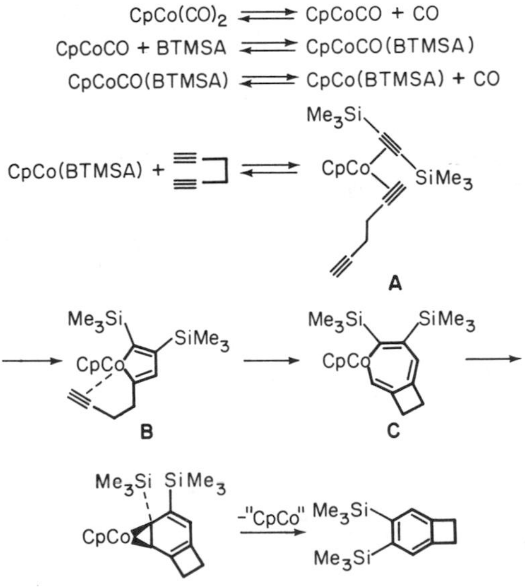

Fig. 7

converting **B** to the product is via an intramolecular Diels–Alder reaction, followed by demetalation. The likelihood that the scheme in Fig. 7 is correct, at least in its broad features, is based on a variety of organometallic studies in the literature.[16]

C. Strategy

Having solved the problem of efficiency without curtailing functional group flexibility, we turned our attention to the ultimate goal of the project, a cobalt-mediated steroid synthesis. A retrosynthetic analysis of the problem (Fig. 8) suggests as a target compound the benzocyclobutene **B**. Using cobalt catalysis, this intermediate could be approached by cocyclization of **C**, in turn probably readily available by alkylation of **E** with **D**. Since 3-substituted diynes of the type **D** were not known, we were presented with the synthetic problem of finding an efficient route to these compounds. On the other hand, we envisaged that an enolate of the type **E** would be accessible from 2-methylcyclopent-2-en-1-one by addition of a vinyl cuprate reagent.

As an alternative approach, it was thought that it might be possible to synthesize **B** through an appropriately functionalized benzocyclobutene in which the functional group would be located on the four-membered ring. This was reduced to a simple question: was it possible to cotrimerize 3-functionalized 1,5-hexadiynes?

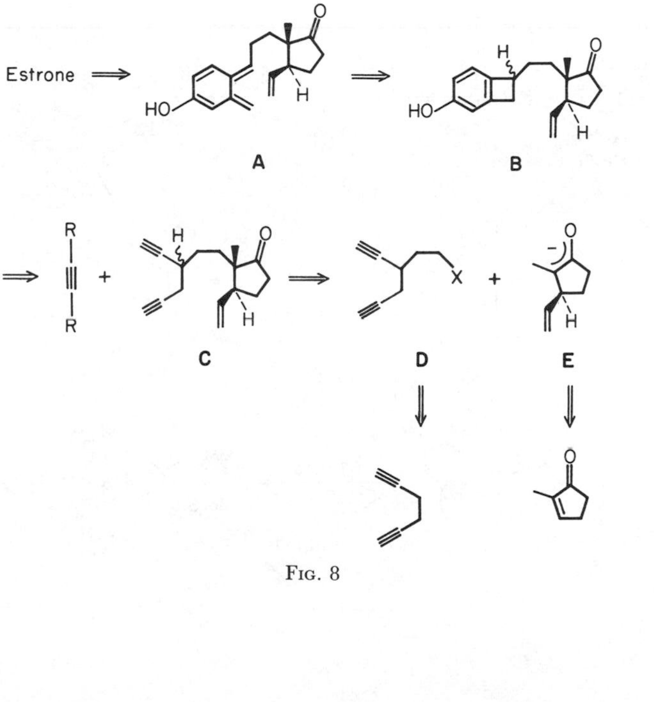

Fig. 8

D. MODEL STUDIES

To probe this question, we prepared the corresponding 1,5-diyne ethers and subjected them to cyclization with BTMSA. On chromatography, a colorless, crystalline compound was obtained, which exhibited spectral properties totally inconsistent with the expected product! The most striking discrepancy was found in the NMR spectrum, which revealed only two sharp singlets in the ratio of 9:1, one typical of a trimethylsilyl group, the other in the aromatic region. Eventually it transpired that the silylated naphthalene shown in Fig. 9 had formed in one remarkable step.

Evidently, the initially formed 3-alkoxybenzocyclobutene is unstable under the reaction conditions with respect to four-ring opening to give the isomeric *o*-xylylene. The latter is trapped by BTMSA, and the cycloadduct is aromatized by loss of an alcohol molecule to give the observed product in about 40% overall yield.[17]

We suspected that the efficiency of the process might be improved if we were to modify the R group such as to incorporate a potential intramolecular dienophile. In principle, such a system should provide rapid access to polycycles, as shown in Fig. 10.

The requisite ethers were prepared by standard techniques and on cyclization indeed provided directly the desired polycyclic systems in fair yields and with generally good stereoselectivity (first four entries in Table I).[18]

Although these results seemed to bode trouble for the synthetic strategy outlined in Fig. 8, they also presented an opportunity possibly to bypass the

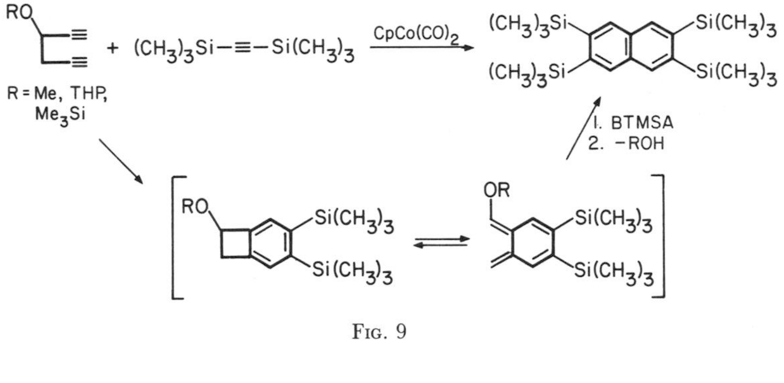

FIG. 9

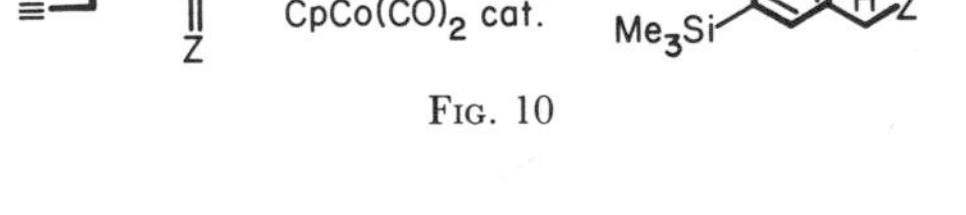

FIG. 10

isolation of intermediate **B** (Fig. 8), and to cyclize **C** directly to the desired steroid. This now made imperative the development of a synthetic method that would provide access to intermediates of type **D**.

We initially explored classical routes to the solution of this problem. To this end, we converted 1,5-hexadiyn-3-ol to the corresponding bromide in an attempt to use this compound as a precursor to organometallic intermediates that might be alkylated at the 3-position. This, however, proved unsuccessful.

At this stage, it occurred to us that it might be possible to metalate 1,5-hexadiyne directly, by employing procedures that had been described in the literature[19] to allow the conversion of a terminal alkyne to the corresponding 1,3-dilithio derivative. Indeed, treatment of our diyne with 3 equivalents of butyllithium and one of TMEDA produced the corresponding trilithio compound. Since the propargylic position in this compound is the most nucleo-

TABLE 1

POLYCYCLES BY ACETYLENE CYCLIZATION

Starting material	Product	Yield (%)
		60
		65
	+ regioisomers (unseparated mixture)	50
		45
	+ cis isomer (ca~15%)	90
	+ cis isomer (<5%)	80

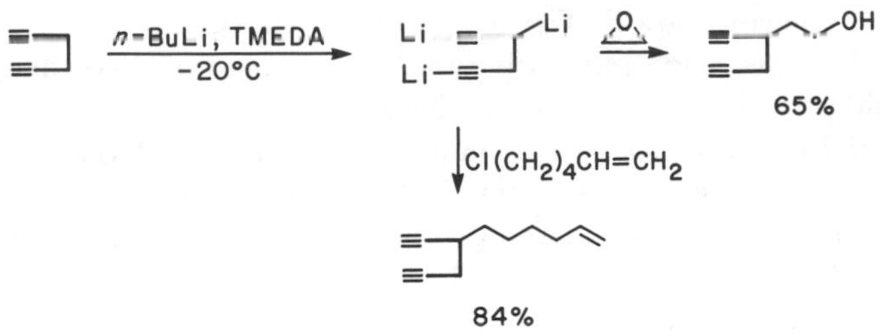

FIG. 11

philic, it could be regiospecifically alkylated to give the desired 3-substituted 1,5-hexadiynes (Fig. 11).[18] Their preparation immediately provided the opportunity of running entries 5 and 6 in Table I.[18]

Although in the case of 3-alkyl substituted 1,5-hexadiynes the intermediate benzocyclobutenes in the cyclization are stable enough to be isolated at shorter reaction times, prolonged heating leads directly to the tricycles. The yields are excellent, particularly considering that five new carbon–carbon bonds are being formed and that the products are generated with now pronounced stereospecificity: the trans-ring juncture is preferred. This suggests that the exo transition state is chosen in the Diels–Alder reaction.

We wondered whether the cobalt catalyst was somehow involved in determining the stereospecificity of this reaction. This notion was readily tested by isolating the intermediate benzocyclobutenes and, after purification, exposing them to the thermolytic conditions. Results identical to those indicated in Table I were obtained, thus making the involvement of CpCo in the Diels–Alder reaction unlikely.

E.   THE COBALT WAY TO STEROIDS: THE CATALYTIC ROUTE

Inspection of the estrone structure shows that the B,C-ring junction also has trans stereochemistry. Thus the result of the last entry in Table I fueled our optimism that a steroid synthesis should be possible. All we had to do was to add on an additional five-membered ring to the enediyne starting material. However, there were still two problems to be overcome. The required starting material **C** (Fig. 8) had to have the stereochemistry shown: namely, the vinyl group had to be positioned cis with respect to the methyl substituent. In addition, we had to worry about the stereochemistry of the B,C-ring junction relative to that of the C,D connection in the intramolecular Diels–Alder step. Even assuming an exo transition state (**A**, in Fig. 8), addition of the vinyl group to the diene from the top would provide the desired trans–anti–trans stereochemistry found in the naturally occurring steroids. However, addition

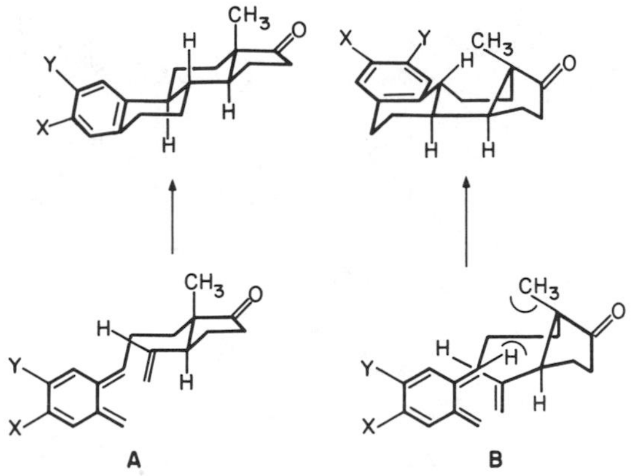

FIG. 12

of the vinyl group from underneath would furnish the trans–syn–trans arrangement.

The two possible exo transition states are depicted in Fig. 12 as **A** and **B**. Molecular models suggested to us that **A** should be preferred over **B**, because the former appeared to adopt an energetically more favorable chair conformation, the latter a boat configuration.

There was one final aspect of concern to the proposed cyclization of **C** (Fig. 8), and that had to do with the relative stereochemistry at the propargylic position. We felt that the configuration at that center was irrelevant because both diastereomers **B** (Fig. 8) should open up by a conrotatory outward movement, giving the same intermediate **A**, in which the carbon in question had become achiral.

The stage was now set to tackle the proposed steroid synthesis. Considering the results shown in Fig. 11, two strategies en route to intermediate **C** (Fig. 8) appeared attractive. The first, which ultimately proved successful, is shown in Fig. 8. However, because we were concerned about alkylating an enolate anion with a relatively unreactive alkyl halide (equilibration of the enolate and products at elevated temperatures could give mixtures), we initially embarked on an alternative approach based on the trapping of enolate **E** (Fig. 8) by a two-carbon synthon, which we hoped could be converted to an alkylating group capable of coupling with the trilithiodiyne (Fig. 11). To this end, 2-methylcyclopenten-2-one was treated with a vinyl cuprate reagent and the resulting enolate was exposed to the relatively reactive ethyl bromoacetate. It was hoped that enolate trapping would occur stereospecifically trans, for steric reasons. Disappointingly, however, an approximately 3:1 mixture of trans and cis isomers was obtained (Fig. 13). Although this mixture

Fɪɢ. 13

was not separable, the indicated product ratio was clearly established by spectroscopic data.

Despite the less than satisfactory alkylation step above, we decided to explore further the potential of this route. The next task was to convert the ester function to a potential electrophilic unit to be used as a connecting point to 1,5-hexadiyne. For this purpose, the ketone was protected and the ester subsequently reduced to give a readily separable mixture of alcohols. Tosylation and displacement with chloride gave the desired β-chloroethylcyclopentane derivative. Unfortunately, alkylation of trilithio-1,5-hexadiyne with this halide was unsuccessful at the low temperatures required to preserve the constitution of the organometallic reagent (Fig. 13). We suspected that it was the hindered nature of the potential alkylating agent that prevented its effective use. It is possible that other leaving groups might give better results. However, because we had 3-ethynyl-5-hexynol in hand (Fig. 11), we decided to employ this intermediate, using the strategy outlined in Fig. 8.

The diynol was first quantitatively converted to the p-toluenesulfonate, which on exposure to sodium iodide in acetone gave the corresponding iodo diyne (Fig. 14). We then envisaged proceeding according to Fig. 8 and using the iodide as the alkylating agent of enolate **E**. However, when the latter, generated as in Fig. 13, was exposed to 4-ethynyl-6-iodohexyne, an intractable mixture ensued. Evidently, the copper present in the mixture (derived from the vinyl cuprate addition to 2-methylcyclopentenone) plays havoc with the unprotected alkyne units of the iodide. In order to circumvent this problem, a procedure of Binkley and Heathcock[20] was called to the rescue (Fig. 14). The intermediate enolate was initially trapped with chlorotrimethylsilane to give a silyl enol ether, which could be purified by distillation. The cyclopentanone enolate could then be regenerated regiospecifically, using lithium amide in liquid ammonia. Alkylation with excess iodo diyne gave a 2:1

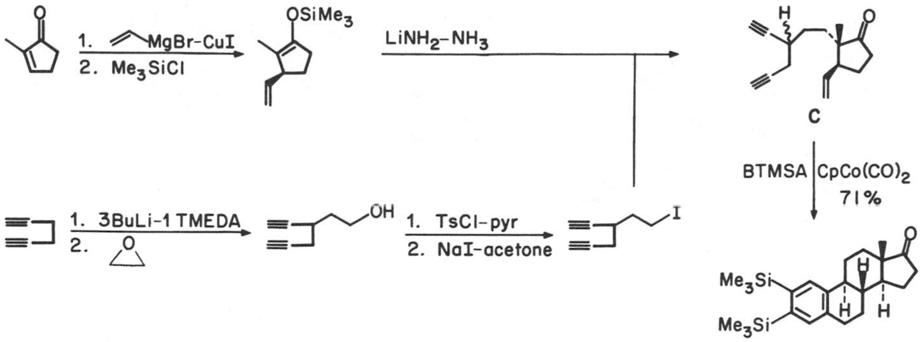

FIG. 14

mixture of the two possible pairs of diastereomers of **C** (Fig. 14). Again, this reaction did not proceed completely stereospecifically, although the major fraction of the diastereomers formed was that in which the stereochemistry at the five-membered ring was as indicated in Fig. 14, e.g., the 2-methyl group is located cis with respect to the vinyl substituent.

We were now ready to execute the exciting experiment of cocyclizing enediyne **C** with BTMSA in the presence of our cobalt catalyst. Imagine our thrill when on column chromatography of the reaction mixture a colorless crystalline material was obtained in 71% yield, exhibiting the spectral properties anticipated for 2,3-bis(trimethylsilyl)estratrien-17-one![21] Chemical structural proof was provided by protodesilylation to give the known parent estratrienone. Thus, a cobalt-mediated steroid synthesis was at hand.

It was gratifying to note that the Diels–Alder step proceeded indeed through the preferred transition state **A** (Fig. 12), as expected. Similar observations were made by other groups, although the stereochemistry of the resulting polycycle in such intramolecular Diels–Alder additions is not always predictable.[9,22] One recognizes that in the cobalt-catalyzed approach, the ABC portion of the steroid is constructed and added on to the preformed D ring in one step with extraordinary specificity.

F.   CONVERSION TO ESTRONE

In order to apply our methods to a bonafide natural product synthesis, it remained to explore its potential with respect to the preparation of A-ring-oxygenated derivatives, such as estrone. This problem raised two questions: was it possible to introduce oxygen directly in the cyclization and, if not, could then perhaps the 2,3-bis(trimethylsilyl) derivative in Fig. 14 be converted regiospecifically to a 3-oxy steroid? We initially addressed the first question and decided to explore the potential of trimethylsilylmethoxyethyne in the cotrimerization with **C** (Fig. 14). Because this monalkyne is now un-symmetrically substituted, the possibility of the formation of regioisomeric mixtures existed. However, at least in theory, one could make a reasonable case that the alkoxy substituent would emerge preferentially located at the 3-position (Fig. 15).

This argument was based on the expectation that utilizing excess mono-alkyne would insure its early incorporation into the cobalt metallacycle. Oxidative coupling to the latter was anticipated to proceed regioselectively. We had observed earlier results consistent with the exclusive formation of intermediate metallacycles in which the trimethylsilyl group was located α to the metal.[15,23] This finding appears to be a reflection of a combination of both steric and electronic effects.[24] Assuming this effect to be operational in our case, one can envisage two possible incorporations of the diyne unit into the metallacycle, according to path a or b in Fig. 15. Utilization of the pre-sumably slightly less hindered end of the diyne (path a) would give eventually the desired estrone derivative. Path b, on the other hand, would lead to the undesired 2-methoxy analog. In addition to the question of regioselectivity, we had to be concerned about the reduced bulk of the monoalkyne to be employed. There was the potential danger that autocyclization would occur, depleting this starting material from the reaction mixture before effective cocyclization could take place.

In the event, it turned out that although the latter worry was unwarranted, another even more detrimental side reaction occurred, namely, CpCo–cyclobutadiene complex formation by cyclodimerization of the alkoxy alkyne. Because this process effectively removed the catalyst from the cocylization

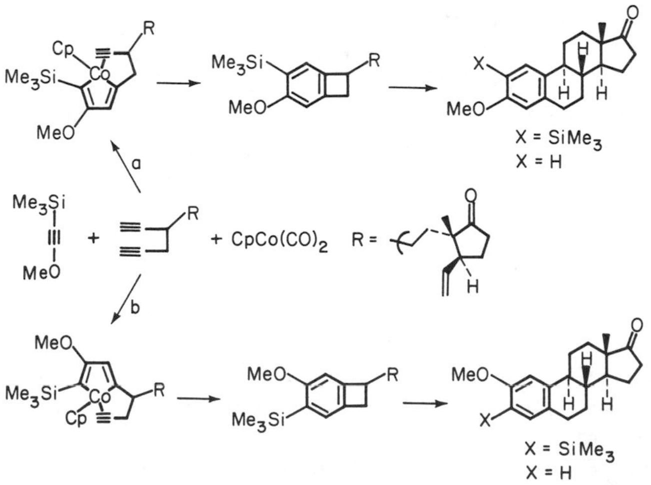

FIG. 15

mixture, yields of cyclized materials were low. Moreover, a mixture of the two possible steroids was formed in a 2:1 ratio. The identity of the products was established by protodesilylation, which showed the major isomer to be the desired one formed by path a (Fig. 15), the minor one being the 2-methoxy derivative formed by path b.

Although the foregoing method held promise, we decided rather to explore the potential of bis(trimethylsilyl)estratrienone as a precursor to estrone. Was it going to be possible to functionalize regioselectively the A ring of this system to achieve the desired goal? In a model compound, 2,3-bis(trimethylsilyl)octahydrophenanthrene (formed in entry 6 of Table I), we had observed earlier that selective bromodesilylation could be achieved, leading to a 4:1 mixture of isomeric bromides. At that time we could not make a definite assignment with respect to which isomer was which. In fact, we felt that it was the 3-position that should be more reactive, being located para to a "pseudo-isopropyl" substituent, as opposed to the 2-carbon, para to a "pseudoethyl" group. On the other hand, spectroscopic arguments favored the opposite assignment.

When 2,3-bis(trimethylsilyl)estratrienone ketal A (Fig. 16) was brominated under carefully controlled conditions, again an approximately 4:1 mixture of bromides was obtained. Structural verification was achieved by converting the products to the corresponding phenols, using a procedure developed by Hawthorne: transmetalation and treatment with trimethoxyboron to form the aryl trimethoxyborate, followed by neutralization with acetic acid to give the arylboronic acid, and finally oxidation with hydrogen peroxide.[25] Protodesilylation of the phenol mixture gave two steroids, the minor component being estrone. Thus electrophilic attack at the A ring appeared to proceed regioselectively at the 2-position. There are similar observations in

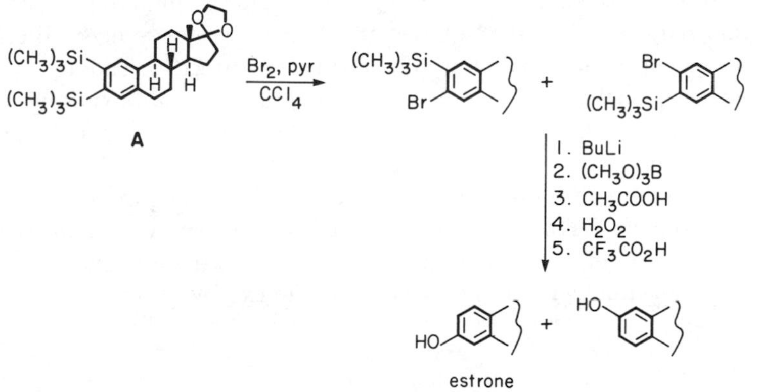

$(CH_3)_3Si$ ⟶ $\xrightarrow[CCl_4]{Br_2, pyr}$ $(CH_3)_3Si$ / Br + Br / $(CH_3)_3Si$

A

1. BuLi
2. $(CH_3O)_3B$
3. CH_3COOH
4. H_2O_2
5. CF_3CO_2H

HO + HO

estrone

FIG. 16

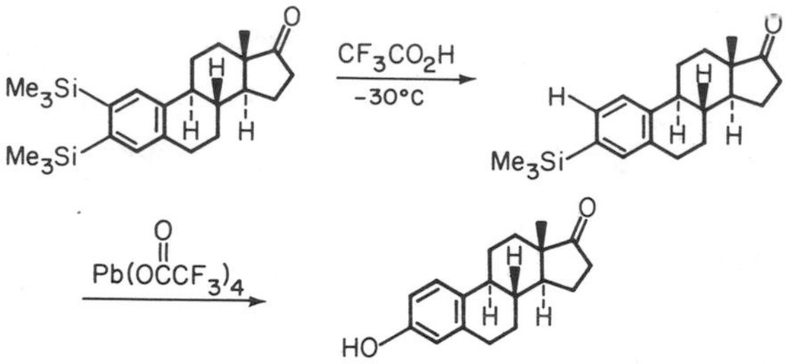

FIG. 17

the literature indicating the increased reactivity of this position in electrophilic aromatic substitution chemistry.[26] An explanation for this behavior might be found in solvation effects on the intermediate pentadienyl cation of electrophilic aromatic substitution. Attack at the 3-position results in a cation in which solvation is made less effective by the "bay region" steric hindrance effect of the C ring. Other long range effects might be operating.

Be that as it may, we realized that this preferential reactivity could be utilizable in a regiospecific introduction of the hydroxy function. If electrophilic substitution occurred more readily at the 2-position, it appeared reasonable first to attempt selectively to protodesilylate 2,3-bis(trimethylsilyl)estratrienone. Indeed, exposure of this compound to a solution of trifluoroacetic acid in carbon tetrachloride at room temperature produced a 4:1 mixture of the monotrimethylsilylated steroids derived from preferential attack at the 2-position. When this reaction was carried out at $-30°C$, the selectivity improved to 9:1. Because protodesilylation is quantitative, a 90% yield of 3-trimethylsilylestratrienone could be obtained in this way (Fig. 17).

The final obstacle in a total synthesis of estrone was now the oxidative cleavage of the phenyl–silicon bond in this product. Although, in principle, this could have been achieved by application of the bromination–boration–oxidation sequence described above, there was a reagent in the literature that was reported to effect this transformation directly: lead tetrakis(trifluoroacetate).[27] We were gratified to discover that careful exposure of this reagent to the 3-silylated steroid gave an excellent yield of the desired natural product estrone.

The cobalt-catalyzed estrone synthesis described in the preceding account constitutes the shortest preparation of racemic estrone known to date, starting from acyclic or monocyclic precursors: 5 steps from 2-methyl-cyclopenten-1-one (21.5% overall yield) and 6 steps from 1,5-hexadiyne (15.1%).[28]

G. OUTLOOK

It will be noted that the foregoing sequences generated racemic steroids. In order to modify them to become enantioselective, one would have to find conditions under which only one stereoisomer is formed in the initial addition of the vinyl cuprate reagent to 2-methylcyclopentenone (Fig. 14), because this is the step in our sequence in which the first chiral center is constructed. Discouraged by past unsuccessful attempts, found in the literature, to carry out such enantioselective cuprate additions, we decided to apply our interest in transition-metal catalysis to this problem. Would it be possible to carry out enantioselective [2 + 2 + 2] cycloadditions? Obviously, this was not viable with alkynes, only because the product of their cyclization is achiral; but what about the cocyclization of alkynes with alkenes? Since, if substituted, the alkene part in such reactions might contain prochiral centers, proper synthesis design might allow a transition metal-mediated cyclization in which achiral starting material is converted to chiral product. It was thought that a successful strategy of this type might allow the use of optically active catalysts to produce optically active products. The next section will deal with our attempts in this area, which have led to a completely new approach to the construction of the steroid nucleus.

III. Cobalt-Mediated [2 + 2 + 2] Cycloadditions en Route to Annulated Cyclohexadienes

A. INTRODUCTION

The [2 + 2 + 2] cycloaddition of three alkyne units, leading to benzene derivatives, provides a powerful addition to the synthetic organic chemist's repertoire in the synthesis of substituted benzenes. On the other hand, it is less useful in the construction of nonaromatic six-membered rings. The question that arises is whether it is possible to cyclize less unsaturated units in a similar way to generate hydroaromatic systems. Such a strategy could become a very attractive alternative to the classical Diels–Alder [4 + 2] cycloadditon: the latter makes only two new carbon–carbon bonds, the former three. Assuming that such reactions could be found, we felt that new retrosynthetic approaches to the steroid nucleus could be formulated.

Inspection of the literature showed that there were several organometallic systems in which two alkyne units had been cocyclized with an alkene to produce complexed or free cyclohexadienes, both catalytically and stoichiometrically.[29] One of those systems was based on the (now familiar) cyclopentadienylcobalt unit (Fig. 18).[30] Utilization of the latter allowed two mechanistic alternatives to be pinpointed. Assuming that the alkynes are the more reactive reaction partners, one may postulate initial formation of metallacyclopentadienes analogous to those encountered in the cyclotrimerization of alkynes. Trapping of these intermediates by the added alkene should then furnish the final product (Fig. 18), either complexed or uncomplexed, the

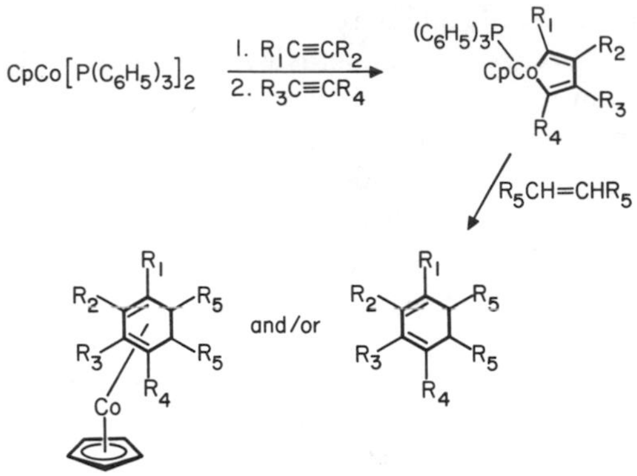

316

VOLLHARDT

FIG. 18

former requiring stoichiometric amounts of metal, the latter being the end product of a catalytic cycle.

It had indeed been shown that cobaltacyclopentadienes could be used as stoichiometric reagents in this transformation (Fig. 18).[30] The exact mode by which the alkene was incorporated into the metallacycle was, however, not known. One possibility is complexation to the metal, followed by a Diels–Alder reaction to the diene unit of the cobaltacycle. If this was correct, the cycloaddition was not stereospecific, because mixtures of both endo and exo complexes were formed with substituted alkenes. An alternative is an insertion reaction of the complexed alkene to give a seven-membered metallacycle, which would then rearrange to a product by an unspecified mechanism.

In addition to metallacyclopentadienes as intermediates, it is possible that

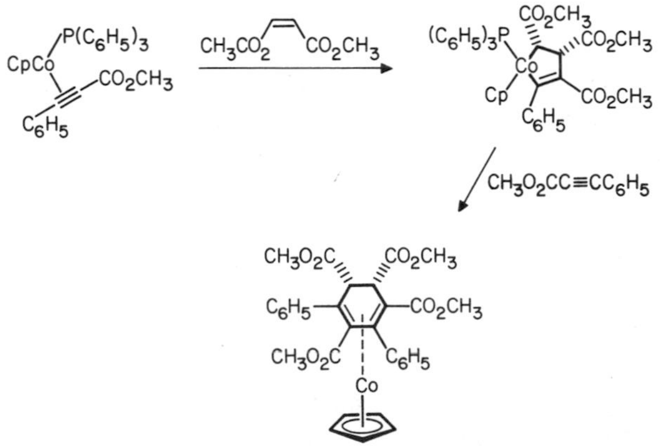

FIG. 19

metallacyclopentenes fulfil this role. Thus it has been shown that CpCo-$[(C_6H_5)_3P]$(alkyne) complexes add to alkenes to give such metallacycles (Fig. 19). Reaction with a second mole of alkyne resulted in the final complexed diene.[31]

Thus, although good precedence existed for the type of reaction in which we were interested, it was clear that mechanistically there was considerable ambiguity. Nevertheless, a new graduate student, Ethan Sternberg, felt sufficiently encouraged by these studies to begin to explore the potential of this reaction in organic synthesis.

B.   INTRAMOLECULAR CYCLIZATIONS: MODEL STUDIES

We felt that a particularly powerful application would be an intramolecular $[2 + 2 + 2]$ cycloaddition to furnish a tricyclic system in one step. For this purpose, we prepared a series of enediynes **A** (Fig. 20) and subjected them to one equivalent of $CpCo(CO)_2$ in boiling isooctane. To our delight, the desired tricycles **B** formed, in most cases in good to excellent yields.[32] The complexes turned out to be easy to handle, being exceedingly nonpolar, readily chromatographed (even by HPLC), and crystallized.

It will be noted that in these cyclizations two new chiral centers are formed: one at cobalt (organometallic compounds with unsymmetrically substituted π ligands are chiral), and one at the tertiary carbon. Consequently, there is the opportunity of forming two diastereomers. Surprisingly, the first two entries in Fig. 20 indicate that stereoselectivity is attainable, at least in certain systems. Although the origin of this effect is obscure, it could become valuable in cases where optically active metals are employed in attempts to induce enantioselectivity.

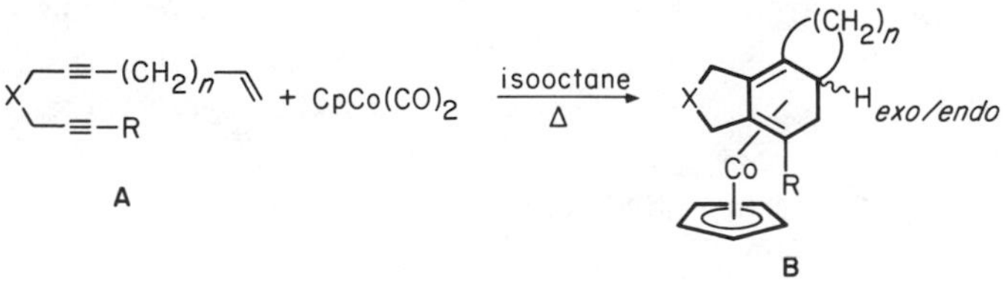

B (yield)

a: X = $(CH_2)_2$, n = 3, R = $SiMe_3$; H_{exo} only, 85%
b: X = O, n = 3, R = $SiMe_3$; H_{exo} only, 35%
c: X = $(CH_2)_2$, n = 4, R = $SiMe_3$; H_{exo}:H_{endo} = 1:1, 93%
d: X = $(CH_2)_2$, n = 3, R = H; H_{exo}:H_{endo} = 1:1, 65%
e: X = $(CH_2)_2$, n = 4, R = H; H_{exo}:(H_{endo} + 3rd isomer) = 1:1, 75%

FIG. 20

The stereochemistry of the products in Fig. 20 could be readily ascertained by NMR spectroscopy, exploiting the anisotropy of the cobalt atom. This effect causes the endo hydrogens in a complexed cyclohexadiene to resonate at relatively low field and their exo counterparts at relatively high field, sometimes above the TMS signal. In addition, X-ray crystallography confirmed the structural assignment of complex *exo*-**Bc**.[32]

Although the intramolecular [2 + 2 + 2] cycloadditions shown in Fig. 20 can be made catalytic by the use of low-valent nickel catalysts, an avenue that is currently being explored more intensively, we were quite happy that the product dienes were formed complexed to cobalt. When free, annulated dienes of the type made in the preceding cyclizations are sensitive compounds. Their instability manifests itself in polymerization, diene rearrangements, and protodesilylation reactions. However, when complexed, the metal functions as an excellent protecting group, which is also readily removed. Oxidative demetalation occurs quantitatively and instantaneously with cupric chloride at room temperature to give the free ligands.

The polycyclic diene systems shown in Fig. 20 were all found to be new compounds. This is not too surprising, since simple synthetic routes to structures of this type are difficult to come by.

C.   Enediyne Cyclizations en Route to Steroids

Although the preceding polycyclization should open up new synthetic strategies en route to a variety of natural products, we were again intrigued by its potential in steroid synthesis. In particular, the first entry in Fig. 20 appeared to us to be a nice model reaction for the construction of the BCD portion of the steroids. Retrosynthetic analysis along these lines reveals a potentially novel construction of the steroid nucleus (Fig. 21) in which the BCD framework is attached in one step to the A ring of the A-ring-aromatic steroids.

We envisaged the construction of precursor **A** by one of two routes, involving retrosynthetic cleavage of either bond a or b (Fig. 21). Bond a could be formed by some kind of metal-mediated coupling of two halides. Alternatively, bond b might be generated by the reaction of an iodo alkyne with a phenethylcopper

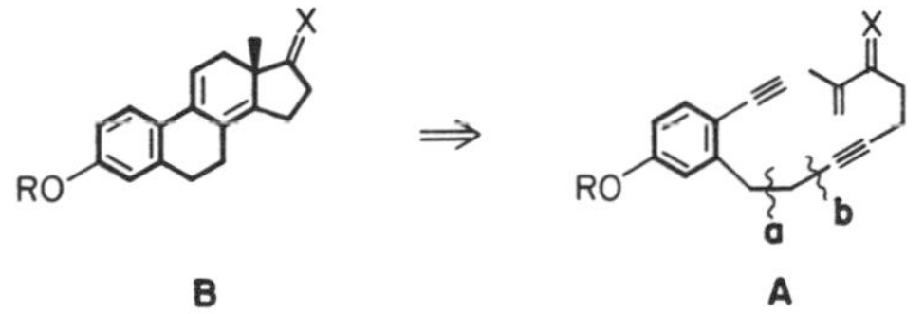

Fig. 21

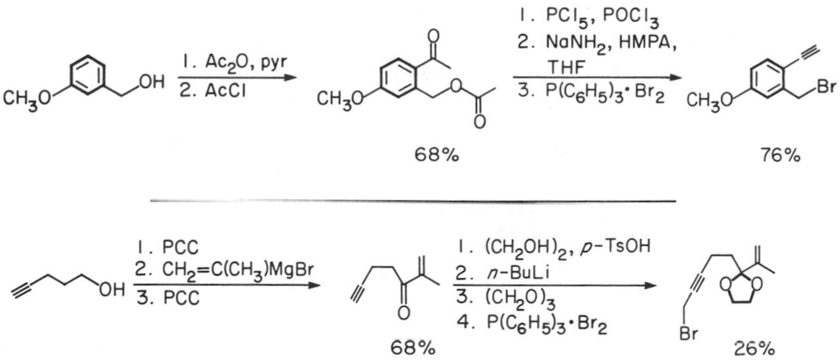

68%   76%

68%   26%

FIG. 22

derivative, exploiting a little-utilized approach to alkyne alkylation developed by Normant.[33]

Our initial investigations were directed toward the bond a approach. The preparation of the required benzylic and propargylic bromides is depicted in Fig. 22 and relies on straightforward methodology.[34] Having the two halves in hand, attempts at their coupling were initiated. Unfortunately, these trials ran into extensive difficulties. Although "brute force" Wurtz-type coupling (*tert*-butyllithium) gave about 25% of the desired enediyne, several other compounds were formed, derived from homocoupling, reduction, and *tert*-butylation. Changing solvents, utilization of sodium instead of lithium, application of other alkyllithium reagents, and stepwise metalation sequences were all unsuccessful.

Disappointed by this failure, we decided to explore route b (Fig. 21). Starting from *m*-methoxyphenethyl alcohol and proceeding through a sequence identical to the one depicted in the top half of Fig. 22 allowed ready access to starting material **A** (Fig. 23). Treatment with Rieke magnesium,[35] followed by cuprate formation, and addition of the iodo alkyne **B** gave, to

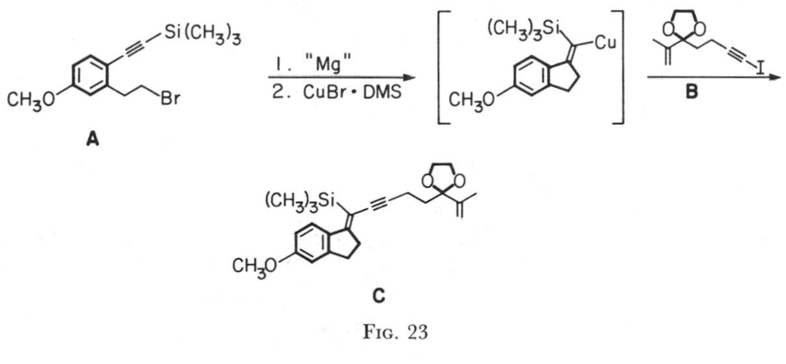

FIG. 23

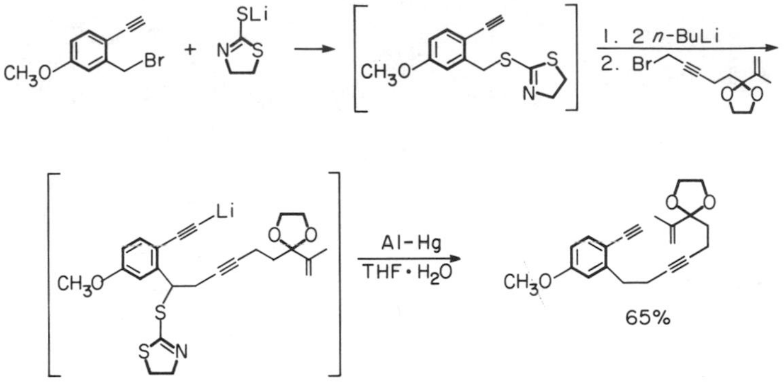

FIG. 24

our disappointment, only the cyclized product **C**. The formation of the latter was strongly indicative of the occurrence of a radical ring closure to an intermediate vinyl copper reagent (shown in brackets).[36]

This failure, which could not be circumvented by modification of the reaction conditions, prompted another go at the type a approach (Fig. 21). For this purpose we sought the help of a sulfur template, which had been demonstrated in the literature to allow coupling reactions of the type desired. This template was based on the thiazoline nucleus,[37] and its utilization in our system lead to the results depicted in Fig. 24. In a one-pot sequence, thiazoline-2-thiolate was benzylated, propargylated, and then desulfurized to give a 65% yield of the desired enediyne. Gratifyingly, on treatment with $CpCo(CO)_2$ this compound cyclized as anticipated to give only one stereoisomer of the steroid complex **A** (Fig. 25).[34] Thus, our desired goal had been reached.

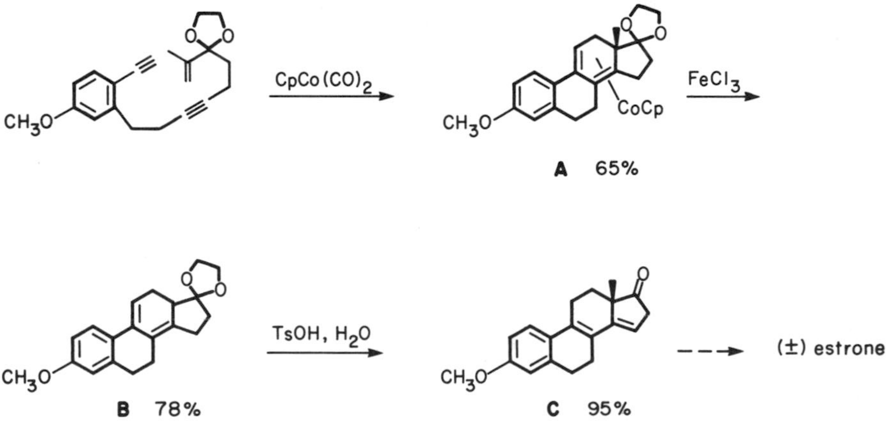

FIG. 25

D.   SYNTHESIS OF THE TORGOV DIENE

We had noticed earlier that the ligand present in **A** (Fig. 25) was an un-known organic entity (at least according to *Chemical Abstracts*). This appeared very surprising since a large number of isomeric estrapentaene deriva-tives were known. The puzzle was resolved when we carried out the oxidative demetalation of complex **A**. A highly air-sensitive, colorless substance was obtained, which visibly decomposed on exposure to air, either neat or in solution, to give intractable white solids. Only when the ligand was liberated under strictly oxygen-free conditions could we obtain a characterizable com-pound, the spectral properties of which left little doubt that we had isolated the cross-conjugated steroid **B** (Fig. 25). This material was thermodynamically unstable with respect to acid-catalyzed rearrangement and deketalization to give the steroid diene **C**, a well-known intermediate in the classical Torgov synthesis of estrone.[38] Thus, the construction of **C** in Fig. 25 provides yet another approach to this substance, employing cobalt-mediated enyne cyclization methodology. This constitutes, to our knowledge, the first success-ful A → BCD construction of A-ring-aromatic steroids. Related acid-induced biomimetic alkene cyclizations en route to steroids have been reported but contrast topologically and mechanistically.[39]

E.   MORE ON STEREOCHEMISTRY: FURTHER MODELS

The complete stereoselectivity in the cyclization leading to **A** (Fig. 25) was again remarkable. In order to probe this question further, we prepared

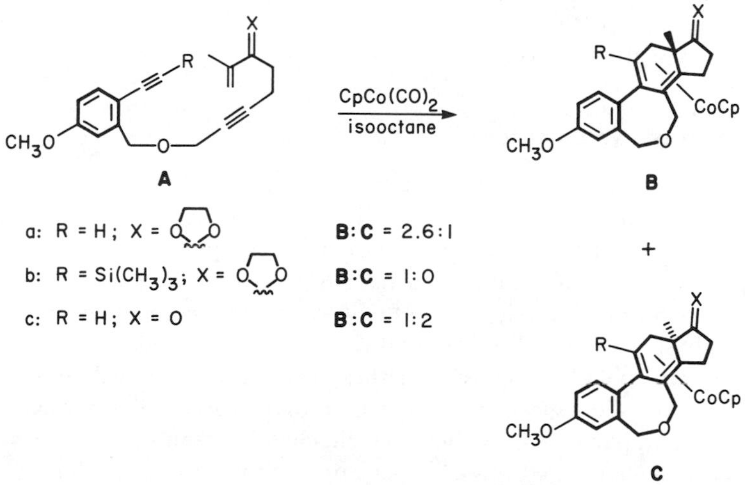

FIG. 26

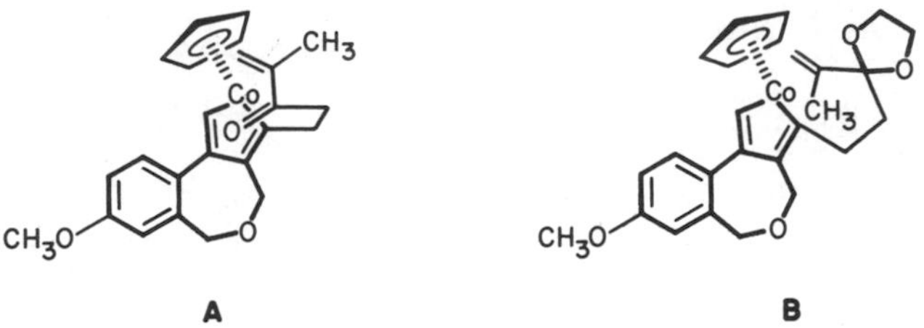

A   B

FIG. 27

the enediyne ethers **A** (Fig. 26), bearing varying R and X groups. On cyclization, the 7-oxa-B homosteroid complexes **B** and **C** were formed in changing
ratios. Interestingly, depending on the presence or absence of the ketal function, either **B** or **C** was seen to predominate. If one assumes that the product
stereoselectivity is determined by an intramolecular Diels–Alder step to a
cobaltacyclopentadiene (Fig. 27), then steric and electronic factors might
dictate predominant formation of **A** (Fig. 27) in the case of entry **c** (Fig. 26),
but more of **B** (Fig. 27) in entries **a** and **b** (Fig. 26).[34] Again, the added trimethylsilyl group appears to enforce stereoselectivity more strictly.

The issue of stereoselectivity is important here not only from a mechanistic
standpoint, but also when considering potentially enantioselective syntheses
using chiral and optically active catalysts. It is encouraging to have found
indications in the foregoing models that complete control of relative stereochemistry might be possible by appropriate manipulation of the organic
substrate. This suggests that one should be able to obtain either enantiomer
of the cyclized product, using the same optically active catalyst but different
protecting groups. Whether this is feasible in practice will have to await the
outcome of further experimentation.

F. SUMMARY AND OUTLOOK

In summary, it is clear that cobalt-mediated $[2 + 2 + 2]$ cycloadditions
have become valuable additions to synthetic organic methodology. Relatively
simple starting materials are converted to rather complex products. Of course,
the appropriate starting compounds have to be readily available. This necessitates the continued development of new synthetic methodology en route to
specifically substituted alkenes and alkynes.

The organometallic chemistry of these cyclizations has led to a further
understanding of the mechanisms by which they proceed. Further expansion
of the method, particularly in heterocyclic chemistry, and in the synthesis of
five- and four-membered rings,[40] opens up exciting synthetic avenues. Ultimately, ways should be found to effect the cyclization of two or even three

alkene units to give the corresponding six-membered-ring derivatives with appropriate stereoselectivity. The reader can readily ascertain the power of the method by picking a complicated natural product containing several fused cyclohexanes and subjecting it to the various possible $[2 + 2 + 2]$ retrocycloadditions. Many molecules unzip to dramatically simplified structures!

ACKNOWLEDGMENTS

I wish to thank my students named in the references pertaining to our work for their enthusiastic collaboration. Permission was granted to reprint Fig. 20 largely from Ref. 32, and Fig. 25–27 from Ref. 34; copyright 1980 and 1982, respectively, American Chemical Society. We are grateful to the New York Academy of Sciences for allowing us to reprint some of the drawings in Ref. 28, and to Pergamon Press for similar permission regarding Ref. 29. This work was supported by NSF and NIH.

REFERENCES

1. R. S. Macomber, *J. Org. Chem.* **38,** 816 (1973).
2. K. C. Bishop, III, *Chem. Rev.* **76,** 461 (1976).
3. C. W. Bird, "Transition Metal Intermediates in Organic Synthesis," Chapter 1. Academic Press, New York, 1967; F. L. Bowden and A. B. P. Lever, *Organomet. Chem. Rev.*, **3,** 227 (1968).
4. E. Müller, *Synthesis* p. 761 (1974).
5. K. P. C. Vollhardt and R. G. Bergman, *J. Am. Chem. Soc.* **96,** 4996 (1974).
6. M. P. Cava and M. J. Mitchell, "Cyclobutadiene and Related Compounds." Academic Press, New York, 1967.
7. I. Klundt, *Chem. Rev.* **70,** 471 (1970).
8. W. Oppolzer and K. Keller, *J. Am. Chem. Soc.* **93,** 3836 (1971); W. Oppolzer, *ibid.* pp. 3833, 3834.
9. See Y. Ito, M. Nakatsuka, and T. Saegusa, *J. Am. Chem. Soc.* **104,** 7609 (1982), and the references therein.
10. A. A. Akhrem and Y. A. Titov, "Total Steroid Synthesis." Plenum, New York, 1970; R. T. Blickenstaff, A. C. Ghosh, and G. C. Wolf, "Total Synthesis of Steroids," Academic Press, New York, 1974.
11. D. Lednicer, ed., "Contraception: The Chemical Control of Fertility." Dekker, New York, 1969; D. Lednicer and L. A. Mitscher, "Organic Chemistry of Drug Synthesis." Wiley, New York, 1977.
12. E. Colvin, "Silicon in Organic Synthesis," Butterworth, London, 1981.
13. See, however, J. R. Fritch, K. P. C. Vollhardt, M. R. Thompson, and V. W. Day, *J. Am. Chem. Soc.* **101,** 2768 (1979).
14. W. G. L. Aalbersberg, A. J. Barkovich, R. L. Funk, R. L. Hillard, III, and K. P. C. Vollhardt, *J. Am. Chem. Soc.* **97,** 5600 (1975); R. L. Hillard, III and K. P. C. Vollhardt, *ibid* **99,** 4058 (1977).
15. K. P. C. Vollhardt, *Acc. Chem. Res.* **10,** 1 (1977); R. L. Hillard, III and K. P. C. Vollhardt, *Tetrahedron* **36,** 2435 (1980); E. R. F. Gesing, J. A. Sinclair, and K. P. C. Vollhardt, *J. Chem. Soc., Chem. Commun.* p. 286 (1980).
16. See, for example, D. R. McAlister, J. E. Bercaw, and R. G. Bergman, *J. Am. Chem. Soc.* **99,** 1666 (1977); H. Yamazaki and Y. Wakatsuki, *J. Organomet. Chem.* **139,** 157 (1977).
17. R. L. Funk and K. P. C. Vollhardt, *J. Chem. Soc., Chem. Commun.* p. 833 (1976).

18. R. L. Funk and K. P. C. Vollhardt, *J. Am. Chem. Soc.* **98,** 6755 (1976); **102,** 5245 (1980).
19. S. Bhanu and F. Scheinmann, *J. Chem. Soc., Perkin Trans.* **1,** p. 1218 (1979).
20. E. S. Binkley and C. H. Heathcock, *J. Org. Chem.* **40,** 2156 (1975).
21. R. L. Funk and K. P. C. Vollhardt, *J. Am. Chem. Soc.* **99,** 5483 (1977).
22. R. L. Funk and K. P. C. Vollhardt, *Chem. Soc. Rev.* **9,** 41 (1980).
23. E. R. F. Gesing, J. P. Tane, and K. P. C. Vollhardt, *Angew. Chem.* **92,** 1057 (1980); *Angew. Chem., Int. Ed. Engl.* **19,** 1023 (1980); C. Chang, J. A. King, Jr., and K. P. C. Vollhardt, *J. Chem. Soc., Chem. Commun.* p. 53 (1981); D. J. Brien, A. Naiman, and K. P. C. Vollhardt, *ibid.* p. 133 (1982).
24. A. Stockis and R. Hoffmann, *J. Am. Chem. Soc.* **102,** 2952 (1980).
25. M. F. Hawthorne, *J. Org. Chem.* **22,** 1001 (1957).
26. T. Nambara, S. Honma, and S. Akiyama, *Chem. Pharm. Bull.* **18,** 474 (1970); T. Nambara, M. Numazawa, and S. Akiyama, *ibid.* **19,** 153 (1971).
27. J. R. Kalman, J. T. Pinhey, and S. Sternhell, *Tetrahedron Lett.* p. 5369 (1972).
28. R. L. Funk and K. P. C. Vollhardt, *J. Am. Chem. Soc.* **101,** 215 (1979); **102,** 5253 (1980); K. P. C. Vollhardt, *Ann. N.Y. Acad. Sci.* **333,** 241 (1980).
29. See C. Chang, C. G. Francisco, T. R. Gadek, J. A. King, Jr., E. D. Sternberg, and K. P. C. Vollhardt, *in* "Organic Synthesis Today and Tomorrow" (B. M. Trost and C. R. Hutchinson, eds.), p. 71, and the references therein. Pergamon, Oxford, 1981.
30. Y. Wakatsuki, T. Kuramitsu, and H. Yamazaki, *Tetrahedron Lett.* p. 4549 (1974); Y. Wakatsuki and H. Yamazaki, *J. Organomet. Chem.* **139,** 169 (1977).
31. Y. Wakatsuki, K. Aoki, and H. Yamazaki, *J. Am. Chem. Soc.* **101,** 1123 (1979).
32. E. D. Sternberg and K. P. C. Vollhardt, *J. Am. Chem. Soc.* **102,** 4839 (1980).
33. A. Commerçon, J. F. Normant, and J. Villeras, *Tetrahedron* **36,** 1215 (1980).
34. E. D. Sternberg and K. P. C. Vollhardt, *J. Org. Chem.* **47,** 3447 (1982).
35. R. D. Rieke and S. E. Bales, *J. Am. Chem. Soc.* **96,** 1775 (1974); R. T. Arnold and S. T. Kulenovic, *Synth. Commun.* **7,** 223 (1977).
36. M. Dagonneau, *Bull. Soc. Chim. Fr.* [2] p. 269 (1982).
37. K. Hirai and Y. Kishida, *Tetrahedron Lett.* p. 2117 (1972); K. Hirai, Y. Iwano, and Y. Kishida, *ibid.* p. 2677 (1977).
38. S. N. Ananchenko and I. V. Torgov, *Tetrahedron Lett.* p. 1553 (1963); A. V. Zakharychev, S. N. Ananchenko, and I. V. Torgov, *Steroids* **4,** 31 (1964); G. H. Douglas, J. M. Graves, G. A. Hughes, B. J. McLoughlin, J. Siddall, and H. Smith, *J. Chem. Soc.* p. 5073 (1963).
39. W. S. Johnson, C. E. Ward, S. G. Boots, M. B. Gravestock, R. L. Markezich, B. E. McCarry, D. A. Okorie, and R. J. Parry, *J. Am. Chem. Soc.* **103,** 88 (1981); see also W. S. Johnson, D. Berner, D. J. Dumas, P. J. R. Nederlof, and J. Welch, *ibid.* **104,** 3508 (1982); W. S. Johnson, D. J. Dumas, and D. Berner, *ibid.* p. 3510.
40. See D. J. Brien, A. Naiman, and K. P. C. Vollhardt, *J. Chem. Soc., Chem. Commun.* p. 133 (1982); J. P. Tane and K. P. C. Vollhardt, *Angew. Chem.* **94,** 642 (1982); *Angew. Chem., Int. Ed. Engl.* **21,** 617 (1982); *Angew. Chem., Suppl.* p. 1360 (1982); B. C. Berris, Y.-H. Lai, and K. P. C. Vollhardt, *J. Chem. Soc., Chem. Commun.* p. 953 (1982).

Chapter 12

EVOLUTION OF A STRATEGY FOR TOTAL SYNTHESIS OF STREPTONIGRIN

Steven M. Weinreb

Department of Chemistry
Pennsylvania State University
University Park, Pennsylvania

I. Introduction

In 1974 we became interested in initiating a research program leading to synthesis of streptonigrin (**1**). This fascinating molecule was first isolated from *Streptomyces flocculus* in 1959,[1] and its structure was subsequently elucidated by Woodward, Biemann, and Rao in 1963, using extensive degradative chemistry coupled with minimal spectral data.[2] We initially felt some trepidation about the correctness of structure **1**, in part because of its rather unprecedented heterocyclic array, and because an obvious biogenetic pathway did not immediately come to mind. These concerns were laid to rest shortly after our synthetic studies had been started when Chiu and Lipscomb[3] determined by X-ray crystallography that the structure of streptonigrin is indeed **1**. In retrospect, one can now fully appreciate the brilliance of the

MeO 6 O 5 4 3
A B 2
NH₂ 7 8 N 6' N 2' CO₂H
O C 3'
NH₂ 5' 4' Me
OH
D
OMe
OMe

1

deductive reasoning that led to the correct formulation for the natural product. It might also be noted that recently Gould and Cane have established a unique biogenesis for streptonigrin.[4] Our interest in developing a synthesis of **1** was heightened by several reports of its clinical effectiveness as a cancer chemotherapy agent. We recently reviewed[5] in detail all aspects of the structure determination, biological properties, synthesis, and biosynthesis of streptonigrin. This chapter will focus exclusively on the evolution of a strategy for its total synthesis, which was finally realized in our laboratories in 1980.

II. Synthetic Strategy

When one inspects streptonigrin (**1**) with an eye toward synthesis, the central pyridine C ring immediately stands out as the pivotal structural feature of the molecule. This heterocyclic ring is fully substituted with five different groups that must in some way be introduced with the correct relationship. A very high degree of functionality is also contained in the remainder of structure **1**. The molecule appears even more challenging from a synthetic chemist's viewpoint by the fact that all functionality is contained entirely in three aromatic systems tightly connected by two biaryl linkages. In addition, one would certainly suspect the stability of the quinolinequinone AB system of streptonigrin, and thus in preliminary synthetic planning it seemed only prudent to leave this functionality in a masked form until the last possible moment.

One rather obvious strategy for creating the tetracyclic framework of **1** might utilize mixed Ullmann couplings, or the equivalent, to form one or both of the biaryl bonds. From a practical point of view, it was felt that this type of coupling might be difficult to achieve in good yield, particularly in regard to generating the quinoline–pyridine linkage. This general approach, however, seems somewhat more reasonable now than it did in the mid-1970s because of recent advances in "Ullmann" methodology. In fact, such an approach has since been successfully tested by Cheng and co-workers in a model CD system.[6]

At a very early stage, it was decided to avoid any route to the C ring of **1** that utilized pyridine electrophilic substitution. Such chemistry did not have aesthetic appeal, and because of the generally poor reactivity that pyridines show toward electrophiles it did not seem that the oxygen-rich benzenoid D ring would allow selectivity.

After ruling out the aforementioned strategies, we were faced with the remaining possibility of constructing the C ring from nonpyridine precursors. In the late 1960s Kametani and co-workers had reported several approaches to the pyridine ring of streptonigrin using classical methods (i.e., Hantzch synthesis) for pyridine or 2-pyridone preparation.[5] Since we did not want to duplicate closely the Japanese work, and because it is obligatory for academic groups to develop novel chemistry, we sought a less ordinary entry into the streptonigrin pyridyl system.

Upon perusal of the literature, we noted a report by Ben-Ishai[7] that methoxyhydantoins such as **2**, upon heating or Lewis acid treatment, eliminate methanol to afford unstable acyl imines (**3**), which can be trapped by various 1,3-dienes in Diels–Alder fashion to produce tetrahydropyridine adducts (**4**). It can readily be seen that **4**, in principle, could serve as a precursor for a substituted pyridine-2-carboxylic acid. On the basis of this

reaction, the first generation (and perhaps naive) strategy shown in Scheme 1 was conceived for total synthesis of streptonigrin. As discussed below, during the years that this project was under investigation a number of key modifications to the original plan became necessary.

It was our intention to combine the trisubstituted unsymmetrical diene **5** with hydantoin **2** to produce adduct **6**. Ideally, a tetrasubstituted diene would have been preferable to **5**, but we felt that it might prove very difficult to synthesize such a diene and, more important, it was not at all clear what the regiochemical outcome of the addition of a complicated tetrasubstituted unsymmetrical diene to **2** would be. Although there was little information available concerning the regiochemistry of imino Diels–Alder reactions[8] with complex dienes when we began, there was enough known to convince us that **6** should be the major adduct of **2** and **5** (see below).

Using the styrene double bond of **6** as a "handle," it seemed reasonable that this compound could be transformed to enone **7**, perhaps via the corresponding epoxide. Conversion of **7** to the imine–enamine **8** would give a system formally in the dihydropyridine oxidation state, which upon further oxidation should readily provide the desired fully substituted pyridine **9**.

Because we wanted to use an unsymmetrical diene in the Diels–Alder step

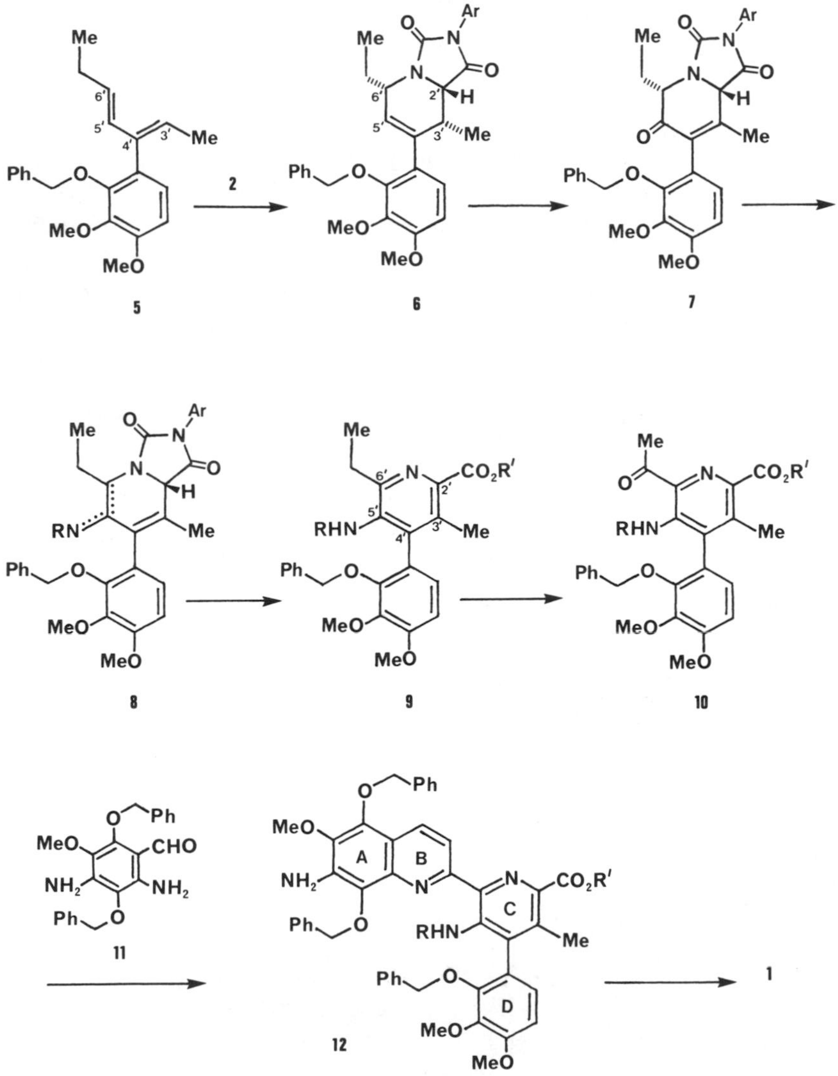

Scheme 1

which would allow for rational control of regiochemistry, it was appealing to introduce an ethyl group at C-6′ of the adduct, which could be later functionalized via a known pyridine N-oxide rearrangement.[9] Thus, it was planned eventually to convert pyridine **9** to acetylpyridine **10**. Construction of the tetracyclic nucleus of streptonigrin was envisioned by Friedlander coupling of pyridine **10** with o-aminobenzaldehyde **11**, affording quinoline **12**. Completion of the synthesis would involve elaboration of the A ring to the quinone and removal of whatever protecting groups proved necessary. It might be noted at this point that it was unclear which functional groups might need protection, but we were rather concerned with the possible instability of the pyridine C-5′ amino substituent in the presence of various oxidizing agents and electrophiles. As it turned out, protection of this group was not required.

III. Preliminary AB-Ring Studies

With this synthetic plan roughly developed on paper, we proceeded to the laboratory to test some of the crucial steps in simplified model systems. Our first goal was to evaluate the feasibility of effecting a Friedlander quinoline synthesis between o-aminobenzaldehyde **11** and a simple 2-acetylpyridine. We quickly found that **11** was not as accessible as first anticipated, and, in fact, we never could synthesize this compound.[10] In an attempt to effect a modified Friedlander sequence,[11] nitro aldehyde **13** was successfully condensed with 2-acetylpyridine to afford chalcone **14**, but it was not possible to

cyclize **14** reductively to pyridylquinoline **15**. After considerable experimentation[10] it was finally concluded that for various reasons protected hydroquinone A-ring precursors were not useful in the proposed Friedlander-type quinoline syntheses.[11] Thus it was decided to use a less substituted starting material to produce the quinolinequinone system, and we turned to the known, easily prepared amino aldehyde **16**.

Condensation of **16** with 2-acetylpyridine, using Triton B as catalyst, gave pyridylquinoline **17**, but only in mediocre yield. Because this was only a model Friedlander reaction, the yield of **17** was not optimized. This poor step did highlight a potential problem that clearly would eventually have to be

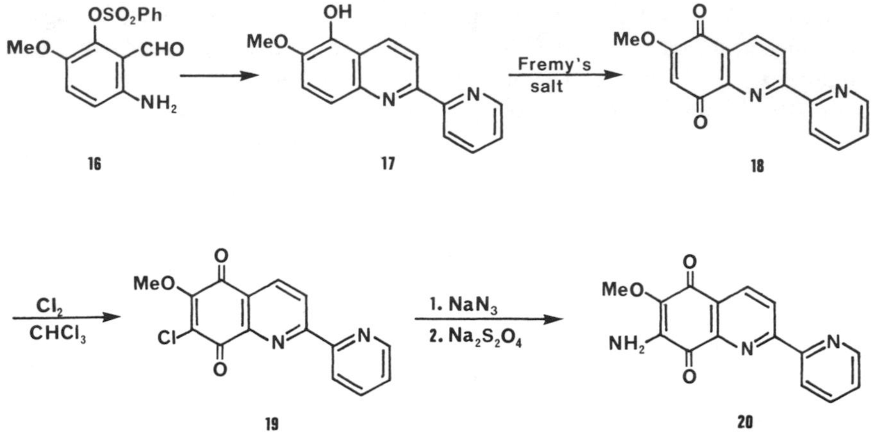

SCHEME 2

addressed. Using methodology borrowed from the studies of several research groups[5] it was possible to convert **17** to quinolinequinone **20** efficiently (Scheme 2).

IV. Preliminary CD-Ring Studies

While this work on the AB-ring system was in progress, we began an extensive series of model studies on construction of the pyridine C ring. As mentioned above, the regiochemistry of cycloaddition of imino dienophiles with unsymmetrical dienes had not been investigated to any great extent. Thus it seemed prudent that we first explore the regiochemistry of the proposed Diels–Alder step shown in Scheme 1. From the sparse information available in the literature,[8] it appeared that, in general, cycloadditions of electron-deficient imines with 1,3-dienes probably proceed via dipolar transition states or intermediates. For example, addition of unsymmetrical diene **21** and hydantoin **2** can provide isomeric transition states **22** and/or **23**. If substituent R is a carbocation-stabilizing group, one would predict that the major regioisomeric adduct would be derived from **22**.

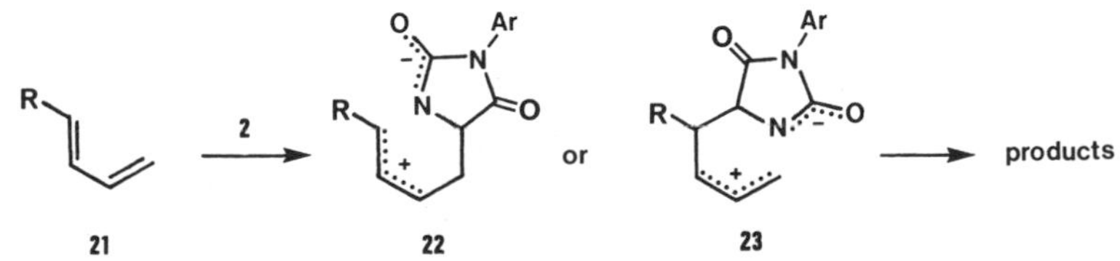

In order to test this theory with a trisubstituted diene such as **5** (Scheme 1), we prepared model compounds **25** and **28** by Wittig reactions of aldehyde **24**.[12] These dienes were mixtures of E–Z-disubstituted double-bond isomers that were not separated. Thermal cycloaddition of **25** with hydantoin **2** in refluxing xylene gave a mixture of chromatographically separable regioisomeric products in a 3:1 ratio with the desired (and predicted) isomer **26** being the major product. The structure and stereochemistry of adducts **26** and **27** were unambiguously established by ^{1}H-NMR decoupling experiments. Not surprisingly, no products were formed from (Z)-diene **25**, which was recovered unchanged. This was a point of concern, because in the actual streptonigrin synthesis we did not want to waste starting materials needlessly, and so we felt that a method would ultimately have to be found to prepare a geometrically pure (E)-diene. Although the regioselectivity of this cycloaddition was not outstanding, it did appear to be sufficiently in our favor to warrant continuation of the approach. Several attempts were also made to use Lewis acid catalysis in the hope that it would improve the regioselectivity, but diene **25** was quite susceptible to polymerization under these conditions. Similarly, the methyl-substituted diene **28** added to **2** thermally to afford a separable 2:1 mixture of adducts **29** and **30**.

Having the ethyl adduct **26** and methyl compound **29** in hand, we began to investigate the strategy shown in Scheme 1 for synthesis of a suitable pentasubstituted pyridine (cf. **6** → **9**).[13,14] Compound **26** was cleanly and stereospecifically epoxidized to produce **31**, but all attempts to rearrange it with acids to ketone **32** failed, giving instead aldehyde **33** (see Scheme 3). The epoxide ring of **31** could be opened with various acids (e.g., HBr, TFA, etc.)

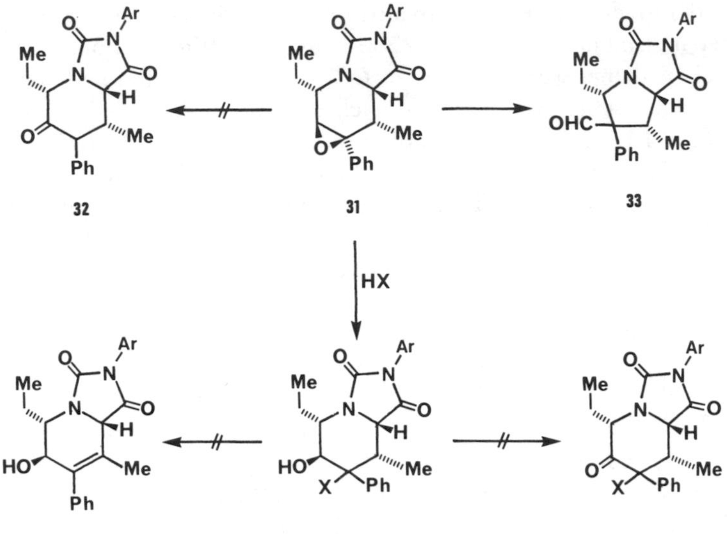

SCHEME 3

to give compounds of type **34**, but all attempts at oxidation to give **36** and elimination to allylic alcohol **35** were totally unpromising. In addition, epoxide **31** was unreactive toward various nitrogen nucleophiles. In all probability, many of the problems encountered with these systems were due to steric congestion near the reacting centers, resulting from the many substituents around the piperidine ring. Several variations of this sort of chemistry were also tried, and we have previously described some of this work.[13,14] These routes were sufficiently unpromising that they are not worth recounting in this chapter.

Although peracid epoxidation of Diels–Alder adduct **26** proceeded nicely, reactions of the styrene double bond with other types of electrophiles were not successful. For example, attempted addition of nitrosyl chloride to **26** to afford the desired unsaturated oxime **37** failed completely.[15]

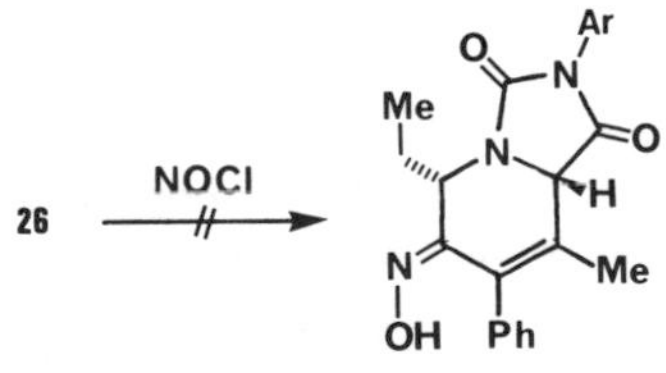

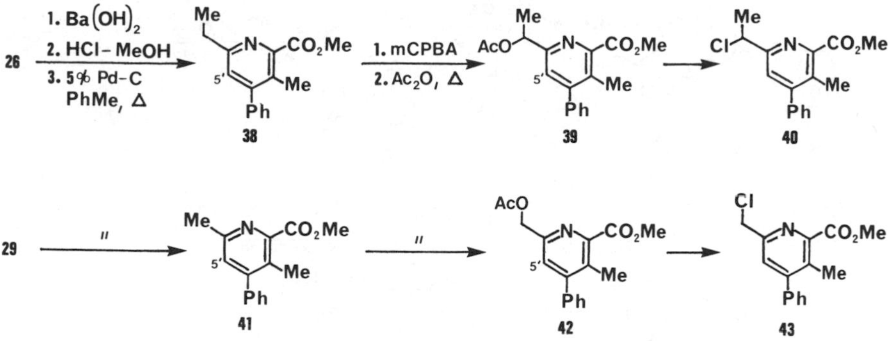

26
1. Ba(OH)$_2$
2. HCl–MeOH
3. 5% Pd–C
PhMe, Δ

→ **38** (Me, Et, N, CO$_2$Me, Me, Ph, 5′)

1. mCPBA
2. Ac$_2$O, Δ

→ **39** (AcO, Me, N, CO$_2$Me, Me, Ph, 5′)

→ **40** (Cl, Me, N, CO$_2$Me, Me, Ph)

29 → ″ → **41** (Me, N, CO$_2$Me, Me, Ph, 5′)

→ ″ → **42** (AcO, N, CO$_2$Me, Me, Ph, 5′)

→ **43** (Cl, N, CO$_2$Me, Me, Ph)

SCHEME 4

Faced with a frustrating inability to introduce the missing C-5′ amino substituent into the tetrahydropyridine ring of imino Diels–Alder adducts, such as **26** and **29**, we were forced to explore the possibility of first aromatizing **26** and/ or **29**, and subsequently introducing the desired substituent into the resulting pyridine(s). Therefore, the hydantoin ring of **26** was cleaved with barium hydroxide (Scheme 4), and the resulting crude amino acid was esterified with methanolic HCl. Aromatization with 5% Pd–C yielded the desired pyridine **38** in acceptable overall yield. Similarly, adduct **29** could be transformed to the dimethylpyridine **41**.[12]

Because we had firmly decided at an early stage to eschew any electrophilic substitution chemistry in synthesis of the pyridine C ring of streptonigrin, it became crucial to find a suitable alternative method to introduce a nitrogen substituent into **38** or **41**. A reasonable approach here seemed to be to introduce a functionalized carbon atom into the vacant C-5′ position of one or both of these pyridines, followed by some type of degradation to produce a primary amino group. A Sommelet–Hauser rearrangement or a related [2,3]-sigmatropic process had some appeal as a solution to this problem, and we decided to test this general idea. Just as we began some preliminary studies, a paper appeared[16] that described exactly the type of [2,3]-sigmatropic rearrangement in the pyridine series that we had in mind. Using this useful work as a guide, we performed the following experiments.

Ethylpyridine **38** was oxidized (Scheme 4) to the corresponding N-oxide, which was treated with hot acetic anhydride to afford rearranged acetate **39** in excellent yield. An identical series of reactions worked equally well on methylpyridine **41**, affording **42**. Both **39** and **42** were converted in a straightforward manner to chlorides **40** and **43**, respectively.

Chloromethylpyridine **43** was alkylated with amino nitrile **44** (Scheme 5) to yield the quaternary ammonium salt **45**. The ammonium salt was then

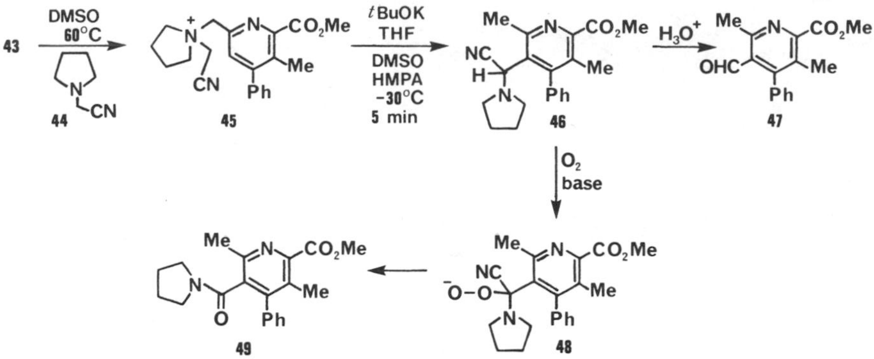

SCHEME 5

treated with potassium *tert*-butoxide in DMSO under nitrogen, and, to our surprise, the product of this reaction was amide **49** (30% yield). Presumably, **45** formed an ylide that underwent a Sommelet–Hauser rearrangement to afford amino nitrile **46**. In the presence of excess base, **46** was probably deprotonated, and the resulting carbanion reacted with traces of oxygen to produce peroxide **48**. Decomposition of **48** would afford the observed amide **49**. However, when the initial [2,3]-sigmatropic rearrangement was run under thoroughly deoxygenated (Fieser's solution) argon under the very carefully defined reaction conditions shown in Scheme 5, it was possible to intercept amino nitrile **46**. Hydrolysis of **46** with aqueous oxalic acid gave the desired pentasubstituted pyridine aldehyde **47** in 43% yield from ammonium salt **45**.

A number of attempts were made to repeat this route with the chloroethyl-pyridine **40**. However, it was not possible to effect the alkylation of amino nitrile **44** with chloride **40** to produce the requisite quaternary salt **50**. It thus

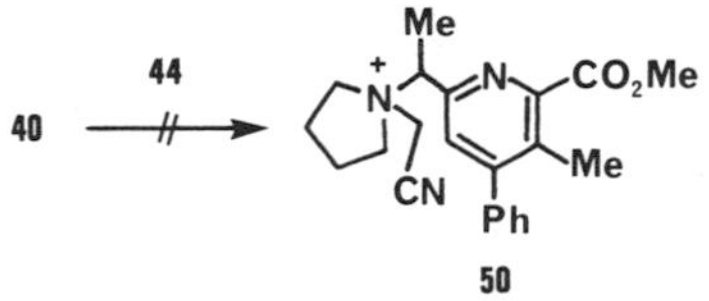

became evident at this point that if we were going to utilize a [2,3]-sigma-tropic rearrangement for introduction of the final pyridine substituent, we would have to use the C-6′ methyl series of Diels–Alder adducts (i.e., **29**). This meant that if a Friedlander-type synthesis of the AB ring system were to be applied as outlined in Scheme 1, it would be necessary to effect a one-carbon homologation of the pyridine fragment at some stage.

A few additional experiments were done on model pyridine aldehyde **47** to establish positively the feasibility of this path to a suitable streptonigrin

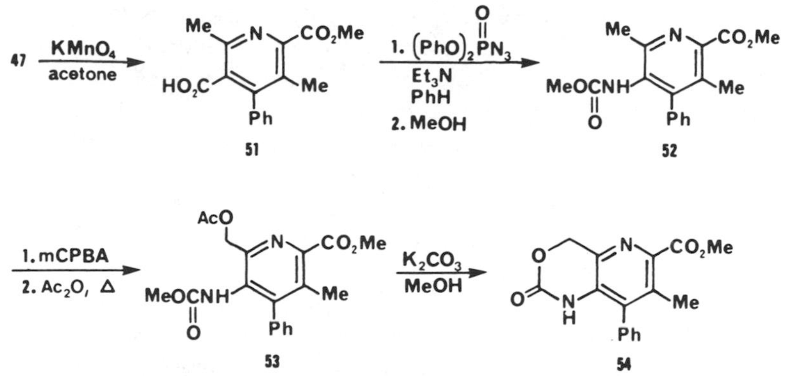

SCHEME 6

C-ring precursor. Toward this end, aldehyde **47** was oxidized to carboxylic acid **51** with neutral potassium permanganate (Scheme 6). To our delight, the Yamada modification[17] of the Curtius rearrangement served to convert acid **51** cleanly to carbamate **52**. At last the final amino substituent had been introduced into the pyridine nucleus! In order to functionalize the C-6' methyl group, compound **52** was oxidized to the N-oxide, which upon treatment with hot acetic anhydride, rearranged to afford acetoxymethylpyridine **53**. In an attempt to convert **53** to the corresponding primary alcohol, it was treated with methanolic potassium carbonate. However, the product of this reaction was the cyclic carbamate **54**. Although this interesting compound was reassuring in that it helped confirm that we had introduced the ring substituents in the correct relationship, it was not particularly useful for further transformations. We therefore explored the possibility of protecting the amino group as an amide rather than as a carbamate.

Curtius rearrangement of acid **51** was performed as shown above, but the intermediate isocyanate was hydrolyzed to afford free amine **55** (Scheme 7).

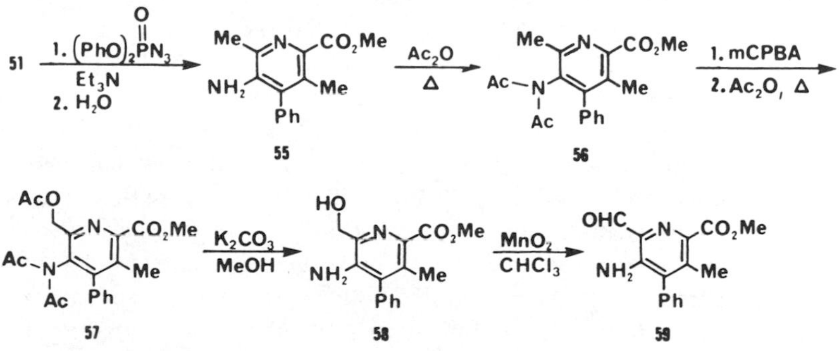

SCHEME 7

This amine was quite unreactive toward acylating agents unless forcing conditions were used. To our surprise, heating **55** with acetic anhydride at 130°C did not give the desired amide but afforded imide **56** instead. Although not anticipated, protection of the amino group as the imide did seem compatible with the planned route. Using the same N-oxide rearrangement pathway as before, **56** was converted to acetoxymethylpyridine **57**. It was our intention now to selectively remove the O-acetyl group of **57** to produce the corresponding alcohol, still having the amino group protected as the imide. However, again to our surprise, treatment of **57** with methanolic potassium carbonate at room temperature removed all acetyl groups, giving amino alcohol **58**. Oxidation of **58** with activated manganese dioxide gave amino aldehyde **59** in high yield. This compound was quite stable and did not show the usual propensity of o-aminobenzaldehydes to undergo self-condensation.

The experiments shown in Scheme 7 indicated to us that the amino substituent in these pyridines is not very basic and is more like a vinylogous amide than an aromatic amine. Thus it appeared that this group might actually be "self-protected," and we could possibly save a step by eliminating the N-acylation. In line with this idea, attempts were made to oxidize amine **55** to N-oxide **60** directly, but to no avail. However, we still believed that by a slight change in the order of steps, N-protection could ultimately be avoided (see below).

V. Synthesis of Streptonigrin

At this stage we were sufficiently confident that the strategy that had been developed would allow synthesis of streptonigrin, so that model studies were discontinued, and we set to work on the natural product. To do so, it was necessary to repeat the successful sequence with a fully oxygenated D-ring precursor. For this purpose we required the α,β-unsaturated aldehyde **64**. This compound proved to be considerably more difficult to prepare than expected.

Scheme 8 outlines the initial approach to **64**, which was patterned after preparation of the model E aldehyde **24**.[18,19] Phenylacetaldehyde derivative **63** was prepared from **61** in the straightforward manner shown. Many attempts were made to effect a mixed aldol condensation between **63** and acetaldehyde, using classical methods (as well as preformed enolate, silyl enol

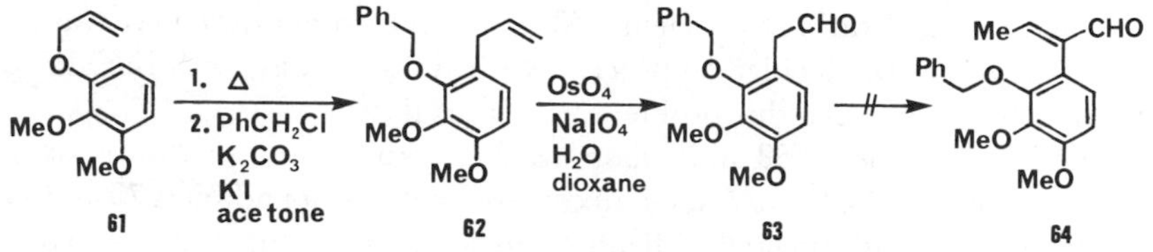

SCHEME 8

ether, and enamine chemistry), without appreciable success. At best, this key starting unsaturated aldehyde could only be produced in low yield from **63**.

Fortunately, we were eventually able to develop an alternate route, which allowed efficient preparation of aldehyde **64** in multigram quantities (Scheme 9). Readily available aldehyde **65** was converted to styrene oxide **66**, using Corey's methylene transfer procedure.[20] Without purification, this epoxide was opened regioselectively with vinylmagnesium bromide to afford homo-allylic alcohol **67** (96% from **65**). Oxidation of **67** with Collins reagent gave a complex mixture of aldehydes. However, when the crude oxidation product was stirred with 10% HCl overnight, isomerization of the various aldehydes to the apparently most stable conjugated E isomer **64** occurred (55%). Using this route, it was easily possible to prepare 15- to 25-g batches of aldehyde **64**.

We next turned to the task of stereoselectively converting this aldehyde to the requisite (E)-diene **68**. As mentioned above, it was quite clear from preliminary studies that the dienes having a (Z)-disubstituted double bond were

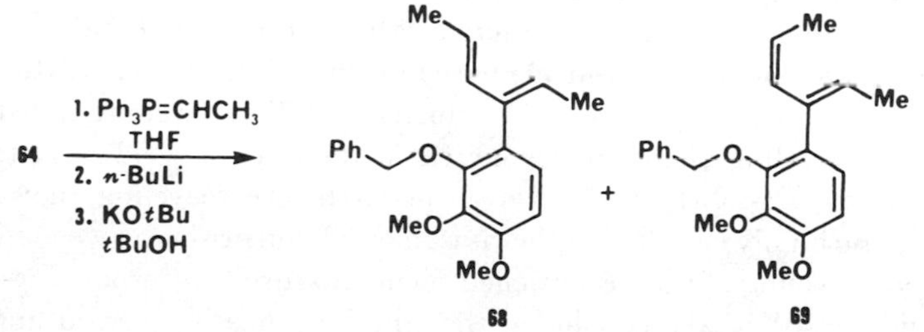

unreactive in imino Diels–Alder cycloadditions with hydantoin **2**. Application of the Schlosser modification[21] of the Wittig reaction to aldehyde **64** allowed formation of an inseparable mixture of the desired (E)-diene **68** and the Z

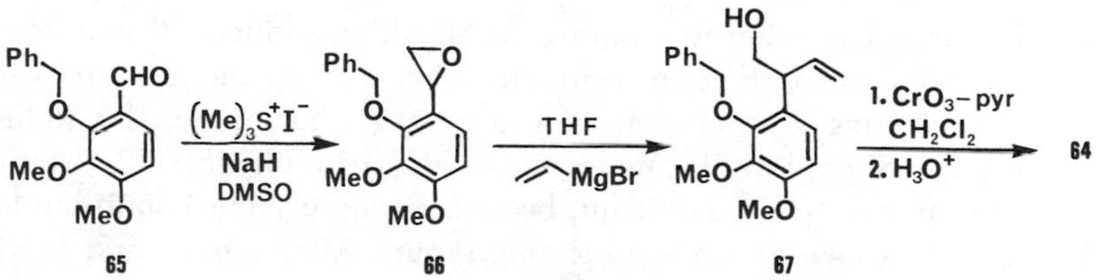

SCHEME 9

isomer **69** in a 2.5/1 ratio (75% yield). Although the isomer ratio here was not at all acceptable, we decided to proceed with the synthesis, with the idea of returning to try to solve this problem at a later date.

The mixture of dienes **68** and **69**, on heating with hydantoin **2** in refluxing xylene for 3 days, gave the expected mixture of Diels–Alder adducts **70** and **71**. However, several unexpected difficulties arose at this point. First, we could not separate the two regioisomeric products, even though in the model series (i.e., **26/27** and **29/30**) isomers could be easily separated by column chromatography. It was estimated by ^{1}H-NMR analysis of the mixture that the ratio of the desired compound **70** to the undesired **71** was about 3:1, as anticipated from model work.

The second problem here was that the Diels–Alder reaction could not be driven to completion, although many variations in experimental conditions were tried. The yield of **70** and **71**, based on the diene isomer mixture, was 39% but could be raised to 56% if the unchanged dienes were recycled once. Continued recycling of the dienes increased the adduct yield even farther. Inspection by ^{1}H-NMR of recovered dienes before recycling showed the mixture to be highly enriched in the cis isomer **69**. Surprisingly, we found that continued recycling of this cis-enriched diene mixture kept producing Diels–Alder adducts. We have concluded that the long heating period and time required for cycloaddition results in slow thermal isomerization of **69** to **68**. Thus some good did come out of the sluggishness of the Diels–Alder step, i.e., we did not have to return to find an improved procedure for stereoselective synthesis of trans diene **68**.

Since it was not possible to separate Diels–Alder adducts **70** and **71**, it became necessary to continue the synthetic scheme using the mixture, with the hope of effecting a separation at a later stage. Therefore, the adduct mixture was hydrolyzed to the amino acids with barium hydroxide. Esterification of this mixture proved difficult, because strongly acidic conditions led to cleavage of the benzyl ether protecting group. After some work, it was found that the amino acids could be converted to a mixture of methyl esters **72** and **73** with thionyl chloride in methanol. This complex mixture undoubtedly consisted of regioisomers and stereoisomers resulting from ester epimerization during the hydantoin hydrolysis.[12] No attempt was made at purification at this point.

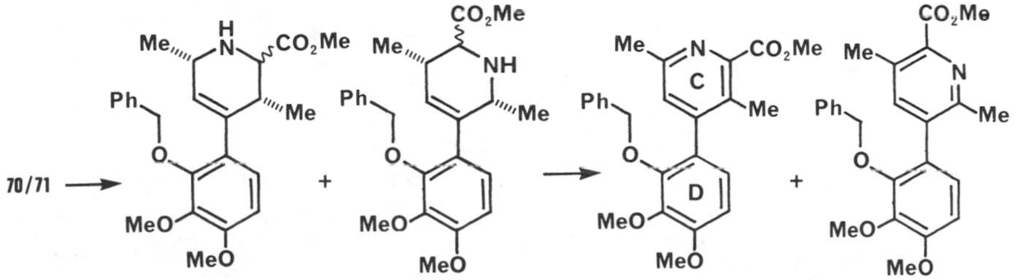

The tetrahydropyridine mixture was heated with 5% Pd–C in toluene overnight to afford a chromatographically separable mixture containing the desired pyridine **74** and only a trace of isomer **75**. The yield of crystalline **74** from the Diels–Alder adducts **70/71** was about 20%. It is not clear where the undesired regioisomer was being lost in this sequence, and because we worked with such bad mixtures it was impossible to analyze the products in individual steps. Using this route, about 5 g of key pyridine **74** was prepared for completion of the total synthesis.

Transformation of **74** to the pentasubstituted pyridine aldehyde **78** (Scheme 10) almost exactly paralleled the conversion of model pyridine **11** to aldehyde **47** (Scheme 5). The only difference in these two series was that the sigmatropic rearrangement of quaternary salt **77** did not seem to require such carefully controlled conditions as the model reaction, and oxidation products were never isolated here. The overall yield from pyridine **74** to aldehyde **78** was about 33%.

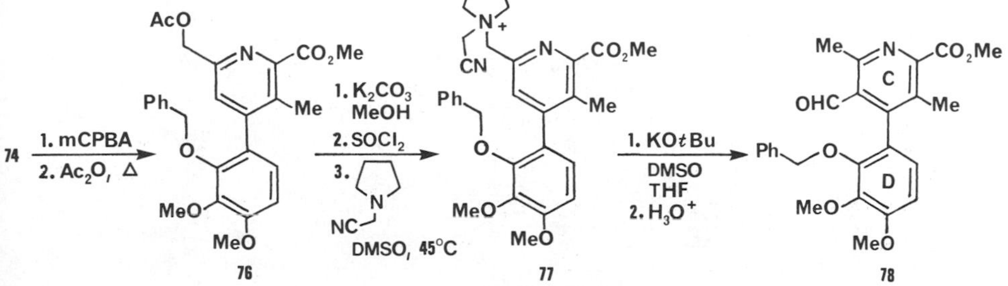

SCHEME 10

Keeping in mind that we wanted to avoid any protection of the C-ring amino substituent, the sequence of reactions shown in Scheme 11 was conceived for converting **78** to amino aldehyde **83**. In light of the fact that model aminopyridine **55** could not be converted to the N-oxide **60**, it was decided to generate an N-oxide before introduction of the amino group. Pyridine **78** was unreactive toward m-chloroperbenzoic acid, probably because of the two electron-withdrawing carbonyl substituents. However, trifluoroperacetic acid did serve to produce the N-oxide aldehyde **79**. This compound was further oxidized to carboxylic acid **80** with potassium permanganate, and Curtius rearrangement led to aminopyridine **81**. Rearrangement of **81** with acetic anhydride, followed by treatment of the crude product with methanolic potassium carbonate, gave the amino alcohol **82**. Oxidation of the alcohol functionality with activated manganese dioxide afforded the stable, crystalline amino alcohol **83**. Yields of the reactions in Scheme 11 averaged about 90%.

We next turned to the problem of devising an efficient method for annulating an AB-quinoline ring system onto CD fragment **83**. One possibility here was to homologate the aldehyde to the methyl ketone and to effect a classical Friedlander quinoline synthesis with o-aminobenzaldehyde **16**, as done in Scheme 2. This idea was discarded in view of the poor yield in our preparation of pyridylquinoline **17**, and the fact that condensations of this type have a generally poor reputation. At this late stage we could certainly not afford to squander our hard-won intermediate **83**.

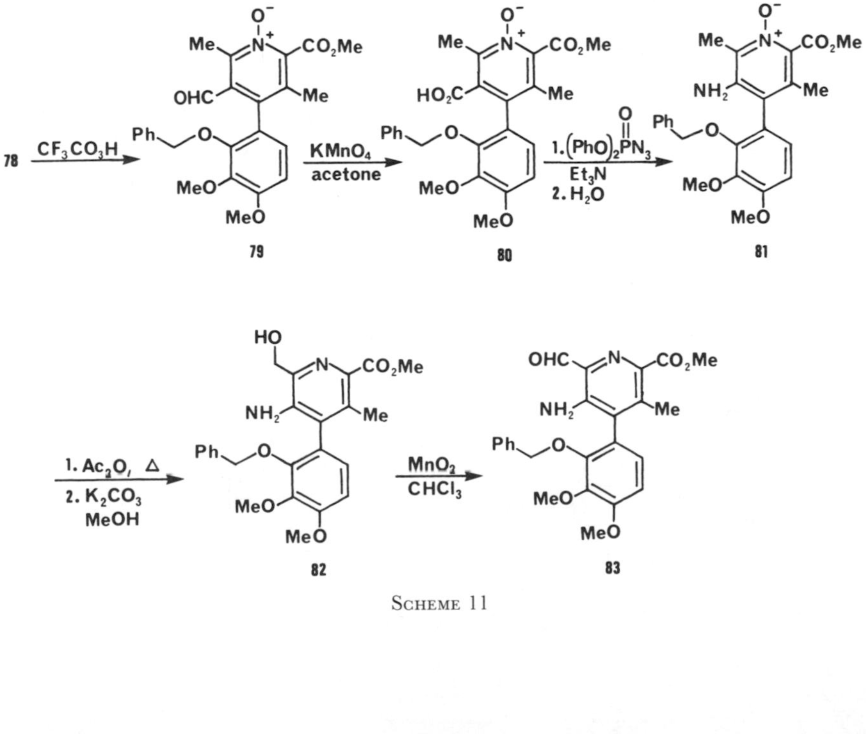

SCHEME 11

After some careful thought, we decided to investigate a new modification of the Friedlander synthesis that would use modern phosphonate carbanion chemistry. This procedure was first tried in the simple system shown in Scheme 12.[14] The required one-carbon homologation of pyridine aldehyde **84** could be achieved with the carbanion derived from dimethyl methylphosphonate, affording hydroxyphosphonate **85**. Oxidation of **85** with activated manganese dioxide produced β-ketophosphonate **86**. As hoped, **86** undergoes a clean Wadsworth–Emmons–Horner reaction with nitrobenzaldehyde **87** to yield the nitro chalcone **88**. Reductive cyclization of **88** afforded the desired pyridylquinoline **89**. This sequence of reactions was easy to perform, and the yields were quite good.

Having successfully tested this approach in the above model, we proceeded to apply it to amino aldehyde **83**. This turned out to be one of the rare instances when a simple model system gave a true indication of how well a method would work in a significantly more complicated case. Conversion of amino aldehyde **83** to tetracyclic pyridylquinoline **93** (Scheme 13) worked at least as well, if not better, than the test system in Scheme 12. Thus homologation of aldehyde **83** to hydroxyphosphonate **90** could be effected in 67% yield, and subsequent alcohol oxidation afforded β-ketophosphonate **91** in 79% yield. The only difficulty experienced in the condensation of **91** to form chalcone **92** was that in a small scale reaction traces of sodium hydroxide in sodium hydride cleaved the sulfonyl protecting group of nitro aldehyde **87**. However, using potassium hydride as base, **92** could be formed in good yield. Reductive cyclization of **92** to tetracyclic compound **93** proceeded as well as in the model system.

The penultimate stage of the total synthesis was now to elaborate the A ring of **93** to the quinolinequinone functionality of streptonigrin. Using the

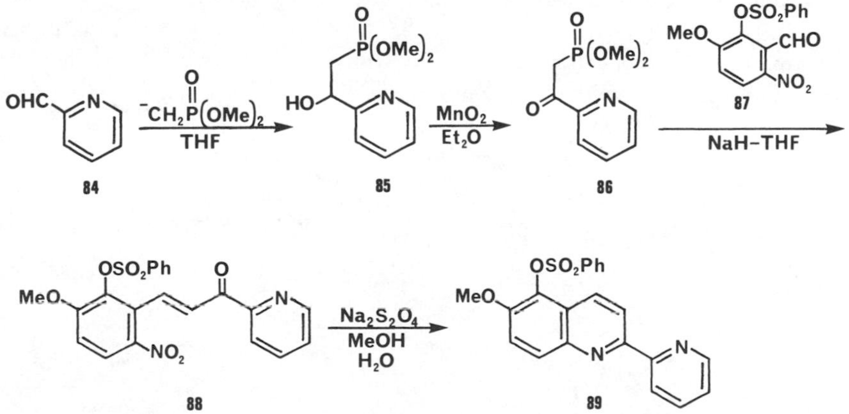

SCHEME 12

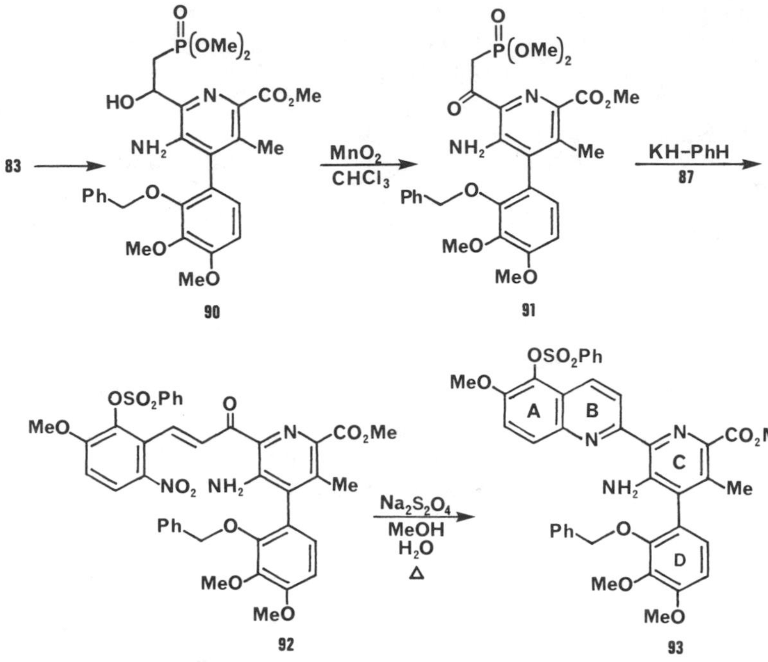

Scheme 13

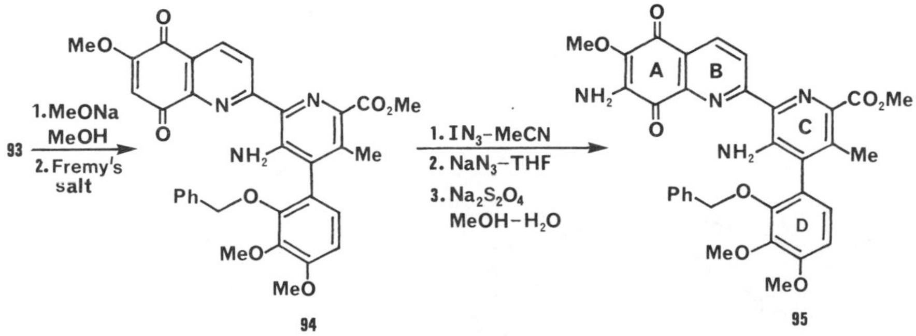

chemistry developed previously in Scheme 2 it was possible to effect these transformations rather efficiently. Cleavage of the sulfonyl protecting group of **93** with sodium methoxide and oxidation of the resulting phenol with Fremy's salt gave the yellow quinone **94** in quantitative yield. We had been worried about this step for quite a while, because it was not clear what sort of selectivity a reactive species like Fremy's salt would show with such a highly functionalized heterocyclic molecule. Treatment of **94** with iodine azide gave the 7-iodoquinone,[22] which was converted to the 7-azidoquinone with sodium azide. We were unable to effect both of these steps in a single reaction, although several attempts were made to do so. Azide group reduction with sodium hydrosulfite gave amino quinone **95**.

Completion of the synthesis required only cleavage of the methyl ester and benzyl ether protecting groups of **95**. In order to test these final steps and to correlate our synthetic material with streptonigrin, we prepared **95** from the natural product kindly supplied by Dr. John Douros of the National Cancer Institute. Upon reaction with methanolic boron trifluoride etherate, streptonigrin (**1**) was converted to its methyl ester **96**. It was interesting (and useful) to note that streptonigrin is stable under strongly acidic conditions. The D-ring phenol of **96** could be benzylated with potassium carbonate–benzyl iodide to produce **95** identical with material prepared by total synthesis.

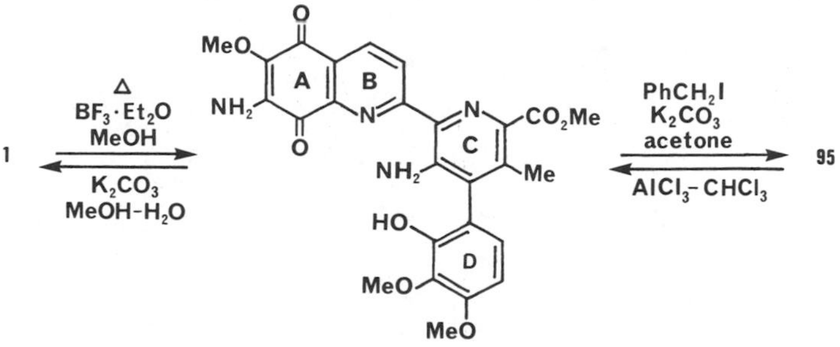

96

Debenzylation of **95** to streptonigrin methyl ester proved much simpler to effect than anticipated. Upon stirring **95** with a hundredfold excess of anhydrous aluminum chloride in chloroform, **96** was formed in 80% yield, again showing the stability of streptonigrin to strong Lewis acids. Cleavage of the methyl ester group of **96** proved considerably more difficult. Whereas **1** is stable to acid, it is rather unstable in the presence of strong base. Treatment of **96** with dilute ammonia did give some streptonigrin, but produced a second compound tentatively assigned quinone imine structure **97**. After much

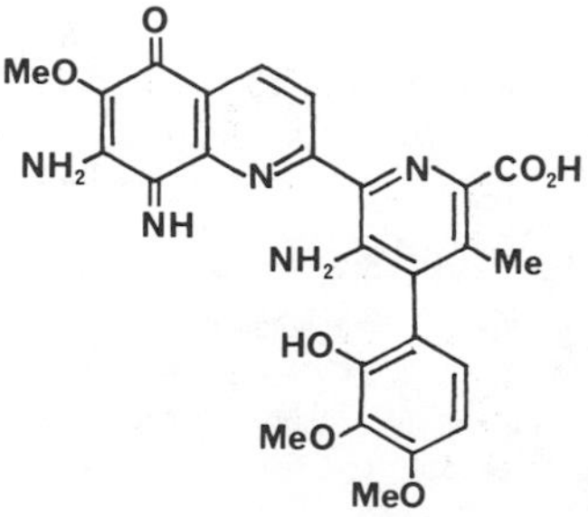

97

experimentation, it was found that ester **96** could best be hydrolyzed with aqueous potassium carbonate at room temperature for two days (65% yield). Repeating the two deprotection steps on synthetic **95** gave streptonigrin (**1**), identical to an authentic sample.

Thus, after about 6 years of sustained effort, the first total synthesis of this challenging natural product had been achieved.[19] The successful route turned out to be somewhat longer than that originally conceived (Scheme 1), although we believe this work paved the way for rather more efficient approaches subsequently described by Kende, [23] Martin,[24] and Boger.[25] In addition, during this project we became deeply interested in further applying imino Diels–Alder chemistry to natural product total synthesis and have been productively engaged in this area during the past few years.[26] This program, although exceptionally frustrating at times, has had both intellectual rewards and worthwhile scientific spinoffs.

ACKNOWLEDGMENTS

Financial support of this work by the National Science Foundation is gratefully appreciated. This difficult project could not have been executed without the skillful collaboration of Drs. Fatima Basha, Satoshi Hibino, Deukjoon Kim, Nazir Khatri, Walter Pye, and Tai-Teh Wu.

REFERENCES

1. K. V. Rao and W. P. Cullen, *Antibiot. Annu.* p. 950 (1959–1960).
2. K. V. Rao, K. Biemann, and R. B. Woodward, *J. Am. Chem. Soc.* **85**, 2532 (1963).
3. Y.-Y. Chiu and W. N. Lipscomb, *J. Am. Chem. Soc.* **97**, 2525 (1975).
4. S. J. Gould and D. E. Cane, *J. Am. Chem. Soc.* **104**, 343 (1982), and references cited therein.
5. S. J. Gould and S. M. Weinreb, *Fortschr. Chem. Org. Naturst.* **41**, 77 (1982).
6. P. J. Wittek, T. K. Liao, and C. C. Cheng, *J. Org. Chem.* **44**, 870 (1979).
7. D. Ben-Ishai and E. Goldstein, *Tetrahedron* **27**, 3119 (1971).
8. For reviews, see S. M. Weinreb and J. I. Levin, *Heterocycles* **12**, 949 (1979); S. M. Weinreb and R. R. Staib, *Tetrahedron* **38**, 3087 (1982).
9. For a review, see S. Oae and K. Orino, *Heterocycles* **6**, 583 (1977).
10. S. Hibino and S. M. Weinreb, *J. Org. Chem.* **42**, 232 (1977).
11. G. Jones, ed., "The Chemistry of Heterocyclic Compounds Quinolines," Vol. 32, Part 1, Chapter 2. Wiley (Interscience), New York, 1977.
12. D. Kim and S. M. Weinreb, *J. Org. Chem.* **43**, 121 (1978).
13. D. Kim and S. M. Weinreb, *J. Org. Chem.* **43**, 125 (1978).
14. D. Kim, Ph.D. Thesis, Fordham University, Bronx, New York (1979).
15. R. Pummerer and F. Graser, *Justus Liebigs Ann. Chem.* **583**, 207 (1953).
16. E. B. Sanders, H. V. Secor, and J. I. Seeman, *J. Org. Chem.* **41**, 2858 (1976); **43**, 324 (1978).
17. K. Ninomiya, T. Shiori, and S. Yamada, *Tetrahedron* **30**, 2151 (1974).
18. R. Kuhn and J. Michel, *Ber. Dtsch. Chem. Ges. A* **67**, 696 (1938).
19. F. Z. Basha, S. Hibino, D. Kim, W. E. Pye, T.-T. Wu, and S. M. Weinreb, *J. Am. Chem. Soc.* **102**, 3962 (1980); S. M. Weinreb, F. Z. Basha, S. Hibino, N. A. Khatri, D. Kim, W. E. Pye, and T.-T. Wu, *ibid.* **104**, 536 (1982).

20. E. J. Corey and M. Chaykovksy, *J. Am. Chem. Soc.* **87**, 1353 (1965).
21. M. Schlosser and K. F. Christmann, *Justus Liebigs Ann. Chem.* **708**, 1 (1967).
22. Cf. A. S. Kende and P. C. Naegely, *Tetrahedron Lett.* p. 4775 (1978).
23. A. S. Kende, D. P. Lorah, and R. J. Boatman, *J. Am. Chem. Soc.* **103**, 1271 (1981).
24. J. C. Martin, *J. Org. Chem.* **47**, 3761 (1982).
25. D. L. Boger and J. S. Panek, *J. Org. Chem.* **47**, 3763 (1982).
26. See, for example, N. A. Khatri, H. F. Schmitthenner, J. Shringarpure, and S. M. Weinreb, *J. Am. Chem. Soc.* **103**, 6387 (1981); B. Nader, T. R. Bailey, R. W. Franck, and S. M. Weinreb, *ibid.* p. 7573; R. A. Gobao, M. L. Bremmer, and S. M. Weinreb, *ibid.* **104**, 7065 (1982).

Chapter 13

METHYNOLIDE AND THE PRELOG–DJERASSI LACTONIC ACID: AN EXERCISE IN STEREOCONTROLLED SYNTHESIS

James D. White

Department of Chemistry
Oregon State University
Corvallis, Oregon

I. Introduction

Like any journey of substantial length, the management of a successful synthesis from its starting point to its destination requires a carefully planned itinerary, the flexibility to accommodate diversions along the way, and the fortitude to see the journey through to a satisfactory conclusion. A future reader of today's chemical journals would probably not be deceived by the impression, conveyed by many publications that report completed syntheses, that success is invariably the result of brilliant strategic analysis coordinated with astute logistics. To be sure, there are instances where the execution of a synthesis adheres with remarkable fidelity to its initial preconception. More

STRATEGIES AND TACTICS
IN ORGANIC SYNTHESIS

347

often, however, the journey is a tortuous and protracted one that continually asserts the unpredictability of travel in all but the most familiar terrain. The synthesis of methynolide (**1**), chronicled below, exemplifies this latter precept; it is chosen as a case study because it represents fairly accurately the modus operandi for a synthesis that is built around a rather specific stratagem.

The naturally occurring lactones, known collectively as macrolides, now comprise a family of well over a hundred substances. Much of the chemistry of this group has been summarized in several reviews,[1] but the very rapid pace at which events in this arena have unfolded makes any contemporary survey impossible. There have, nevertheless, been several accomplishments that stand out from the broad tapestry of work in this field. One of these is the synthesis of methymycin (**2**) by Masamune in 1975.[2] Another is the recognition, due originally to Celmer, of functional group and configurational regularity in the macrolides.[3] The latter is a consequence of the consistency associated with the enzyme-mediated assembly of propionate subunits, and gives rise to a structural complementarity that, in principle, should permit synthesis of different macrolides from similar segments. Practical plans that employ this concept have already been described.[4]

In developing a strategy for construction of methynolide (**1**), the aglycone of **2**, we sought a route that could ultimately be extended beyond this prototypical structure to "homologous" members of the polyoxo series, such as pikromycin (**3**).[5] Our plan from the outset focused on a segment A corresponding to carbons 1–7 of **1**, in the knowledge that this moiety had been removed (with configuration intact) from methynolide in the course of degradative studies to yield the familiar Prelog–Djerassi lactonic acid (**4**).[6]

Comparison of synthesized material with **4** derived from the natural source would provide an obvious checkpoint for stereochemical accuracy.

In parallel with the synthesis of **4**, the approach to **1** called for elaboration of a fragment (**5**), in optically active form, corresponding to carbons 9–11. It was envisaged that this piece would be united with an equivalent of **4**, after homologation of the latter to incorporate carbon 8, and the derived seco acid **6** (in some protected form) would, upon final lactonization, yield **1**.

It is surely not by happenstance that the syntheses of **1**, completed by Masamune,[2] by Grieco,[7] and by Yamaguchi,[8] are all based on the strategy outlined in Scheme 1. Stereocontrol is most easily accomplished when chiral elements are in proximity to each other, and the contiguity of centers at carbons 2, 3, 4, and 6 points clearly to a dissection of **1** in which the configurations at carbons 10 and 11 are generated separately from those in the larger segment. Our attention was therefore focused first on the section of the methynolide perimeter which incorporates carbons 1–7 and which is embodied in the Prelog–Djerassi lactonic acid (**4**).

It was felt from the outset that the stereochemical outcome of our planned reaction sequence would be more readily predictable if it were carried out on a rigid frame, and the bicyclic structure **7** was chosen as the initial target in the expectation that it could be converted easily to **4**. This oxabicyclo[3.2.1]octane possesses the same relative configuration of all three methyl groups as is found in **4**; however, an inversion at C-5 is required to set the correct configuration at the δ carbon of the Prelog–Djerassi lactone. Oxidative scission at the indicated bond in **7** opens this bicycle to a close facsimile of **4**.

A further motive for selecting **7** as our synthetic objective lay in the opportunity that it afforded for exploiting the novel and intriguing chemistry of

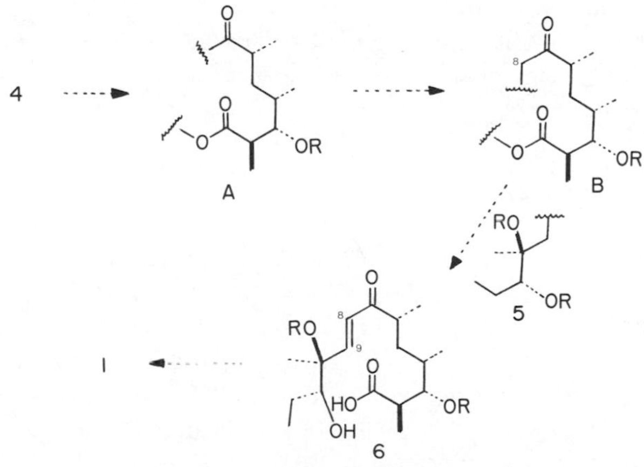

SCHEME 1

oxoallyl cations, which had been discovered independently by Hoffmann[9] and Noyori[10] a few years earlier. These workers reported that cationic species such as **8**, produced by dehalogenation of α,α′-dibromo ketones, undergo cycloaddition to furans to yield 8-oxabicyclo[3.2.1]octanes (**9**), in which the predominant, if not exclusive, isomer possesses a cis (diequatorial) orientation of the methyl groups. It was clear that this convenient tactic for fixing the relative configuration of methyl substituents occupying a 1,3 relationship is ideally suited for establishing the centers at C-4 and C-6 of methynolide. In addition, the ketone function of **9**, although it must be replaced by a methylene unit for **1**, would permit the introduction of a hydroxyl group at an "oxygen" site if such were needed for a different macrolide.

Our opening gambit thus became cycloaddition of **8** to a suitably substituted furan. The selection of appropriate substituent(s) for the furan turned out to be one of the more critical elements in this plan, and several permutations were explored before a viable bicyclic precursor to **4** could be fabricated. Nevertheless, some interesting chemistry evolved from what were ultimately fruitless digressions, and this forms the basis of much of the discussion that follows.

II. Excursions from 2,4-Dimethylfuran

Tempted by the prospect of a route that could lead directly to an intermediate containing all of the carbons and the three methyl substituents of segment B (Scheme 1), we elected first to examine the cycloaddition of **8** with 2,4-dimethylfuran (**10**). The oxoallyl cation (**8**), generated from 2,4-dibromo-3-pentanone with zinc–copper couple in glyme,[11] underwent an exothermic reaction with **10** to give a 56% yield of cycloadducts **11** and **12**. Although a clean separation of these isomers could not be achieved, configurational assignment was easily made from the chemical shifts of the methyl doublets in the NMR spectra. This analysis, which accorded well with studies of similar systems carried out by Hoffmann,[12] established the ratio of cis (11) to trans (12) isomers as 5:1.

Since the keto function of 11 was not needed beyond this point, the next task undertaken was reduction of this function to a methylene unit. All attempts to accomplish the transformation by direct means met with failure, and it quickly became apparent that reversible processes that involved nucleophilic attack at the carbonyl group (as, for example, by hydrazine and its derivatives) would be doomed. A molecular model of 11 shows that the steric compression that develops as the hydroxyl substituent at C-3 is forced into an endo orientation is substantial and probably accounts for this failure. It was, nevertheless, possible to reduce the carbonyl groups of 11 and 12 with sodium borohydride to give, in 75% yield, a 2:1 mixture of endo and exo alcohols 13 and 14, respectively (minor quantities of the alcohols resulting from reduction of 12 were removed chromatographically at this stage). Distinction between 13 and 14 was easily made by NMR spectroscopy, which displayed the endo C-3 proton of 14 as an unusually high-field signal (δ 2.88) due to shielding by the π system of the double bond. Unfortunately, although 14 could readily be acetylated to give 15 or converted to its mesylate 16, the major alcohol 13 was quite resistant toward a derivatization sequence that might have made reductive displacement or trans (diaxial) elimination a feasible entry to the desired methylene system. Likewise, derivatives of exo alcohol 14 (e.g., 16) were also unreactive toward displacement, presumably because attack at C-3 from the hindered endo side is sterically prohibited. An interesting facet of the reactivity associated with the bicyclic system present in 13 came to light when this alcohol was treated with methanesulfonyl chloride under forcing conditions. Exo mesylate 16 was obtained, identical with the product from 14. A possible explanation for the inversion process involves participation by the π bond in an S_N1 type of solvolysis, leading to a non-classical species **17**. The latter has precedent in solvolytic studies carried out by Winstein on bicyclooctyl systems[13] and, indeed, the inversion from **13** to **16** is not observed in derivatives where the double bond is saturated. The C-6 methyl substituent also assists this reaction, since the analogous inversion does not occur in similar systems originating from 2-alkyl furans.

Having failed to remove the oxygen functionality at C-3, it was decided to postpone this operation to a later stage, where interference from the 6,7 π bond might be avoided. The first approach along these lines entailed treatment of **11** with *m*-chloroperbenzoic acid, which led to an epoxide in 96% yield. On the reasonable presumption that attack at the double bond of **11** took place from the exo side, the stereochemistry of this product was formulated as **18**. However, a complication emerged at the very next step for, although reduction of **18** with sodium borohydride furnished a substance that contained a secondary hydroxyl group, this compound was clearly not the expected endo alcohol. First, the product, in contrast to **13**, readily formed an acetate and a mesylate; more informatively, the signal due to the proton at C-7 in epoxide **18** had vanished from the NMR spectrum and a new singlet for one proton had appeared at δ 3.76. The data pointed strongly toward structure **19** as our reduction product, and this was confirmed by the observation that derivatives of **19**, including the chloride **20**, were recovered unchanged after attempted elimination. The same product (**19**) was obtained upon epoxidation of **13** with *m*-chloroperbenzoic acid, reaffirming that proximity of the endo hydroxyl at C-3 to the initially formed exo 6,7-oxide provides the opportunity for an intramolecular opening of the latter.

In constructing **18** we had envisaged a Lewis acid-catalyzed rearrangement of the epoxide at some later stage to install a keto group at C-7. This would simultaneously have put the C-6 methyl group in the correct (exo) configuration and would have provided a substituent at C-7 that could facilitate the desired oxidative scission of the 1,7 bond (cf. **7**). When exposure of **18** to boron trifluoride etherate and other acids failed to yield a tractable product, we returned to **11** in the hope that a suitable intermediate might be accessible via a hydroboration–oxidation sequence. The reaction of **11** with diborane, followed by oxidation with sodium dichromate, gave in 50% yield a 2:3 mixture of diketones **21** and **22**. Without separation, this mixture was treated with sodium methoxide, resulting in clean epimerization of **21** to the more stable exo isomer **22**. Several attempts were made to (a) cleave the tetra-hydrofuranone ring of **22** via Baeyer–Villiger oxidation and (b) reduce the C-3 keto group. To our disappointment, **22** was quite unreactive toward trifluoroperacetic acid, with which we had hoped to effect a selective oxidation at the furanone carbonyl. Also, all attempts to differentiate the two ketones of **22** by reductive methods were fruitless.

In a last-ditch effort to save this route, the acetate **15** was subjected to hydroboration–oxidation. Again a mixture of epimeric ketones **23** and **24** was produced (53%), which was treated with 2 N sodium hydroxide to afford a single hydroxy ketone (**25**). The latter was converted to both its mesylate (**26**) and acetate (**27**), but neither substance was of any avail for reductive removal of the functional group at C-3 or for a Baeyer–Villiger oxidation of the ketone.

The lack of reactivity of the C-7 ketone in this bicyclic system (a fact that was reinforced by its failure to ketalize) may perhaps be another manifestation of the resistance of sp^2 carbons in this framework to undergo rehybridization to tetrahedral configuration. In parallel with observations made in regard to the ketone at C-3, the increased steric compression that accrues from forcing a developing hydroxyl group at position 7 into the endo cavity is sufficient to prohibit reaction. In light of these results, it was of paramount importance that we find an alternative means for opening the bicyclic framework of the furan cycloadduct, and our plans were therefore realigned to avoid the obstacles that had blocked putative routes from 2,4-dimethylfuran.

III.   Further Expeditions from 8-Oxabicyclo[3.2.1]octanes

Although the oxabicyclooctane systems described in the foregoing section failed to provide us with an avenue to the methynolide segment B, we nevertheless felt that this strategy deserved further examination in the context of other furans bearing different substitution. The intrinsic merits of an approach that assembles a sizable fraction of the target structure via a few straightforward and stereocontrolled processes persuaded us to revise our original scheme by replacing **10** with a simple furfuryl alcohol derivative. The omission of a methyl substituent at C-6 of the bicyclic nucleus was not considered a serious obstacle because our plan envisaged a means for introducing this at a later stage. Furthermore, it had already been demonstrated that the configuration at this center of the bicyclic nucleus can be nicely controlled through thermodynamic epimerization. The presence of a functionalized substituent at C-7 of the bicyclooctane seemed especially advantageous when the possibility of a Wittig olefination or similar reaction for connecting segments B and **5** came to mind. Finally, a compelling practical reason for changing the nature of the furan was the discovery that cycloaddition with oxoallyl cation **8** was much more efficient when the furan was present in excess.

Furfuryl benzyl ether (**28**), prepared in 75% yield by alkylation of furfuryl alcohol with benzyl bromide in the presence of sodium hydride, was the first of several 2-alkyl furan substrates surveyed and was found to give **29** in 58% yield upon exposure to **8**. Traces of unwanted methyl epimers were readily removed from **29** by chromatography at this stage. Reduction of **29** with sodium borohydride afforded a 93% yield of endo alcohol **30**, and several forays were made from this substance toward an intermediate in which the C-3 oxygen function was removed.

First, **30** was converted to its mesylate **31** in 75% yield, in the hope that a reductive displacement could be effected at C-3. As befits an axial mesylate, however, elimination was the dominant process under all conditions, including treatment with lithium triethylborohydride, and only a pair of dienes (**32**) was obtained from **31**. Although we considered it likely that the newly produced olefinic bond would be hydrogenated selectively from the exo direction to give the desired cis methyl configuration, there appeared to be no way to forestall concomitant reduction of the 6,7 double bond. Because this would remove functionality indispensable to introduction of both a methyl group and oxygen substitution at these sites, it was clearly important to devise a sequence that differentiated the olefinic bonds. Epoxidation was one possibility for accomplishing this.

28

29

30: R = H
31: R = Ms

32

33: R = H
34: R = Ms

In view of previous findings with **18**, it was not surprising to discover that treatment of **30** with *m*-chloroperbenzoic acid in dichloroethane gave, not the 6,7-epoxide, but the ether **33** resulting from a further, intramolecular attack by the endo, C-3 alcohol. The regiochemistry of the epoxide opening is suggested by the presence of a strongly hydrogen-bonded hydroxyl absorption in the IR spectrum of alcohol **33** and was confirmed by the NMR spectrum of the derived mesylate **34**, which displayed a singlet for the C-7 proton at δ 4.99 and which therefore required the absence of a vicinal, bridgehead proton (the H—C—C—H dihedral angle across C-6,7 in **34** is approximately 90°). The remarkable selectivity exhibited in this epoxide opening (98% yield) could be the result of close attendance of the oxygen of the benzyloxy group in protonation of the epoxide, but experiments designed to test this mechanism were not pursued.

The facile synthesis of **33** appeared to foreclose an approach that brings an alcohol and epoxide into confluence within this bicyclic framework, and we therefore explored options based on mesylate **31**. Epoxidation of **31** with *m*-chloroperbenzoic acid furnished a single product in 75% yield, which, on steric grounds, was presumed to be the exo epoxide **35**. As before, attempted displacement of the axial mesylate of **35** resulted only in a mixture of isomeric olefins **36** (89%); however, we were gratified to find that hydrogenation of this mixture over palladium-on-carbon produced **37** in 95% yield. Thus, reduction of the C-3 oxygen functionality to a methylene unit, although it must be effected indirectly, can be accomplished in good yield to give a stereochemically homogeneous product. The accompanying removal of the benzyl protecting group was regarded as an asset, it being our hope that the primary alcohol might facilitate regioselective opening of the epoxide to install the C-6 methyl substituent via a cuprate complex. Unfortunately, epoxide **37** proved totally resistant toward nucleophiles, including lithium dimethylcuprate, demonstrating again the inaccessibility of the endo side of this bicycle. Lewis acid-catalyzed rearrangement of the epoxide was also attempted without success and, although the aldehyde **38** and chloride **39** could be prepared, these too were sterile intermediates.

Having failed to find a tactic for cleaving the oxabicyclooctane system at the 1,7 bond, we turned to a plan that hinged upon fragmentation of the adjacent 6,7 bond. It was recognized that this necessitated a more circuitous approach to the skeleton of B because a carbon must ultimately be removed from C-1 and added at C-6. Balanced against this, however, was the attractive prospect of employing the 6,7 double bond in a direct oxidative scission that bypassed the uncooperative epoxide intermediates of previous approaches.

Hydroxylation of **29** with catalytic osmium tetroxide and potassium perchlorate gave, in 82% yield, the crystalline diol **40**. This was converted to its acetonide **41** with isopropenyl acetate containing a trace of acid, and the ketone was reduced with sodium borohydride to provide the endo alcohol **42** (65%) along with a lesser quantity of the exo isomer **43**. Without separation, these alcohols were subjected to mesyl chloride in pyridine for 3 days, a protocol that afforded 78% of the pair of olefin isomers **44** along with the exo mesylate **45**. Hydrogenation of **44** over palladium-on-carbon proceeded quantitatively to yield the crystalline alcohol **46**. Acetylation of **46** gave **47** in 81% yield.

At this point, a sequence, which appeared initially promising, came to an abrupt halt. No conditions could be found for removal of the acetonide from **47**, and the glycol that was the intended substrate for oxidative cleavage was therefore out of reach. Vigorous treatment of **47** with acidic reagents resulted in loss of the acetate before hydrolysis of the ketal, and it was decided that this exceptionally stable acetonide, perhaps a consequence of the favorable geometry in the system, should be replaced by a diol protecting group that could, in principle, be removed during the hydrogenation step used to generate the cis dimethyl substituents. In order to provide a blocked C-1 substituent that would survive this sequence, we began our next approach from furfuryl acetate (**48**).

Cycloaddition of **48** with **8** gave keto acetate **49** in a disappointing 30% yield. However, hydroxylation of **49** under the conditions employed for **29** afforded the crystalline diol **50** in good yield (77%) and this was smoothly converted to the crystalline benzylidene derivative **51** (66%) with benzaldehyde. Reduction of **51** was carried out with several hydride reagents in an attempt to optimize stereoselectivity in favor of the endo alcohol. Sodium borohydride gave no better than a 2:1 mixture of endo and exo alcohols **52** and **53**, respectively; however, diisobutylaluminum hydride improved this ratio to 9:1, although the primary acetate was also reduced in this case.

Fortunately, selective acetylation of the crystalline diol **54** presented no difficulty, the endo secondary hydroxyl group being much the more hindered site. Treatment of **52** with mesyl chloride in pyridine led directly to the olefin mixture **55**, which was hydrogenated without purification to give a 66% yield of the desired diol **56** as a beautifully crystalline solid. Our high hopes for this intermediate were soon shattered, however, by the finding that the only isolable product from oxidation of **56** with sodium metaperiodate or lead tetracetate was the hydroxy ketone **57**. It can be presumed from this result that the coordinated species, which must intervene for carbon–carbon bond fragmentation to occur, fails to activate the 7-hydroxyl group, again pointing up the sterically hindered environment of this center. There appeared to be no obvious detour around this impasse, and we therefore abandoned the concept of a 6,7 bond cleavage for an approach that would open the oxabicyclooctane framework via rupture of the external bond to the C-1 bridgehead.

IV. The Prelog–Djerassi Lactonic Acid (4)

The foregoing attempts to convert an oxabicyclooctane into a useful segment of the methynolide structure had, for the most part, foundered on our inability to open the bicyclic nucleus. To circumvent this problem, we decided to aim for a variant of this bicycle, containing a hydroxyl group at the bridgehead, in the expectation that a simple retroketalization would open the oxide bridge. If a further rupture of the derived cycloheptane could be contrived, a structure closely resembling both segment A and **4** could be envisaged.

Initially, two objections to this plan were raised. First, omission of the bridgehead carbon substituent necessitates a subsequent and somewhat untidy homologation of A by one carbon. Second, the 2-alkoxy furan needed for this approach was considered to be a risky partner in cycloaddition with a species as electrophilic as the oxoallyl cation **8**. A conceivable solution to this latter difficulty lay in concealing the bridgehead hydroxyl function within an acetyl substituent, so that a Baeyer–Villiger oxidation of the latter could be employed at a later stage to establish the precursor for ring scission. This strategy is formulated in the generalized sequence **58** → **59** → **60**.

The ketal **61**, prepared from 2-acetylfuran with ethylene glycol in the presence of trimethyl orthoformate and *p*-toluenesulfonic acid, underwent cycloaddition with **8** to give the desired adduct **62** as a crystalline solid in 53% yield. As in previous cycloadditions with **8**, NMR evidence pointed convincingly to the stereochemistry shown, which is in accord with the known preference for **8** to add in the "W" conformation and endo mode.[9] Reduction of **62** with diisobutylaluminum hydride afforded crystalline, endo alcohol **63** in 85% yield.

At this stage, two options presented themselves. In the first, dehydration of **63** (through its mesylate) could be projected as leading to a diene (**64**), hydrogenation of which would lead to simultaneous saturation of both olefinic bonds. We were fearful, however, that this route might come to grief at the diallyl ether **64**, which would be susceptible to hydrogenolysis and which could result in an untimely fission of the oxide bridge to give **65**. For this reason, a slightly longer sequence, beginning with hydrogenation of **63** to **66**, was pursued.

The crystalline alcohol **66**, prepared quantitatively from **63**, was treated with methanesulfonyl chloride in pyridine to give the expected mixture of regioisomeric olefins **67** in 78% yield. Hydrolysis of the ketal function with aqueous acetic acid produced ketone **68** (90%), and this was followed by hydrogenation over palladium-on-carbon to give **69** in 96% yield after chromatographic purification. As was found in previous instances involving this bicyclic nucleus, the very high selectivity for hydrogen addition to the exo side provides a convenient means for restoring the cisoid relationship of the two methyl substituents.

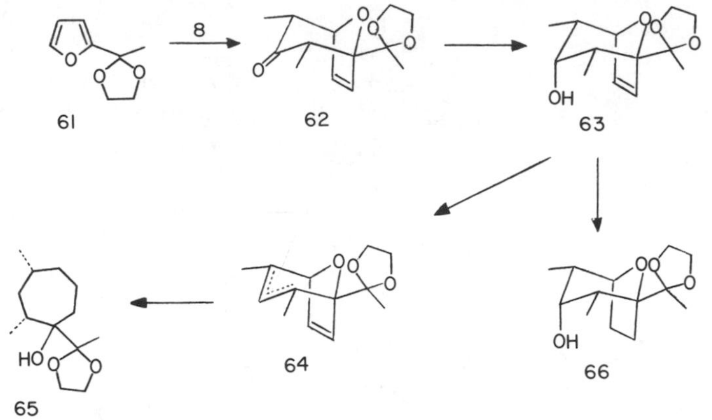

61   62   63

65   64   66

From **69**, unlatching the oxide bridge proved to be straightforward. Baeyer–Villiger oxidation of this ketone with *m*-chloroperbenzoic acid furnished an acetate (**70**), which was treated with potassium carbonate in methanol to give the cycloheptanone **71** as a low-melting solid in 60% yield. Comparison of the latter with **4** and segment A of **1** indicates that, in order for the 5-hydroxy group of **71** to become the C-3 oxygen substituent of **1**, an inversion must be performed at this center, and our next task was this configurational reversal. The mesylate **72** was prepared with methanesulfonyl

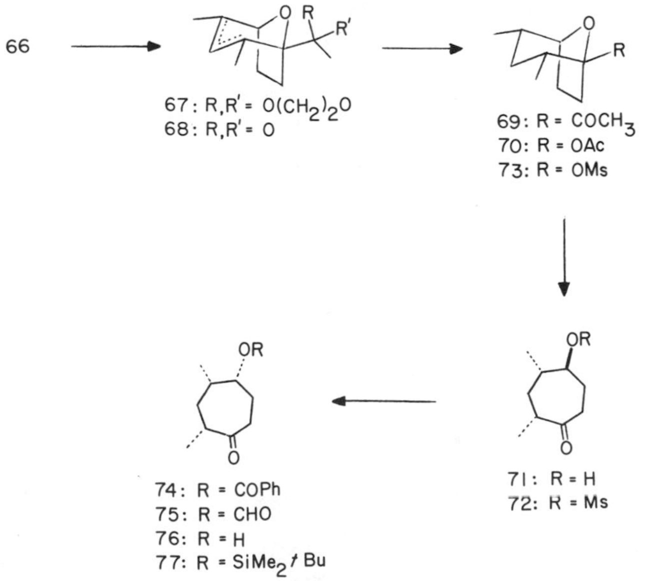

66

67: R,R' = O(CH₂)₂O
68: R,R' = O

69: R = COCH₃
70: R = OAc
73: R = OMs

74: R = COPh
75: R = CHO
76: R = H
77: R = SiMe₂ *t* Bu

71: R = H
72: R = Ms

chloride in pyridine (when triethylamine was used in this reaction, the major product was **73**, indicating that an equilibrium exists between **71** and the bicyclic hemiketal), and displacement was effected with either potassium benzoate or with tetra-*n*-butylammonium formate. The inverted benzoate **74** and formate **75** were each converted to **76** upon methanolysis in the presence of potassium carbonate, with no apparent complication from epimerization α to the carbonyl group. In practical terms, inversion through the formate **75** was more convenient, proceeding in 77% yield from **72** to the alcohol **76**.

In preparation for introduction of the 6-β-methyl substituent, along with functionality that would facilitate oxidative scission of the cycloheptane ring, the hydroxyl group of **76** was blocked as its *tert*-butyldimethylsilyl ether. The ether **77**, acquired in 97% yield, was actually designed to serve a steric as well as protective purpose, in that the bulky silyl derivative was expected to guide the incoming methyl group to the β side of the ring and thus trans to the preexisting substituents. First, however, it was necessary to incorporate α,β unsaturation into the cycloheptanone **77** in a regioselective manner, and this led to examination of methodology by which a substituent could be established at C-6 without epimerization at C-2. Kinetic deprotonation of **77** with lithium 2,2,6,6-tetramethylpiperidide proved to be the key and, when the resulting enolate was allowed to react with phenylselenenyl chloride and then with hydrogen peroxide, the desired enone **78** was produced in 57% yield. Anticipating that this substance might be configurationally labile at both C-2 and C-5 (as, indeed, it proved to be), **78** was treated promptly with lithium dimethylcuprate in ether at 0°C. It was expected, on the basis of results obtained with cuprate additions to other α,β-unsaturated ketones, that the intermediate enolate from **78** would be regiostable, and, in fact, trapping with trimethylsilyl chloride gave a single enol ether (**79**).[14]

As mentioned above, the Prelog–Djerassi lactonic acid (**4**) provides a useful benchmark for confirming configurational fidelity in our sequence and, having **79** in hand, the stage was now set for our route to converge upon this key substance. Ozonolysis of **79**, followed by a reductive workup with sodium borohydride, yielded the hydroxy acid **80** and, when the latter was exposed to dilute hydrochloric acid to remove the silyl blocking group, the resulting δ-hydroxy acid underwent spontaneous lactonization to give **81**. A final oxidation of **81** with Jones's reagent led to (±)-**4** in an overall yield for the 6 steps from **78** of 26%.[15] Comparison of **4** with a racemic sample kindly provided by Professor Satoru Masamune indicated that the two materials corresponded in every detail. This identity allowed us to proceed confidently forward with our plan to merge a segment derived from **79** with an equivalent of **5** to obtain methynolide seco acid **6**.

V. A Cycloheptanoid Route to Methynolide

In planning for the linkage of **79**, representing subunit A of methynolide (**1**), with **5**, we chose to prepare the latter in the absolute configuration corresponding to C-10 and -11 of **1**, i.e., $10(S),11(R)$. Although the confluence of an optically active segment with a racemic derivative of **79** would, in principle, lead to two diastereomers of **6**, this approach was clearly preferable to a scheme involving two racemic components when transformations beyond the macrolide itself, such as attachment of the sugar residue, were envisioned. Thus, the acquisition of a version of **5**, suitable for coupling to **79** and bearing the appropriate absolute configuration, was our next goal.

The E isomer of 2-methyl-2-pentenal (**82**), obtained from aldol condensation of propionaldehyde,[16] was oxidized to the corresponding carboxylic acid **83**. The latter, upon treatment with 70% hydrogen peroxide in acetic acid containing sulfuric acid, afforded *erythro*-2,3-dihydroxy-2-methylvaleric acid (**84**) in 50% yield. Resolution of this acid with brucine proved to be exceptionally facile, producing both (+)- and (−)-**84** with high efficiency. Thus two crystallizations of the brucine salt of (±)-**84** from acetone gave optically pure dextrorotatory acid, and crystallization of the residual salt from ethyl acetate yielded the levo enantiomer.

Assignment of absolute configuration to enantiomers of **84** was made by the application of Horeau's rule.[17] The antipodal acids were first converted to the corresponding esters (+)- and (−)-**85** with diazomethane, and these were treated separately with racemic 1-(2-phenylbutyroyl)imidazole in benzene.[18] The optical rotation of the 2-phenylbutyric acid recovered from the reaction mixture was then measured. From the enantiomeric excess of 2-phenylbutyric acid in this kinetic resolution (dextro **85** gave recovered acid with $[\alpha]^{25}$ $+0.56°$, and levo **85** resulted in acid with $[\alpha]^{25}$ $-0.48°$), Horeau's rule specifies that $(+)$-**85** has R configuration at the secondary alcohol and therefore possesses the desired $2(R),3(R)$ stereochemistry for methynolide.

82: R = CHO
83: R = CO₂H

84: R = H
85: R = CH₃

87: R = CO₂H
88: R = CHO
89: R = CH₂OH

92

90: R = OMs
91: R = SPh
86: R = SPh(O)

Our first design for connecting the two synthetic segments envisioned homologation from **79** to an intermediate that could be used for alkylation of a nucleophile derived from $(+)$-**84**. The sulfoxide **86** appeared to be an ideal candidate for the latter purpose because, in addition to providing activation of the adjacent methylene, this functionality afforded a means for installation of the 8,9 double bond of methynolide through an elimination reaction. Accordingly, **84** was treated with 2,2-dimethoxypropane in the presence of p-toluenesulfonic acid to give acetonide **87** in quantitative yield, and this was reduced with lithium aluminum hydride to **89** (90%). Conversion of this alcohol to its mesylate **90** with methanesulfonyl chloride in pyridine, followed by a displacement with sodium thiophenoxide in dimethylformamide, produced the sulfide **91** (55%), which underwent oxidation with sodium metaperiodate to the crystalline sulfoxide **86** in 92% yield. However, when attempts were made to test the efficacy of the anion from **86** in alkylation with 1-bromomethylcycloheptene, it became apparent that β elimination to a vinyl sulfoxide was the dominant pathway.

The inescapable flaw in this scenario persuaded us to adopt an alternative scheme for uniting the two segments, in which the electrophile and nucleophile roles were reversed. Specifically, it was decided to homologate a derivative of **5** rather than **79**, and the epoxide **92** became our surrogate for the C-8–11 section of structure **1**. This substance was readily obtained from aldehyde **88** (prepared by oxidation of **89** with pyridinium chlorochromate in 87% yield) by treatment with dimethylsulfonium methylide. A 95% yield of a single epoxide was formed, which, on the assumption that Cram's Rule prevails in this alkylation, probably has R configuration at the new chiral center. Since this center ultimately becomes trigonal, this fortuitous stereoselectivity is, of course, inconsequential, and a rigorous proof of configuration was not undertaken.

Attention was now turned to the preparation of a suitable nucleophile from **79** for union with **92**. A cycloheptenyl anion, in which the trimethylsiloxy function of **79** was replaced by a lithium atom, was designated for this purpose and, fortunately, this species proved to be readily accessible from the cycloheptanone **93**, derived by hydrolysis of **79**. Condensation of **93** with 2,4,6-triisopropylbenzenesulfonylhydrazide furnished **94** in 82% yield and, when this was treated with butyllithium in tetramethylethylenediamine at −70°C, the intermediate vinyllithium **95** was produced with apparent high regioselectivity. Addition of **92** to this intermediate resulted in a 63% yield of a 1:1 mixture of the alkylation product **96** and its diastereomer containing unnatural configuration at C-2, -3, -4, and -6. This mixture was acetylated with acetic anhydride in pyridine to give **97**, and the latter was subjected to the reductive ozonolysis procedure previously employed in the synthesis of the Prelog–Djerassi lactone (**4**) from **79**. Chromatography of the ozonolysis product mixture on alumina resulted in elimination of acetic acid from the acetoxy ketone to give the trans enone **98** (1 : 1 mixture of isomers) in 56% yield. The same diastereomeric mixture had been obtained by Grieco in an earlier synthesis of **1**[7] and, following that protocol, **98** was carried forward to methynolide seco acid **100**. Thus oxidation of **98** with Jones's reagent afforded the carboxylic acid **99**, from which the acetonide was removed by treatment with dilute hydrochloric acid in tetrahydrofuran. The dihydroxy acid **100** was separated from its undesired stereoisomer by thin layer chromatography and its IR and NMR spectra were found to be identical with those of **100** prepared by Grieco et al.[7] This seco acid has been previously converted to methynolide (**1**) by Masamune[2] and, hence, the route described above constitutes a formal, total synthesis of the aglycone.

In conclusion, it has to be admitted that a conceptually attractive route to methynolide turned out to be laden with unforeseen difficulties. These necessitated numerous deviations from our original plan, which undoubtedly delayed completion of the project. Nonetheless, some intriguing observations were made in the course of this work that bear directly on our future plans in the macrolide area. These will be reported in due course.

ACKNOWLEDGMENTS

I am greatly indebted to the several collaborators who participated in this adventure. In particular, Dr. Yoshiyasu Fukuyama is recognized for the immense quantity of dedicated effort that he gave to this project. Dr. Martin Demuth, Dr. Balasubramania Gopalan, and Palaykotai Raghavan also contributed in many significant ways to the successful outcome of the work. Thanks are due as well to Prof. Paul Grieco for providing spectra of compound **100** and to Prof. Satoru Masamune for a sample of the Prelog–Djerassi lactone. Financial support for this work was provided by the National Institute for Allergy and Infectious Diseases.

REFERENCES

1. S. Masamune, G. S. Bates, and J. W. Corcoran, *Angew. Chem., Int. Ed. Engl.* **16**, 585 (1977); K. C. Nicolaou, *Tetrahedron* **33**, 683 (1977); T. G. Back, *ibid.* p. 3041.

2. S. Masamune, C. U. Kim, K. E. Wilson, G. O. Spessard, P. E. Georghiou, and G. S. Bates, *J. Am. Chem. Soc.* **97**, 3512 (1975); S. Masamune, H. Yamamoto, S. Kamata, and A. Fukuzawa, *ibid.* p. 3513.

3. W. D. Celmer, *in* "Biogenesis of Antibiotic Substances" (Z. Vanek and Z. Hostalek, eds.), pp. 99–129. Academic Press, New York, 1965; W. D. Celmer, *Pure Appl. Chem.* **28**, 413 (1971).

4. G. Stork, I. Paterson, and F. K. C. Lee, *J. Am. Chem. Soc.* **104**, 4686 (1982).

5. H. Brockmann and W. Henkel, *Chem. Ber.* **84**, 284 (1951).

6. C. Djerassi and J. A. Zderic, *J. Am. Chem. Soc.* **78**, 6390 (1956); R. Anliker, D. Dvornik, K. Gubler, H. Heusser, and V. Prelog, *Helv. Chim. Acta* **39**, 1785 (1956).

7. P. A. Grieco, Y. Ohfune, Y. Yokoyama, and W. Owens, *J. Am. Chem. Soc.* **101**, 4749 (1979).

8. J. Inanaga, T. Katsuki, S. Takimoto, S. Ouchida, K. Inoue, A. Nakano, N. Okukado, and M. Yamaguchi, *Chem. Lett.* p. 1021 (1979).

9. H. M. R. Hoffmann, K. E. Clemens, and R. H. Smithers, *J. Am. Chem. Soc.* **94,** 3940 (1972).

10. R. Noyori, Y. Baba, S. Makino, and H. Takaya, *Tetrahedron Lett.* p. 1741 (1973).

11. A. P. Cowling and J. Mann, *J. Chem. Soc., Perkin Trans.* **1,** p. 1564 (1978).

12. J. G. Vinter and H. M. R. Hoffmann, *J. Am. Chem. Soc.* **96,** 5466 (1974).

13. S. Winstein and P. Carter, *J. Am. Chem. Soc.* **83,** 4485 (1961).

14. E. S. Binkley and C. H. Heathcock, *J. Org. Chem.* **40,** 2156 (1975).

15. A preliminary account of this work has been published: [J. D. White and Y. Fukuyama, *J. Am. Chem. Soc.* **101,** 226 (1979)].

16. M. Häusermann, *Helv. Chem. Acta* **34,** 1482 (1951).

17. A. Horeau and H. B. Kagan, *Tetrahedron* **20,** 2431 (1964).

18. H. Brockmann, Jr. and N. Risch, *Angew. Chem., Int. Ed. Engl.* **13,** 664 (1974).

INDEX